MODELLING AND APPLICATION OF STOCHASTIC PROCESSES

MODELLING AND APPLICATION OF STOCHASTIC PROCESSES

edited by

Uday B. Desai
Washington State University
Pullman, Washington

KLUWER ACADEMIC PUBLISHERS
Boston/Dordrecht/Lancaster

Distributors for North America:
Kluwer Academic Publishers
101 Philip Drive
Assinippi Park
Norwell, Massachusetts 02061, USA

Distributors for the UK and Ireland:
Kluwer Academic Publishers
MTP Press Limited
Falcon House, Queen Square
Lancaster LA1 1RN, UNITED KINGDOM

Distributors for all other countries:
Kluwer Academic Publishers Group
Distribution Centre
Post Office Box 322
3300 AH Dordrecht, THE NETHERLANDS

Consulting Editor: Jonathan Allen

Library of Congress Cataloging-in-Publication Data

Modelling and application of stochastic processes.

1. Stochastic processes. I. Desai, Uday B.
T57.32.M63 1986 519.2 86-15185
ISBN 0-89838-177-0

Printed in the United States of America

CONTENTS

LIST OF CONTRIBUTORS

B. D. O. Anderson
Systems Engineering Department
National Australian University
Canberra
GPO Box 4, ACT 2600
Australia

K. S. Arun
University of Illinois at Urbana-Champaign
Coordinated Science Laboratory
Urbana, IL 61801

Y. Avniel
Department of Electrical and Computer Engineering
Drexel University
Philadelphia, PA 19104

E. S. Chornoboy
Department of Biomedical Engineering
The Johns Hopkins University
Baltimore, MD 21205

F. S. Cohen
Department of Electrical Engineering
University of Rhode Island
Kingston, RI 02881

U. B. Desai
Department of Electrical and Computer Engineering
Washington State University
Pullman, WA 99164

B. W. Dickinson
Department of Electrical Engineering
Princeton University
Princeton, NJ 08544

A. E. Frazho
School of Aeronautics and Astronautics
Purdue University
West Lafayette, IN 47907

J. D. Klein
Jet Propulsion Laboratory
4800 Oak Grove Drive
Pasadena, CA 91109

W. J. Kolodziej
Department of Electrical and Computer Engineering
Oregon State University
Corvallis, OR 97331

S. Y. Kung
Signal and Image Processing Institute
University of Southern California
Los Angeles, CA 90089

R. R. Mohler
Department of Electrical and Computer Engineering
Oregon State University
Corvallis, OR 97331

R. E. Skelton
School of Aeronautics and Astronautics
Purdue University
West Lafayette, IN 47907

J. K. Tugnait
Exxon Production Research Company
P. O. Box 2189
Houston, TX 77252

R. J. Vaccaro
Department of Electrical Engineering
University of Rhode Island
Kingston, RI 02881

E. I. Verriest
School of Electrical Engineering
Georgia Institue of Technology
Atlanta, GA 30332

H. L. Weinert
Department of Electrical Engineering and Computer Science
The Johns Hopkins University
Baltimore, MD 21218

PREFACE

The subject of modelling and application of stochastic processes is too vast to be exhausted in a single volume. In this book, attention is focused on a small subset of this vast subject. The primary emphasis is on realization and approximation of stochastic systems. Recently there has been considerable interest in the stochastic realization problem, and hence, an attempt has been made here to collect in one place some of the more recent approaches and algorithms for solving the stochastic realization problem. Various different approaches for realizing linear minimum-phase systems, linear nonminimum-phase systems, and bilinear systems are presented. These approaches range from time-domain methods to spectral-domain methods. An overview of the chapter contents briefly describes these approaches. Also, in most of these chapters special attention is given to the problem of developing numerically efficient algorithms for obtaining reduced-order (approximate) stochastic realizations. On the application side, chapters on use of Markov random fields for modelling and analyzing image signals, use of complementary models for the smoothing problem with missing data, and nonlinear estimation are included.

Chapter 1 by Klein and Dickinson develops the nested orthogonal state space realization for ARMA processes. As suggested by the name, nested orthogonal realizations possess two key properties; (i) the state variables are orthogonal, and (ii) the system matrices for the $(n+1)$st order realization contain as their "upper" n-th order blocks the system matrices from the n-th order realization (nesting property). This realization is achieved by appropriately selecting a basis in the predictor space obtained by projecting the "future" onto the "past." The nested orthogonal realization can be viewed as an extension of the AR ladder form to ARMA processes.

Chapter 2 by Skelton and Anderson develops an algorithm for obtaining reduced-order stochastic realizations which match not only the first q output covariance lags of a higher order model, but also match the first q Markov parameters. As such they refer to the reduced model as the q-Markov covariance equivalent realization.

The next four chapters deal with the use of "balancing" concepts to develop reduced-order stochastic models.

Chapter 3 by Desai uses the theory of canonical correlation analysis to develop a realization approach, as well as a transformation approach for obtaining continuous-time balanced stochastic realizations. Minimum-phase reduced-order models are obtained as "subsystems" of the balanced stochastic realization. Because of the minimum-phase property, a reduced-order (approximate) Kalman-Bucy filter is obtained by simply inverting the reduced-order model. This approach is then extended to develop reduced-order hierarchical estimators.

Chapter 4 by Arun and Kung generalizes the principal component analysis approach to the problem of obtaining reduced-order stochastic realizations. Relationship of this approach to deterministic as well as stochastic balancing methods is also presented.

Chapter 5 by Vaccaro develops algorithms for obtaining approximate stochastic realizations from a finite set of covariance lags. These algorithms are based on the ideas of deterministic balancing and some interesting scaling ideas.

Chapter 6 by Verriest presents a brief overview of the open-loop deterministic balancing scheme, and shows how it leads to the LQG-balanced realizations. Using the RV-coefficient concept, a common framework is provided for different stochastic realization algorithms. The chapter also presents an interesting viewpoint on the notion of modelling and model reduction.

An alternate approach towards unifying different balancing schemes is to view the canonical correlation analysis problem as a generalized singular value decomposition problem: $Hv_i = \sigma_i Ru_i$, $H'u_i = \sigma_i Qv_i$. Then if H is the cross covariance between the past and the future, and R and Q are respectively the covariances of the past and the future, then we obtain the balanced stochastic realization. On the other hand, if H and R remain as above but $Q = I$, then we obtain the stochastic realization based on principal component analysis. Thus by picking different weighting matrices R and Q, one can obtain different kinds of "balanced realizations."

Chapter 7 by Avniel solves the stochastic realization problem in the spectral domain by associating a scattering matrix with a multivariate stationary stochastic process. It is an interesting and a useful fact that this scattering matrix can be viewed as the generalized phase function for the spectral density. The main thrust of this work is to show that the scattering matrix for stationary sequences can be viewed as the analog of the frequency response function for deterministic systems. It is this analog which brings stationary sequences into the domain of classical systems theory.

Chapter 8 by Tugnait considers the problem of realizing nonminimum phase stochastic systems. The realization algorithms discussed in Chapters 1 through 7 use only the second-order statistics of the process under consideration, and, since these statistics are phase insensitive, one essentially obtains a minimum-phase realization. In order to extract the phase information, the fourth-order statistics (or cummulants) need to be used. In this chapter, using both the second-order and the fourth-order statistics, an algorithm for realizing nonminimum phase stochastic systems is developed.

Chapter 9 by Frazho first very briefly reviews the use of noncommutative dilation theory for solving the linear stochastic realization problem as well as the extension (partial realization) problem. The solution to the extension problem is the well-known maximum entropy extension characterized in terms of the reflection coefficients (Schur contractions), or equivalently in terms of an auto-regressive realization. Noncummutative dilation theory is then extended to solve the bilinear stochastic realization problem as well as the bilinear extension problem. Once again the solution to the extensive problem is provided in terms of Schur contractions.

Chapter 10 by Cohen considers the problem of modelling image signals by two dimensional Markov random fields. Essentially, the statistics of the Markov random field model the dependence of an image pixel on its neighboring pixels. Special attention is given to the multilevel Gibbs model and the Gaussian Markov random field.

Both the parameter estimation problem and the image synthesis problems are discussed and illustrated through various examples.

Chapter 11 by Weinert and Chornoboy uses complementary models for solving the smoothing problem with blackouts. The interesting thing about this approach is that it is simpler, requires the solution of only one Riccati equation, and generalizes easily to the case of multiple blackout intervals. Also, the smoothed estimate can be easily updated without reprocessing the data if the initial state variance is changed. This last feature usually follows when a smoothing problem is solved using the method of complementary models. Parallel implementation of the algorithm is also discussed.

Chapter 12 by Mohler and Kolodziej considers the modelling and estimation problems for processes generated by bilinear stochastic systems which have observation feedback. It is shown how problems in neutron kinetics, heat exchange, and immunology can be modeled by bilinear stochastic systems. The solution to the state estimation problem for processes generated by such models is presented and illustrated through a tracking example.

This work would never have been completed without the help of a number of people. Foremost among them are the contributors of the various chapters, to whom I am very grateful. I would like to thank the Publisher, Mr. Carl Harris, for encouraging me to edit this book and, above all, for his immense patience with the unavoidable delays in putting this volume together. I am also grateful to Professors Robert E. Skelton, Richard Vaccaro and Chin Hsu for their valuable assistance in editing this book.

Uday B. Desai

MODELLING AND APPLICATION OF STOCHASTIC PROCESSES

1

NESTED ORTHOGONAL REALIZATIONS FOR LINEAR PREDICTION OF ARMA PROCESSES

JEFFREY D. KLEIN
Jet Propulsion Laboratory, 4800 Oak Grove Drive, Pasadena, CA 91109, USA

BRADLEY W. DICKINSON
Department of Electrical Engineering, Princeton University, Princeton, NJ 08544, USA

Abstract

In this chapter we introduce a nested set of partial whitening filters for ARMA processes having orthogonal state variables. This provides a generalization of the tapped delay line realization for an ARMA process. The nested orthogonal sequence of realizations is obtained by similarity transformation using the inverse Cholesky factorization of its state covariance matrix. The resulting realization may be viewed as an extension of the AR ladder form to ARMA processes, and an algorithm for performing the factorization which generalizes the Levinson algorithm is described.

Introduction

It is a standard engineering assumption that a stationary discrete-time Gaussian process $\mathbf{x} \triangleq \{x_t \; ; \infty < t < \infty\}$ either has or is well-approximated by a process with a rational power spectral density $f(z)$. That is

$$f(z) = T(z)\,\overline{T(\overline{z}^{-1})} = \frac{B(z)\,\overline{B(\overline{z}^{-1})}}{A(z)\,\overline{A(\overline{z}^{-1})}} \tag{1}$$

Here $T(z) = B(z)/A(z)$; $B(z) = \sum_{i=0}^{n} b_i z^i$, $A(z) = \sum_{i=0}^{p} a_i z^i$, $a_0 = 1$ and $A(z) \neq 0$ inside the unit circle. (We use z to denote the unit delay and overbar to denote conjugation.) Also, it is assumed that $A(z)$ and $B(z)$ have no common factors and $B(z) \neq 0$ for $|z| \leq 1$, so that $T(z)$ is causally invertible.

There are two main reasons for assuming rationality. First, assuming zero mean, the number of parameters needed to describe the probability structure of $\mathbf{x}$ is finite (i.e. $n + p + 1$), whereas in general this description requires the entire covariance sequence $\{r_i \; ; 0 \leq i < \infty\}$. Second, $T(z)$ defines a "model" for $\mathbf{x}$. That is, realizations of $\mathbf{x}$ satisfy the autoregressive-moving average (ARMA) difference equation

$$x_t = -\sum_{i=0}^{p} a_i x_{t-i} + \sum_{i=0}^{n} b_i u_{t-i} \tag{2}$$

where **u** is a stationary white Gaussian "input" process with zero mean and unit variance. Thus **x** may be viewed as the output of a linear system with impulse response function $T(z)$ driven by the input **u**. Furthermore, the model is causally invertible so that the input process **u** can be recovered from **x** by a system of the same form having impulse response function $1/T(z)$:

$$u_t = (-\sum_{i=1}^{n} b_i u_{t-i} + \sum_{i=0}^{p} a_i x_{t-i})/b_0$$

These features are important when an economical, structural representation of a random process is required for applications, for instance:

Prediction of time series. The linear least-squares estimate of x_t based on past inputs $\{u_{t-j}, 1 \le j \le n\}$ and past outputs $\{x_{t-i}, 1 \le i \le p\}$ is

$$\hat{x}_t = -\sum_{i=1}^{p} a_i x_{t-i} + \sum_{i=1}^{n} b_i u_{t-i}$$

and the variance of the prediction error $x_t - \hat{x}_t$ is b_0^2.

Signal modeling. The problem is that of estimating the ARMA parameters of **x**, $\{p; a_i, 1 \le i \le p; n; b_j, 0 \le j \le n\}$, from a finite realization $\{x_t, 1 \le t \le N\}$ and any prior information about **x**. This is important in a number of applications such as: signal analysis/synthesis systems, where the model parameters serve to encode the information in the observation record; and spectral estimation, where the ARMA parameter estimates may be used in Eq. (1) to give an estimate of the power spectral density of **x**.

The signal modeling applications have motivated a great deal of research on estimation of ARMA parameters. However, it turns out that the problem is much more difficult when both $B(z)$ and $A(z)$ must be estimated than when $B(z)$ is known. For this reason, we will restrict our attention to situations where $B(z)$ is given. We refer the interested reader to [1] for more information on the general problem.

Considerably more progress has been made in the case where $B(z)$ is taken as given. The simplest case, when $B(z) = b_0$ (the autoregressive or AR model) has been widely used, but only recently has it been realized that the more general case leads to similar structural features that allow a complete analysis paralleling the AR case. Roughly speaking, the reason for this is that information contained in the process which is necessary for obtaining linear least-squares predictions is contained in a finite-dimensional *predictor space* $P^{\mathbf{x}}$ [2] determined by $B(z)$. The orthogonal projection operator with range $P^{\mathbf{x}}$ then determines $A(z)$, and can be obtained by solving a set of linear equations (commonly called the normal equations).

The difficulty with estimating $B(z)$ as well as $A(z)$ is that the predictor space must be chosen before the projection determining $A(z)$ can be specified. Since the known methods of estimating $B(z)$ from data depend upon $A(z)$, the problem jointly estimating $B(z)$ and $A(z)$ become nonlinearly coupled. By fixing $B(z)$, this coupling is eliminated.

This may seem to be a gross oversimplification of the modeling problem. However, in some applications prior information about $\mathbf{x}$ may be available to allow a reasonable guess of $B(z)$, independent of the data. We would then hope that this information could be used to find a model for $\mathbf{x}$ in which $P^{\mathbf{x}}$ has smaller dimension than the one associated with the often-used AR model. In other words, we hope that an ARMA predictor will have much smaller dimension than an AR predictor for equal prediction error energy. We expect this to be the case for signals with very narrow bandwidth zeros, for example.

Now consider the approximate modeling filter. In terms of least-squares, the performance of the approximate filter will be a function of the dimension of the associated predictor space. Since the actual dimension of $P^{\mathbf{x}}$ is not known (and may be infinite), a reasonable way to proceed with the estimation is to compute a sequence of approximants whose predictor spaces are of successively higher dimensions. One way to do this is to introduce additional zeros into the modeling filter, one at a time. Then the "best" approximant can be chosen from this sequence according to some criterion. (However, we will not be concerned with this order selection problem in this chapter.)

To employ this strategy to greatest benefit, we will require a nested basis for $P^{\mathbf{x}}$; such a basis is introduced in the next section. We say the the modeling (or whitening) filter associated with $P_i^{\mathbf{x}}$, the i-dimensional approximating predictor space, is "i^{th}-order". Obviously, it will be desirable if the $(i+1)^{th}$-order approximant can be easily computed from that of ith-order. The well-known ladder filter structure for the modeling (or whitening) filters of AR form has this property. It is natural to ask if a similar nested filter structure exists for the ARMA case.

This question is considered in [3,4]. By computing a sequence of rational functions of increasing degree which approximate the spectral density of a given process $\mathbf{x}$, $f(z) = \sum_{i=-\infty}^{\infty} r_i z^i$, where $r_i = E(x_{t+i}\overline{x}_t)$, a sequence of transfer functions for approximate modeling filters for $\mathbf{x}$ are obtained. A number of roughly equivalent methods following this approach are presented in [3]. One based on the *Schur algorithm* is discussed in detail in [4] and will be outlined here.

In this approach the locations $\{\beta_i^{-1}; 0 \le i \le l, \ |\beta_i| < 1\}$ and corresponding multiplicities m_i of the zeros of the approximate modeling filter are chosen first. Then the Schur algorithm is used recursively to obtain a sequence of rational approximants, $\hat{c}_k(z)$, for $0 \le k \le \sum_{i=0}^{l} m_i$, to the Carathéodory function $c(z)$ which is defined as

$$c(z) = \frac{1}{2} r_0 + \sum_{i=1}^{\infty} r_i z^i$$

so that $f(z) = c(z) + \overline{c(\overline{z}^{-1})}$. Each $\hat{c}_k(z)$ satisfies one more of the constraints $\{\hat{c}^{(j)}(\beta_i) = c^{(j)}(\beta_i), 0 \le i \le l, 0 \le j \le m_i - 1\}$ than its predecessor. The approximants are solutions to the Nevanlinna-Pick interpolation problem in the unit disk for $c(z)$ [5]. From $\hat{c}_k(z)$, the approximate modeling filter $\hat{T}_k(z)$ is obtained, corresponding to the approximate power spectral density function $f_k(z) = \hat{c}_k(z) + \hat{c}_k(\overline{z}^{-1})$. It is shown in [4] that the divergence of $\sum_i^{\infty}(1 - |\beta_i|)$ implies convergence of $\hat{T}_k(z)$ to $T(z)$ on the unit circle, i.e.

$$\int_{-\pi}^{\pi} |\hat{T}_k(e^{j\theta}) - T(e^{j\theta})|^2 d\theta \to 0 \text{ as } k \to \infty$$

We note that the AR modeling problem is given by the special case of one zero of appropriate multiplicity at infinity, i.e. $\beta_0 = 0$. The interpolation constraints in this case imply that the first $k + 1$ terms in the power series for $\hat{c}_k(z)$ match those of $c(z)$. This is just the covariance-matching property of AR spectral estimates [6].

For the AR case, the corresponding order-recursive procedure for computing the whitening filters from the covariance sequence is the Levinson algorithm [7]. In this case the Levinson and Schur algorithms both require a finite number of covariance lags [8] and have roughly the same complexity. Until recently there has been no analog to the Levinson algorithm when $\beta_i \neq 0$. In this chapter we will present a generalization for this case [9] and apply it to the problem of finding order-recursive whitening filters for $\mathbf{x}$. As in the AR case we obtain a nested family of realizations of the whitening filters (and modeling filters).

The Shift Operator and State Equations for the Predictor Space

First we define some notation. Consider a zero-mean, stationary process $\mathbf{x} \triangleq \{x_t\,;\, t \in \mathbf{Z}\}$ with finite variance. It generates the Hilbert space $H_\infty = \mathbf{sp}\,\{x_t\,;\, t \in \mathbf{Z}\}$, the closed linear span of finite linear combinations of the random variables, with inner product $<\alpha_1,\alpha_2> \triangleq E(\alpha_1\alpha_2)$. The closure is with respect to the associated norm which we indicate as $\|\alpha\| \triangleq (E\,|\alpha|^2)^{1/2}$. We also define the "present and future subspace" and the "past subspace" as

$$H_+ = \mathbf{sp}\,\{x_t\,;\, t \geq 0\}$$

$$H_- = \mathbf{sp}\,\{x_t\,;\, t < 0\}$$

respectively. (We are using stationarity to choose $t = 0$ as the time origin.) The *predictor space* of $\mathbf{x}$ is then given by $P^{\mathbf{x}} \triangleq P_- H_+$, where P_- is the orthogonal projection operator with range H_-.

In the "frequency domain" we employ the Kolmogorov isomorphism which associates x_i with z^{-i}, mapping H_∞ to $L_2(f)$, the Hilbert space of functions of the complex variable z square-integrable on the unit circle with respect to the spectral density function $f(z)$. The inner product in $L_2(f)$ is given by $<z^{-i},z^{-k}> = r_{i-k}$. As before, the covariances $\{r_i\}$ are related to the spectral density $f(z)$ by

$$f(z) = \sum_{i=-\infty}^{\infty} r_i z^i \;;\; r_i = \frac{1}{2\pi j} \oint z^{-i} \frac{f(z)dz}{z}$$

For example, the function $1 \in L_2(f)$ is the image of the random variable $x_0 \in H_\infty$; similarly $\sum_{i=-\infty}^{\infty} d_i x_{-i} \in H_\infty$ corresponds to $\sum_{i=-\infty}^{\infty} d_i z^i \in L_2(f)$. By abuse of notation, we use the same symbol to denote subspaces of H_∞ and of $L_2(f)$ that correspond under the Kolmogorov isomorphism, e.g. $P^{\mathbf{x}}$ (and $M^{\mathbf{x}}$, introduced below), relying on context for the appropriate interpretation. Thus the predictor space is denoted by

$$P^{\mathbf{x}} = \{h(z) \mid h(z) = P_{\mathbf{sp}\{z^i:\, i \geq 1\}} g(z)\}$$

where $g(z) \in L_2(f)$ is a power series in negative powers of z.

For linear prediction of $\mathbf{x}$ to be practical, we require that $p \triangleq \dim P^{\mathbf{x}}$ be finite. Equivalently, we assume that $f(z)$ is rational. Hence $\mathbf{x}$ can be modeled as the response of a finite-dimensional time-invariant linear system to a white input signal.

In a prediction filter, once x_0 becomes available we want to incorporate the new information in x_0 into estimates of x_s for $s \geq 1$. That is, we want to "time-shift" $P^{\mathbf{x}}$, moving the time origin from time 0 to time 1. For this purpose, we exploit the fact that $P^{\mathbf{x}}$ is "almost" shift-invariant. Specifically, in order to get shift invariance we enlarge $P^{\mathbf{x}}$ by defining the space $M^{\mathbf{x}} \triangleq \mathbf{sp}\,\{1\} \oplus P^{\mathbf{x}}$. More concretely, for finite dimensional $P^{\mathbf{x}}$, take a basis of $P^{\mathbf{x}}$, say $\{h_1, \cdots, h_p\}$ and form the $(p+1)$-dimensional space $M^{\mathbf{x}} \triangleq \mathbf{sp}\,\{x_0 - P_- x_0, h_1, \cdots, h_p\}$.

The time origin shift is conveniently described in a subspace of l_2, the space of square-summable sequences, which is isomorphic to $M^{\mathbf{x}}$, denoted $D \subseteq l_2$. This isomorphism takes a function in $M^{\mathbf{x}}$, $h(z) = \sum_{i=0}^{\infty} d_i z^i$, to the infinite sequence of its coefficients, $(d_0, d_1, \cdots) \in D$. On D, the left shift operator of l_2 acts as follows:

$$S(d_0, d_1, d_2, \cdots) = (d_1, d_2, \cdots)$$

so that on $M^{\mathbf{x}}$, S is given by abuse of notation:

$$S\, h(z) = z^{-1}(h(z) - h(0))$$

The following result shows that D, and hence $M^{\mathbf{x}}$, is shift-invariant. (It does not depend on finite dimensionality of $P^{\mathbf{x}}$.)

Lemma. The subspace D (as well as $M^{\mathbf{x}}$) is an invariant subspace of S.

Proof. Any nonzero element of D can be uniquely decomposed as $h_{1,0}\,\delta_0 + h_1$ where $\delta_0 = (1, 0, 0, \cdots)$ and $h_1 = (0, h_{1,1}, h_{1,2}, \cdots)$. In obvious functional notation, $h(z) = h_{1,0}\,\delta_0(z) + h_1(z)$ where $\delta_0(z) = 1$, $h_{1,0} = h(0)$ and $h_1(0) = 0$. Clearly $S\,\delta_0 = 0 \in D$. For the second term, define $w_1 \in H_+$ so that $P_- w_1 = h_1$. Then if $P_-^{(1)}$ denotes the orthogonal projection operator whose range in the subspace of H_∞ given by $\mathbf{sp}\,\{x_t;\ t \leq 0\}$, stationarity of $\mathbf{x}$ gives

$$z^{-1}(h_1(z) - h_1(0)) = z^{-1}h_1(z) = P_-^{(1)}\ z^{-1}w_1(z) = P_-\ z^{-1}w_1(z) + k\,(1 - P_-1) \in M^\times$$

where k is a constant, so that $S\ h_1 \in D$.

It can be shown, see Appendix 2, that any S-invariant $(p+1)$-dimensional subspace of l_2 is the kernel of some $(p+1)^{th}$-degree polynomial, say $\Phi(S) = \Phi_p(S) \triangleq \prod_{i=0}^{p}(S - \beta_i I) = \sum_{i=0}^{p+1} a_{p+1-i} S^i$, $a_0 = 1$, whose zeros all have modulus less than 1. (This result should be distinguished from the well-known Beurling Theorem of Hardy space theory, see [10] for example, which characterizes all closed invariant subspaces of the *right* shift operator on l_2.) Thus any element of the subspace $M^\times \subset l_2$ satisfies the recursion

$$a_{p+1}d_j + a_p\,d_{j+1} + \cdots + a_1 d_{j+p} + d_{j+p+1} = 0\,,\quad j \geq 0 \tag{3}$$

or in terms of functions, the power series coefficients of all elements of $M^\times$ satisfy (3). Thus to specify $M^\times$ we fix the polynomial $\Phi(\cdot)$.

Note that we have $\ker \Phi_{j-1}(S) \subset \ker \Phi_j(S)$. Thus $\{\ker \Phi_j(S);\ 0 \leq j \leq p\}$ is a *nested* sequence of subspaces, so any basis of $\ker \Phi_{j-1}(S)$, e.g. $\{\delta_0, \cdots, \delta_{j-1}\}$, can be augmented to provide a basis for $\ker \Phi_j(S)$; we only require that $\delta_j \notin \ker \Phi_{j-1}(S)$. We say that such a basis is a *nested basis* for $\ker \Phi(S)$.

For the construction of a whitening (or innovations) filter for $\mathbf{x}$ based on this characterization of $M^\times$, we will find it most convenient to consider bases whose equivalents in D are chosen to reflect the direct sum structure of $M^\times$ as $\mathbf{sp}\,\{1\} \oplus P^\times$. We do this by taking at least one zero of $\Phi(\cdot)$ to be 0, since δ_0 spans $\ker S$. A basis of this form is not necessary (see [9] for examples). However, it simplifies the discussion that follows and leads to a nice state space representation of the resulting whitening filter. Thus we will always take $\beta_0 = 0$ in $\Phi(\cdot)$.

We obtain the state equations for $P^\times$ by noting that if $\chi_{i,t}$ is the ith basis element of $P^\times$ and corresponds to the function $h_i(z)$, then $\chi_{i,t+1}$ corresponds to $z^{-1}\,h_i(z)$. Denoting $\chi_{i,t}^{S}$ as the element of $P^\times$ corresponding to the function $S\ h_i(z) = z^{-1}\,(h_i(z) - h_i(0))$, we have

$$\chi_{i,t+1} = \chi_{i,t}^{S} + h_i(0)\,x_t$$

Thus the action of S acting on a chosen basis determines the form of the state equations. In fact, since $M^\times$ is S-invariant, we may as well work with the restriction of S to $M^\times$, denoted $S \mid \ker \Phi(S)$.

As an example, consider the AR case where $\Phi(S) = S^{p+1}$. The obvious choice of basis for $P^\times$ is $\{z, z^2, \cdots, z^p\}$ in terms of functions, or $\{x_{-1}, x_{-2}, \cdots, x_{-p}\}$ in terms of random variables. Thus the state at time $t = 0$ is given by $\chi_0 = (x_{-1}, x_{-2}, \cdots, x_{-p})'$. (We use the prime to denote transposition and adopt the convention of writing all vectors as columns.) Now consider the equation

$$S \mid \ker S^{p+1} \begin{bmatrix} 1 \\ z \\ \cdot \\ \cdot \\ z^p \end{bmatrix} = Z \begin{bmatrix} 1 \\ z \\ \cdot \\ \cdot \\ z^p \end{bmatrix} = \begin{bmatrix} 0 & 0 & \cdot & 0 & 0 \\ 1 & 0 & \cdot & \cdot & 0 \\ 0 & 1 & \cdot & \cdot & \cdot \\ \cdot & \cdot & \cdot & \cdot & \cdot \\ 0 & 0 & \cdot & 1 & 0 \end{bmatrix} \begin{bmatrix} 1 \\ z \\ \cdot \\ \cdot \\ z^p \end{bmatrix} \tag{4}$$

(In what follows, we use the symbol Z for any "downshift matrix", leaving the dimension to be understood by context, as is the usual custom for identity matrices.)

The time domain version of the state equations can now be obtained by inspection:

$$\chi_{t+1} = Z\,\chi_t + e_1 x_t \tag{5a}$$

where e_1 is the unit vector $(1, 0, 0, \cdots, 0)'$. The output equation

$$u_t = c'\chi_t + x_t \tag{5b}$$

then gives the whitening filter for the AR process $\mathbf{x}$ when the vector $c = (c_{1p}, c_{2p}, \cdots, c_{pp})'$ is determined by the normal equations

$$R\,\bar{c} = -v\,, \quad R = E\,\chi_t \chi_t^*, \quad v = E\,\chi_t \bar{x}_t$$

where the * denotes conjugate transpose.

For the general case with arbitrary β_i (each with modulus less than 1), there are several useful bases for $\ker \Phi(S)$ [9], two of which are given here. First consider the set

$$\Delta_p \triangleq \{\delta_i \mid \delta_0 = 1;\ \delta_i = z^i / \Phi^r(z),\ 1 \leq i \leq p\}$$

where $\Phi^r(z) = z^{p+1}\,\Phi(z^{-1})\}$.
Following the same procedure that led to (4), we obtain the matrix associated with the operator $S \mid \ker \Phi(S)$

$$\begin{bmatrix} 0 & 0 & \cdot & \cdot & 0 \\ 1 & -a_1 & \cdot & \cdot & -a_p \\ 0 & 1 & \cdot & \cdot & \cdot \\ & \cdot & \cdot & \cdot & \cdot \\ & & \cdot & 1 & 0 \end{bmatrix} \triangleq \begin{bmatrix} 0 & \vec{0}\,' \\ e_1 & A_p^{\Delta} \end{bmatrix}$$

where the symbol $^{\rightarrow}$ is used to explicitly indicate a column vector quantity. This basis is of some interest because it provides a state equation in "controller form"

$$\chi_{t+1}^{\Delta} = A_p^{\Delta}\,\chi_t^{\Delta} + e_1\,x_t$$

The second basis, which we consider in greater detail, is suggested by the following result.

Theorem [11]. For a pth-degree polynomial $\Phi(\cdot)$ having all its zeros inside the unit circle and a $p \times p$ matrix V, the following are equivalent:

a) $\Phi(V) = 0$, and rank $(I - V^* V) = 1$.

b) V is unitarily similar to $S \mid \ker \Phi(S)$.

c) $\Phi(V) = 0$, and V is unitarily similar to $M_B(Z)$.

Here $M_B(Z) = \Gamma^{-1}(Z + B)(I + \bar{B}Z)^{-1}\Gamma$ with $B = \mathrm{diag}\{\beta_0, \cdots, \beta_p\}$ and $\Gamma = \mathrm{diag}\{\gamma_0, \cdots, \gamma_p\}$ with $\gamma_i = (1 - |\beta_i|^2)^{1/2}$. The operator $M_{(\cdot)}(\cdot)$ is known as a Möbius transform [11,12]; with B and Z as arguments, it has the following form:

$$M_B(Z) = \begin{bmatrix} \beta_0 & & & & \\ \gamma_0\gamma_1 & \beta_1 & & & \\ \gamma_0\bar{\beta}_1\gamma_2 & \gamma_1\gamma_2 & \beta_2 & & \\ \cdot & \cdot & \cdot & \cdot & \cdot \\ \gamma_0\prod_{i=1}^{p-1}(-\bar{\beta}_i)\gamma_p & \cdot & \cdot & \cdot & \beta_p \end{bmatrix}$$

The theorem implies that we may think of $M_B(Z)$ as a canonical representative of $S \mid \ker \Phi(S)$. Notice that just as in the case of the first basis described above, $M_B(Z)$ reduces to the downshift matrix Z in the case where $\beta_i = 0$ for all i. In addition, it is nested with respect to upper (and lower) principal submatrices. This fact will be quite useful when we consider the ARMA generalization of the AR ladder form. Incidentally, the rank one property in statement a) of the Theorem (with $V = M_B(Z)$) plays a crucial role in generalizing the theory of *low shift-rank* matrices; for a discussion, see [9].

Because we require $\beta_0 = 0$, the basis associated with the representation of $S \mid \ker \Phi(S)$ by $M_B(Z)$ is [11]

$$\Theta_p \triangleq \{\theta_i \mid \theta_i(z) = \frac{\gamma_i(z - \bar{\beta}_{i-1})}{\gamma_{i-1}(1 - \beta_i z)}\,\theta_{i-1}(z),\ 1 \le i \le p\ ;\ \theta_0(z) = 1\}$$

More explicitly, for $1 \le i \le p$,

$$\theta_i(z) = \frac{\gamma_i \prod_{j=0}^{i-1}(z - \bar{\beta}_j)}{\prod_{j=0}^{i}(1 - \beta_j z)}$$

Using the basis Θ_p, the resulting state equations are

$$\chi_{t+1}^{\Theta} = M_{B_p}(Z)\,\chi_t^{\Theta} + b_{\Theta}\, x_t \tag{6a}$$

where $B_p = \mathrm{diag}(\beta_1, \cdots, \beta_p)$. As before, the output equation is

$$u_t = c'_\Theta \chi_t^\Theta + x_t \tag{6b}$$

where $c_\Theta = -\overline{R_\Theta^{-1} \; v_\Theta}$, and $R_\Theta = E\chi_t^\Theta \chi_t^{\Theta *}$ and $v_\Theta = E\chi_t^\Theta \overline{x}_t$.

This type of basis and the associated realization have the following properties:

a) The set $\{\beta_i, 1 \le i \le p\}$ determines the set of poles of the whitening filter, $\{\beta_i^{-1}, 1 \le i \le p\}$, since the $\{\beta_i\}$ are the eigenvalues of the state transition matrix $M_{B_p}(Z)$.

b) Unlike Δ_p, Θ_p is a nested basis for ker $\Phi(S)$. Each basis element contains one more of the β_i than its predecessor. Note that the θ_i can be visualized as the sections of a generalized tapped delay line.

c) The state equations associated with Θ are nested also. This is reflected by the structure of $M_B(Z)$ and b_Θ.

d) The realization in (6) is controllable. We prove more than this in Appendix 1 by constructing the similarity transformation between the realizations associated with Δ_p and Θ_p.

Nested Realizations Via State Orthogonalization

In this section we consider the problem of nested realizations of the whitening filter for **x**. That is, given a realization $Q_n = \{A_n, b_n, c_n\}$ associated with the projection of x_0 onto $H_{1,n} \triangleq \mathbf{sp}\{h_i; 1 \le i \le n\}$, we want to construct Q_{n+1}, the realization associated with $P_{H_{1,n+1}} x_0$, which takes the form

$$A_{n+1} = \begin{bmatrix} A_n & \vec{0} \\ \# & \# \end{bmatrix}, \quad b_{n+1} = \begin{bmatrix} b_n \\ \# \end{bmatrix}, \quad c_{n+1} = \begin{bmatrix} c_n \\ \# \end{bmatrix}$$

(where # denotes an entry of no importance); if $P^{\times}$ has dimension p, we want this nesting to hold for all $n < p$. This structure means that Q_n is *nested* within Q_{n+1} in the sense that Q_{n+1} if formed simply by augmenting Q_n without otherwise changing it. For AR processes, the ladder form is an example of this structure [14]. We now extend this idea to ARMA processes.

Consider the realization of the whitening filter for x using Θ_p as given in the previous section, (6). In this case the state equations are nested and determined by the choice of $\{\beta_i\}$. The statistics of the process enter only into the equation for the output vector c_Θ. In general, if we wish to add another pole to the whitening filter, we must recompute the output vector, even though the state equations are nested. We now show that a modified realization for which this is unnecessary is obtained by orthogonalizing the state vector with respect to the spectral density of **x**.

We can transform the realization (6) into one with an uncorrelated state vector if we have the inverse Cholesky factorization of the pth-order state covariance matrix, $R_p^{-1} = E\chi_t \chi_t^*$, i.e.

$$R_p^{-1} = U_p \Gamma_p^{-1} U_p^*$$

where $\Gamma_p = \mathrm{diag}\{\sigma_1^2, \cdots, \sigma_p^2\}$ and U_p is unit upper triangular. Since $\Gamma_p = U_p^* R_p U_p$, the system

$$\tilde{\chi}_{t+1} = \tilde{A}\ \tilde{\chi}_t = \tilde{b}\, x_t\ ,\quad u_t = \tilde{c}\,'\tilde{\chi}_t + x_t \tag{7}$$

has state covariance Γ_p if $\tilde{\chi} = U_p^* \chi_t$. We will see that this similarity transformation leads to a completely nested realization. The Cholesky factor U_p has the following nested structure:

$$R_p^{-1} = U_p \Gamma_p^{-1} U_p^* = \begin{bmatrix} U_{p-1} & \vec{u}_p \\ \vec{0}\,' & 1 \end{bmatrix} \begin{bmatrix} \Gamma_{p-1}^{-1} & \vec{0} \\ \vec{0}\,' & \sigma_p^{-2} \end{bmatrix} \begin{bmatrix} U_{p-1}^* & \vec{0} \\ \vec{u}_p^* & 1 \end{bmatrix}$$

where $R_{p-1}^{-1} = U_{p-1}\Gamma_{p-1}^{-1}U_{p-1}^*$. Since A and b are nested and U_p is upper triangular, $\tilde{A}$ and $\tilde{b}$ are also nested. The normal equations determining $\tilde{c}$ take the form

$$\tilde{c} = -\overline{\tilde{\Gamma}_p^{-1}\, \tilde{v}} \tag{8}$$

where $\tilde{\Gamma}_p = E\,\tilde{\chi}_t \tilde{\chi}_t^* = \mathrm{diag}\{\sigma_1^2, \cdots, \sigma_p^2\}$ and $\tilde{v} = E\,\tilde{\chi}_t\, \overline{x}_t$. Since $\tilde{A}$ and $\tilde{b}$ are nested, so is $\tilde{v}$. Then (8) shows that $\tilde{c}$ is nested as well. Thus we have obtained a nested realization for the whitening filter for **x** with the additional property that its state variables are uncorrelated.

Returning to the example of the AR case, with $\Phi(S) = S^{p+1}$, and with realization (5), the inverse Cholesky factors of R_p are given by

$$U_p = \begin{bmatrix} 1 & c_{1,1} & \cdot & c_{p-1,p-1} \\ & 1 & \cdot & \cdot \\ & & \cdot & c_{1,p-1} \\ & & & 1 \end{bmatrix} \tag{9}$$

and $\Gamma_p = \mathrm{diag}\{E_0, \cdots, E_{p-1}\}$ where E_j is the jth-order prediction error energy. The Levinson algorithm [7] recursively computes the columns of U_p and Γ_p via

$$c_{i,j+1} = c_{i,j} - k_{j+1}\, \overline{c}_{j+1-i,j}\ ,\quad 1 \le i \le j+1\ ;\ 0 \le j \le p-2 \tag{10}$$

where $c_{0,j} = 1$ and $c_{j+1,j} = 0$ for all j. The $\{k_j\}$ coefficients appearing in (10) are the PARCOR (or reflection) coefficients obtained from the covariance sequence by

$$k_j = \sum_{i=0}^{j-1} c_{i,j-1} r_{j-i} / E_{j-1}\ ,\quad E_j = (1 - |k_j|^2)\, E_{j-1} \tag{11}$$

for $1 \le j \le p-1$ with $E_0 = r_0$. Using these relations, it can be shown [14] that $\{\tilde{A}, \tilde{b}, \tilde{c}\}$ are given by

$$\tilde{A} = \begin{bmatrix} 0 & & & & & \\ 1 & 0 & & & & \\ \bar{k}_2 k_1 & 1 & 0 & & & \\ \bar{k}_3 k_1 & \bar{k}_3 k_2 & 1 & & & \\ \cdot & \cdot & \cdot\ \cdot & 1 & 0 & \\ \bar{k}_{p-1} k_1 & \bar{k}_{p-1} k_2 & \cdot\ \cdot & \bar{k}_{p-1} k_{p-2} & 1 & 0 \end{bmatrix}, \quad \tilde{b} = \begin{bmatrix} 1 \\ -\bar{k}_1 \\ \cdot \\ \cdot \\ \cdot \\ -\bar{k}_{p-1} \end{bmatrix}, \quad \tilde{c} = \begin{bmatrix} -k_1 \\ -k_2 \\ \cdot \\ \cdot \\ \cdot \\ -k_p \end{bmatrix} \tag{12}$$

which defines the ladder form of the whitening filter. Addition of a stage to this filter does not affect the preceding stages because of the nested structure of the realization.

Notice also that since $\{\tilde{A}, \tilde{b}, \tilde{c}\}$ is nested, so is $\{\tilde{A} - \tilde{b}\,\tilde{c}, \tilde{b}, \tilde{c}\}$, which is a realization of the modeling filter for **x**. In the AR example, we have

$$\tilde{A} - \tilde{b}\,\tilde{c} = \begin{bmatrix} k_1 & k_2 & \cdot\ \cdot & k_{p-1} & k_p \\ q_1 & -\bar{k}_1 k_2 & \cdot\ \cdot & -\bar{k}_1 k_{p-1} & -\bar{k}_1 k_p \\ 0 & q_2 & \cdot\ \cdot & \cdot & \cdot \\ & 0 & \cdot\ \cdot & \cdot & \cdot \\ & & \cdot\ q_{p-2} & -\bar{k}_{p-2} k_{p-1} & -\bar{k}_{p-2} k_p \\ & & 0 & q_{p-1} & -\bar{k}_{p-1} k_p \end{bmatrix}$$

where $q_j = 1 - |k_j|^2$.

We can obtain a natural generalization of the ladder form to ARMA processes by using the nested structure of the state equations in the Θ basis. For this we need the inverse Cholesky factorization of the state covariance of (6) as a similarity transformation to orthogonalize the state vector. In the next section we give an algorithm for this computation.

Inverse Cholesky Factorization of R_p

We have seen that by an appropriate choice of basis we can obtain a nested whitening filter for **x** if we know the inverse Cholesky factorization of its state covariance matrix. In the AR case with state equations (5), this is efficiently computed (with $O(p^2)$ operations) using the Levinson algorithm. Furthermore, the quantities arising in the algorithm, the PARCOR coefficients, are important in the ladder filter implementation of the realization, even though this is not "obvious" from the state equations themselves (12). The Levinson algorithm relies on the Toeplitz structure of the state covariance matrix.

$$R_{Toep} = \frac{1}{2\pi j} \oint \left\{ \begin{bmatrix} z \\ z^2 \\ \cdot \\ z^p \end{bmatrix} [\bar{z}\ \bar{z}^2 \cdot \bar{z}^p] \right\} \frac{f(z)dz}{z}$$

In this section we extend this idea to matrices such as R_p which, while not Toeplitz, have structure that is inherited from the predictor space and the choice of basis. The state covariance of (6) is given by

$$R_p = \frac{1}{2\pi j} \oint \Theta_{1p}(z)\Theta_{1p}(z)^* \frac{f(z)dz}{z}$$

where $\Theta_{1p}(z) \triangleq z^{-1}[\theta_1(z), \cdots, \theta_p(z)]'$ is obtained from the functions comprising the Θ basis. In fact, the elements of the vector $\Theta_{1p}(z)$ comprise a basis for $\ker \Phi_{1p}(S)$, where $\Phi_{1p}(S) = \prod_{i=1}^{p}(S - \beta_i)$. We will compute the inverse Cholesky factorization of R_p by finding a nested basis for $\ker \Phi_{1p}(S)$, orthogonal with respect to $f(z)$. This can be done recursively.

For this purpose we define the reproducing kernel for $\ker \Phi_{1p}(S)$ with respect to $f(z)$:

$$K_p(z,y) = \Theta_{1p}(z)' \, \overline{R}_p^{-1} \, \overline{\Theta}_{1p}(y) = \Theta_{1p}(z)' \, \overline{U}_p \Gamma_p^{-1} U_p' \, \overline{\Theta}_{1p}(y) \tag{13}$$

$K_p(z,y)$ satisfies

$$\frac{1}{2\pi j} \oint K_p(z,y)\overline{g}(z)\frac{f(z)dz}{z} = \overline{g}(y), \quad |y| = 1, \quad g(z) \in \ker \Phi_{1p}(S) \tag{14}$$

Also, define a set of functions spanning $\ker \Phi_{1p}(S)$, orthogonal with respect to $f(z)$, $\{\phi_i(z), 1 \le i \le p\}$ which satisfies

$$\frac{1}{2\pi j} \oint \phi_l(z)\overline{\phi}_m(z)\frac{f(z)dz}{z} = \delta_{l,m}$$

This gives

$$K_p(z,y) = \sum_{i=1}^{p} \phi_i(z)\overline{\phi}_i(y)$$

From (13) we see that one possible definition of $\{\phi_i(z), 1 \le i \le p\}$ is $[\phi_1(z), \cdots, \phi_p(z)]' = \Theta_{1p}(z)' \, \overline{U}_p \Gamma_p^{-1/2}$. That is, the ith column of $\overline{U}_p \Gamma_p^{-1/2}$ is the *coordinate representation* of $\phi_i(z)$ with respect to the Θ_{1p} basis of $\ker \Phi_{1p}(S)$. While this is not the only possible choice for the orthogonal basis functions, this choice has certain advantages because of the existence of the order-recursive formulas for it which we now derive.

The generalized inverse Cholesky algorithm is based on recursions for the reproducing kernels $\{K_n(z,y), 1 \le n \le p\}$ which can be derived from the generalized Christoffel-Darboux formulas [4,15]:

$$K_n(z,y) = \frac{\phi_{n+1}^b(z)\overline{\phi}_{n+1}^b(y) - \phi_{n+1}(z)\overline{\phi}_{n+1}(y)}{1 - \xi_{n+1}(z)\,\overline{\xi}_{n+1}(y)} \tag{15a}$$

$$K_n(z,y) = \frac{\phi_n^b(z)\overline{\phi}_n^b(y) - \xi_n(z)\,\overline{\xi}_n(y)\phi_n(z)\overline{\phi}_n(y)}{1 - \xi_n(z)\,\overline{\xi}_n(y)} \tag{15b}$$

where $\xi_i(z) = (z - \overline{\beta}_i)/(1 - \beta_i z)$ and $\phi_n^b(z) = \prod_{i=1}^{n} \xi_i(z)\overline{\phi}_n(\overline{z}^{-1})$ is a generalized "reverse" of $\phi_n(z)$. Note that (15b) yields $K_n(z, \overline{\beta}_n) = \overline{\phi}_n^b(\overline{\beta}_n)\phi_n^b(z)$ and $K_n(\overline{\beta}_n, \overline{\beta}_n) = |\overline{\phi}_n^b(\overline{\beta}_n)|^2$. (15a) and (15b) can be combined to give

$$\begin{bmatrix} K_{n+1}^b(z,y) \\ K_{n+1}(z,y) \end{bmatrix} = q_{n+1}\Omega(\rho_{n+1})\Gamma_{\xi_{n+1}}(z)\Omega(-\xi_{n+1}(y)\rho_{n+1}) \begin{bmatrix} K_n^b(z,y) \\ K_n(z,y) \end{bmatrix} \tag{16}$$

where

$$\Omega(a) = (1 - |a|^2)^{-1/2} \begin{bmatrix} 1 & a \\ \overline{a} & 1 \end{bmatrix}, \quad q_{n+1} = q_{n+1}(y) = \frac{(1 - |\xi_{n+1}(y)\rho_{n+1}|^2)^{1/2}}{(1 - |\rho_{n+1}|^2)^{1/2}}$$

$$K_n^b(z,y) = \prod_{i=1}^{n} \xi_i(z)\overline{K}_n(\overline{z}^{-1},y), \quad \Gamma_{\xi_{n+1}}(z) = \mathrm{diag}\{\xi_{n+1}(z), 1\}$$

$$\rho_{n+1} = \rho_{n+1}(y) = \phi_n(y)/\overline{\phi}_{n+1}(y)$$

The substitution $y = \overline{\beta}_{n+1}$ yields

$$\begin{bmatrix} \alpha_{n+1}\phi_{n+1}(z) \\ \overline{\alpha}_{n+1}\phi_{n+1}^b(z) \end{bmatrix} = C_n \begin{bmatrix} \xi_{n+1}(z)K_n^b(z, \overline{\beta}_{n+1}) \\ K_n(z, \overline{\beta}_{n+1}) \end{bmatrix} \tag{17}$$

where

$$C_n = (1 - |\mu_n|^2)^{-1} \begin{bmatrix} 1 & \mu_n \\ \overline{\mu}_n & 1 \end{bmatrix}$$

$\alpha_{n+1} = \phi_{n+1}^b(\overline{\beta}_{n+1})$ and $\mu_n = \rho_{n+1}(\overline{\beta}_{n+1})$. The $\{\mu_n, 1 \le n \le p-1\}$ are known as the *Schur parameters* of **x**.

We can eliminate the factors of α_{n+1} and $\overline{\alpha}_{n+1}$ to within a multiplier of unit magnitude by normalizing $K_n(z,y)$. Defining $\tilde{K}_n(z,y) = K_n^{-1/2}(y,y)K_n(z,y)$, we have from (16) that

$$K_{n+1}(y,y) = \frac{(1 - |\xi_{n+1}(y)\rho_{n+1}(y)|^2)}{(1 - |\rho_{n+1}(y)|^2)} K_n(y,y)$$

and

$$\begin{bmatrix} \tilde{K}_{n+1}^b(z,y) \\ \tilde{K}_{n+1}(z,y) \end{bmatrix} = \Omega(\rho_{n+1}(y))\Gamma_{\xi_{n+1}}(z)\Omega(-\xi_{n+1}(y)\rho_{n+1}(y)) \begin{bmatrix} \tilde{K}_n^b(z,y) \\ \tilde{K}_n(z,y) \end{bmatrix}$$

Letting $y = \overline{\beta}_{n+1}$ and noting that $K_n(y,y)$ is real, we get

$$\begin{bmatrix} \epsilon_{n+1}\phi_{n+1}(z) \\ \overline{\epsilon}_{n+1}\phi_{n+1}^b(z) \end{bmatrix} = \Omega(\mu_n)\Gamma_{\xi_{n+1}}(z)K_n^{-1/2}(\overline{\beta}_{n+1}, \overline{\beta}_{n+1}) \begin{bmatrix} K_n^b(z, \overline{\beta}_{n+1}) \\ K_n(z, \overline{\beta}_{n+1}) \end{bmatrix}$$

where $\epsilon_{n+1} = \phi_{n+1}^{b}(\overline{\beta}_{n+1}) / \mid \phi_{n+1}^{b}(\overline{\beta}_{n+1}) \mid$. Letting $y = \overline{\beta}_{n+1}$ in (15b), we obtain

$$\begin{bmatrix} \epsilon_{n+1}\phi_{n+1}(z) \\ \overline{\epsilon}_{n+1}\phi_{n+1}^{b}(z) \end{bmatrix} = \tilde{C}_n \begin{bmatrix} \xi_{n+1}(z) K_n^b(z, \overline{\beta}_{n+1}) \\ K_n(z, \overline{\beta}_{n+1}) \end{bmatrix}$$

where

$$\tilde{C}_n = \left\{ \frac{\mid \phi_{n+1}^{b}(\overline{\beta}_{n+1}) \mid^2 - \mid \xi_n(\overline{\beta}_{n+1})\phi_n(\overline{\beta}_{n+1}) \mid^2}{1 - \mid \xi_n(\overline{\beta}_{n+1}) \mid^2} \right\}^{-1/2} \Omega(\mu_n)$$

Note that since the orthonormal functions defined by (14) are unique up to a constant factor of unit magnitude, we can take $\epsilon_{n+1}\phi_{n+1}(z)$ to be the $(n+1)^{st}$ one. Therefore we will drop the factor of ϵ_{n+1} since it is irrelevant for our purposes.

Now we can obtain an order update formula for ϕ and ϕ_b by using (15b). Since $\overline{\xi}_n(\overline{z}^{-1}) = \xi_n^{-1}(z)$, we have that

$$\begin{bmatrix} \xi_{n+1}(z) K_n^b(z, \overline{\beta}_{n+1}) \\ K_n(z, \overline{\beta}_{n+1}) \end{bmatrix} = W \begin{bmatrix} \phi_n(z) \\ \phi_n^b(z) \end{bmatrix} \tag{18}$$

where

$$W = \Gamma_1 \begin{bmatrix} w_1 & w_2 \\ \overline{w}_2 & \overline{w}_1 \end{bmatrix} \Gamma_{\xi_n}(z)$$

and where $\Gamma_1 = \text{diag}\,\{\xi_{n+1}(z)/(\xi_n(z) - \xi_n(\overline{\beta}_{n+1}))\,,\, 1/(1 - \xi_n(z)\overline{\xi}_n(\overline{\beta}_{n+1}))\}$ and with $w_1 = \phi_n^b(\overline{\beta}_{n+1})$ and $w_2 = -\xi_n(\overline{\beta}_{n+1})\phi_n(\overline{\beta}_{n+1})$. Some algebra then yields

$$\begin{bmatrix} \phi_{n+1}(z) \\ \phi_{n+1}^{b}(z) \end{bmatrix} = \tilde{C}_n \begin{bmatrix} \xi_{n+1}(z) K_n^b(z, \overline{\beta}_{n+1}) \\ K_n(z, \overline{\beta}_{n+1}) \end{bmatrix}$$

$$= \tilde{C}_n D_n \begin{bmatrix} \phi_n(z)(z - \overline{\beta}_n)/(1 - \beta_{n+1} z) \\ \phi_n^b(z)(1 - \beta_n z)/(1 - \beta_{n+1} z) \end{bmatrix} \tag{19}$$

where

$$D_n = (1 - \mid \beta_n \mid^2)^{-1} \begin{bmatrix} a & -b \\ -\overline{b} & \overline{a} \end{bmatrix}$$

and $a = (1 - \beta_n \overline{\beta}_{n+1}\phi_n^b(\overline{\beta}_{n+1})$ and $b = (\overline{\beta}_{n+1} - \overline{\beta}_n)\phi_n(\overline{\beta}_{n+1})$.

We can get the coordinates of the orthogonal functions directly from R_p as follows. It is easily shown that when the $\{\beta_i\}$ are given, $\phi_n(z)$ determines $\phi_n^b(z)$ (see Appendix 6 of [9]) and hence also determines $K_n^b(z, \overline{\beta}_{n+1})$ and $K_n(z, \overline{\beta}_{n+1})$. The only free parameter in (19) is μ_n. This can be obtained from R_{n+1} by choosing μ_n so that $\vec{\phi}_{n+1}$, the coordinate vector of $\phi_{n+1}(z) = \vec{\phi}_{n+1}'\Theta_{1,n+1}(z)$, satisfies the equation

$$R_{n+1}\vec{\phi}_{n+1} = \begin{bmatrix} 0 \\ \cdot \\ \cdot \\ 0 \\ \sigma_{n+1}^2 \end{bmatrix}$$

where σ_{n+1}^2 is a positive constant. This is exactly what is done in the Levinson algorithm.

Let $\vec{k}_n^{\xi}$ and $\vec{k}_n$ be the component vectors with respect to $\Theta_{1,n+1}$ for $\xi_{n+1}(z)K_n^b(z, \overline{\beta}_{n+1})$ and $K_n(z, \overline{\beta}_{n+1})$ respectively. It is shown in [15] that $|\mu_n| < 1$ for nondeterministic $\mathbf{x}$. Thus μ_n may be determined by forcing

$$R_{n+1}(\vec{k}_n^{\xi} + \mu_n \vec{k}_n) = \begin{bmatrix} 0 \\ \cdot \\ \cdot \\ 0 \\ \varsigma \end{bmatrix} \tag{20}$$

where ς is a nonzero constant. At first glance this looks like too many equations for μ_n. However, the recursion (17) guarantees that $\vec{k}_{n+1} \triangleq (\vec{k}_n^{\xi} + \mu_n \vec{k}_n)$ is of this form. Thus only one of the equations $e_i' R_{n+1}\vec{k}_{n+1} = 0$, $1 \leq i \leq n+1$ need be solved for μ_n. The question is, which one?

In order that μ_n be uniquely determined, it is only necessary that $e_i' R_{n+1}\vec{k}_{n+1} \neq 0$ for some i. Since

$$K_n(z, \overline{\beta}_{n+1}) = \Theta_{1,n+1}(z)' \begin{bmatrix} R_n^{-1} & 0 \\ 0 & 0 \end{bmatrix} \overline{\Theta}_{1,n+1}(\overline{\beta}_{n+1})$$

we have that $\vec{k}_n = [(R_n^{-1}\vec{d})', 0]'$ where $\vec{d} = \overline{\Theta}_n(\overline{\beta}_{n+1})$. Hence

$$R_{n+1}\vec{k}_n = \begin{bmatrix} R_n & r_{n+1} \\ r_{n+1}^* & \rho_{n+1} \end{bmatrix} \begin{bmatrix} R_n^{-1}\vec{d} \\ 0 \end{bmatrix} = \begin{bmatrix} \vec{d} \\ \# \end{bmatrix}$$

Since the top entry of $\vec{d}$ is $\gamma_1/(1 - \beta_1\overline{\beta}_{n+1}) \neq 0$, it is sufficient to solve the top row of (20) for μ_n once $\vec{k}_n^{\xi}$ and $\vec{k}_n$ are available.

$\vec{k}_n^{\xi}$ and $\vec{k}_n$ can be computed using (19) since we have $\vec{\phi}_n$ and $\vec{\phi}_n^b$ from the solution of the nth-order problem. $\vec{\phi}_n$, $\vec{\phi}_n^b$, $\xi_{n+1}(z)K_n^b(z, \overline{\beta}_{n+1})$, and $K_n(z, \overline{\beta}_{n+1})$ are elements of $\ker \Phi_{1,n}(S)$, hence the multiplication by functions of z in (19) may be computed in terms of coordinates using the results of Appendix 1.

To be more precise, define Z_{n+1}^{Θ} to be the matrix representation of multiplication by z with respect to $\Theta_{1,n+1}$ obtained using Appendix 1. Then if

$P^{\Theta}_{n+1} = (Z^{\Theta}_{n+1} - \overline{\beta}_n I)(I - \beta_{n+1} Z^{\Theta}_{n+1})^{-1}$ and $Q^{\Theta}_{n+1} = (I - \beta_n Z^{\Theta}_{n+1})(I - \beta_{n+1} Z^{\Theta}_{n+1})^{-1}$, we have that

$$\begin{bmatrix} \vec{k}_n^{\,\xi} \\ \vec{k}_n \end{bmatrix} = [D_n \otimes I] \begin{bmatrix} P^{\Theta}_{n+1} \vec{\phi}_{n,n+1} \\ Q^{\Theta}_{n+1} \vec{\phi}^{\,b}_{n,n+1} \end{bmatrix}$$

where $\vec{\phi}_{n,n+1} = [\vec{\phi}_n{}', 0]'$, $\vec{\phi}^{\,b}_{n,n+1} = [\vec{\phi}_n^{\,b\prime}, 0]'$ and $\otimes$ denotes the Kronecker product. Finally,

$$\begin{bmatrix} \vec{\phi}_{n+1} \\ \vec{\phi}^{\,b}_{n+1} \end{bmatrix} = [\tilde{C}_n \otimes I][D_n \otimes I] \begin{bmatrix} P^{\Theta}_{n+1} \vec{\phi}_{n,n+1} \\ Q_{n+1_\Theta} \vec{\phi}^{\,b}_{n,n+1} \end{bmatrix} \tag{21}$$

For example, consider the AR case, with all $\beta_i = 0$ and take the order equal to p. Then the covariance matrix R_{n+1} is Toeplitz. To demonstrate that the algorithm above reduces to the Levinson algorithm in this situation, we note that $z\,\phi_n(z)$ has coordinate vector $[0, \vec{\phi}_n{}']'$ with respect to the basis $\Theta_{1,n}$ since $\phi_n(z)$ has coordinates $[\vec{\phi}_n{}', 0]'$. Thus $P^{\Theta}_{n+1} = Z$, $Q^{\Theta}_{n+1} = I$ and (21) becomes

$$\begin{bmatrix} \vec{\phi}_{n+1} \\ \vec{\phi}^{\,b}_{n+1} \end{bmatrix} = \Omega(\mu_n) \begin{bmatrix} \omega_n Z \vec{\phi}_{n,n+1} \\ \overline{\omega}_n \vec{\phi}^{\,b}_{n,n+1} \end{bmatrix}$$

with $\omega_n = \phi_n^b(0)/\,|\,\phi_n^b(0)\,|$. Ignoring this unit magnitude factor as before, the top equation become

$$\vec{\phi}_{n+1} = (1 - |\,\mu_n\,|^2)^{-1/2}(Z\vec{\phi}_{n,n+1} + \mu_n \vec{\phi}^{\,b}_{n,n+1}) \tag{22}$$

Noting from (9) and the corresponding definition of Γ_p that $\vec{\phi}_i = E_{i-1}^{-1/2}\,[\overline{c}_{i-1,i-1}, \cdots, \overline{c}_{1,i-1}, 1]'$, (22) gives

$$\vec{\phi}_{n+1} = \begin{bmatrix} \overline{c}_{nn} \\ \cdot \\ \cdot \\ \overline{c}_{1n} \\ 1 \end{bmatrix} = E_n^{1/2}(1 - |\,\mu_n\,|^2)^{-1/2} E_{n-1}^{-1/2} \left\{ \begin{bmatrix} 0 \\ \overline{c}_{n-1,n-1} \\ \cdot \\ \overline{c}_{1,n-1} \\ 1 \end{bmatrix} + \mu_n \begin{bmatrix} 1 \\ c_{1,n-1} \\ \cdot \\ c_{n-1,n-1} \\ 0 \end{bmatrix} \right\}$$

which is seen to be the conjugate of the update formula (10) with $\mu_n = -\overline{k}_n$ and $E_n = (1 - |\,k_n\,|^2)E_{n-1}$.

We leave the general case for Appendix 1, since things are considerably more complicated when the β_i are nonzero and not identical. In general the slowest operations involved in the order update, i.e. the multiplications by $(z - \overline{\beta}_n)/(1 - \beta_{n+1}z)$ and $(1 - \beta_n z)/(1 - \beta_{n+1}z)$ in (19), may not be easy to perform in terms of coordinates. Efficient implementation of the algorithm remains as a topic for further research.

Conclusion

We have given a family of realizations for the whitening and modeling filters of ARMA processes which have a nesting property analogous to that of the ladder realization of AR processes. We can thus treat processes with zeros not necessarily at ∞. We began by characterizing the predictor space, $P^{\times}$, and giving two useful bases for it. $P^{\times}$ was shown to be a function of the zeros of $\mathbf{x}$, $\{\beta_i^{-1}, 1 \leq i \leq p\}$. The basis chosen for $P^{\times}$ determines the form of the state equations for the whitening filter of $\mathbf{x}$. We then showed how the nested basis we introduced, Θ_{1p}, leads to state equations in which successively larger subspaces of $P^{\times}$ will yield state equations which are "nested". After noting that nested *output* equations result as well if such a basis is transformed by similarity into a nested *orthogonal* one, we gave a generalization of the Levinson algorithm which yields the necessary transformation, the inverse Cholesky factorization of R_p. The resulting realization is a generalization of the AR ladder, although its form is much more complicated.

Our tool in obtaining state equations from the shift-invariant augmented predictor space, $M^{\times}$, is the left shift operator (sometimes called the backward shift operator). Originally introduced in the system theory literature for the study of the realization problem for infinite dimensional systems, the shift operator approach has been applied to the study of linear and nonlinear systems problems by Helton, Fuhrmann, Frazho, Gilbert, Rugh, and others. For finite dimensional problems, the shift operator approach is essentially an algebraic one, although it provides certain insights that differ from those obtained by the formal algebraic approach. For a comparison, see Fuhrmann [16] and Kalman [17].

There are many remaining open problems concerning the application of the structural results presented in this chapter to signal modeling problems. Before discussing some of these, we would like to describe a point of contact between the theory of linear prediction and the theory of statistical inference, in particular ARMA model parameter estimation. For this purpose, we limit attention to stationary Gaussian processes. We mentioned in the Introduction that the assumption of a rational spectral density function leads to a finitely-parametrized process model in ARMA form. In technical terms, this can be stated in a much stronger way: a rational spectral density is both necessary and sufficient for existence of a *finite-dimensional* process model, where the class of possible models includes both linear and nonlinear ones (with suitable smoothness properties) and where finite-dimensional means that the model has a *realization* with a finite-dimensional state space [18]. Adopting the viewpoint of nonlinear filtering theory, we can consider another, more stringent, way to categorize ARMA models as "finite" or not; in terms of the notion of *sufficiency*, as developed in the theory of statistical inference, we may ask whether a class of ARMA models with fixed numbers of poles and zeros admits a sufficient statistic which can be generated as the output process of a finite dimensional system driven by the observations [19]. This turns out to be the case only when the zeros of the models are not only fixed in number but also take a set of known values [20]. In the terms of this chapter, this means that the models have a known predictor space, $P^{\mathbf{x}}$. Evidently, this endows the class of models discussed in this chapter with some added importance, at least from a theoretical point of view, although the practical significance is not yet completely understood.

We now turn to a discussion of "practical" issues related to parameter estimation for the models described in this chapter. We cannot expect to use simple (linear) methods to determine the zeros of $\mathbf{x}$, but nonlinear and iterative methods may be feasible. In connection with iterative methods, or simply under the assumption of known, perhaps approximate, zeros, an obvious question is: "If the zeros of $\mathbf{x}$ are known, why not just prefilter $\mathbf{x}$ with the appropriate all-pole filter to remove them *first* and then estimate the poles of $\mathbf{x}$ by standard methods for AR processes?" The answer is that by using the order-recursive method we describe, poles can be added to the whitening filter one at a time, while prefiltering adds them all at the beginning. By introducing these poles one at a time, we need not recompute say, the $(i-1)^{th}$-order approximant if we decide to use a different β_i than originally planned. Thus we have the option of selecting β_i based on, e.g., minimization of the ith-order prediction error with respect to β_i. The best way to do this is an open question. However, we are assured of asymptotic convergence of the approximate filter to the ideal if $\sum_{i=0}^{N}(1 - |\beta_i|) \to \infty$.

Suppose that we have reason to believe that the zeros of $\mathbf{x}$ lie in certain regions of the complex plane, but we don't know the actual values for β_i. Since the guesses we use play a role in determining the computational efficiency of the generalized Levinson algorithm, an obvious approach is to pick the set of β_i so that the assumed zeros of $\mathbf{x}$ lie in the desired parts of the complex plane and that P_n^{Θ} and Q_n^{Θ} are as sparse as possible for all n. We have not examined this in detail, but it appears that one way to ensure the desired sparseness is to choose the sequence $\{\beta_i\}$ to be periodic, i.e., $\beta_{i+l} = \beta_i$ with $l << p$. This and other possibilities remain to be explored.

Finally, another question we have not considered is: what is the actual form of the ARMA ladder realization? We have not shown any direct relation between the structure we have derived (by requiring a nested realization with diagonal state covariance) and the structure treated in [8] (see also [21]). It would be interesting to relate these explicitly.

Acknowledgments

We would like to thank A. Frazho for suggesting the proof in Appendix 2. We would also like to thank U. Desai for the invitation to contribute this chapter to his book. This work was supported at Princeton University through National Science Foundation Grants ECS82-05772 and ECS84-05460.

Appendix 1

In this appendix we derive the matrix representation of $S \mid \ker \Phi(S)$ with respect to an *arbitrary* basis when $\Phi(z)$ is the degree $p+1$ polynomial $\Phi(z) = \prod_{i=0}^{p} (z - \beta_i) = \sum_{i=0}^{p+1} a_{p+1-i} z^i$ where $|\beta_i| < 1$ for $1 \leq i \leq p$. We will use the canonical basis $\Delta \triangleq \{\delta_i \mid \delta_i = z^i / \Phi^r(z),\ 0 \leq i \leq p\}$ where $\Phi^r(z) = \sum_{i=0}^{p+1} a_i z^i$ with $a_0 = 1$. Corresponding to (4) we have

$$S \mid \ker \Phi(S) \begin{bmatrix} \delta_0 \\ \cdot \\ \cdot \\ \delta_p \end{bmatrix} = A_\Delta \begin{bmatrix} \delta_0 \\ \cdot \\ \cdot \\ \delta_p \end{bmatrix} = \begin{bmatrix} -a_1 & \cdot & \cdot & -a_p \\ 1 & \cdot & \cdot & \cdot \\ \cdot & \cdot & \cdot & \cdot \\ 0 & \cdot & 1 & 0 \end{bmatrix} \begin{bmatrix} \delta_0 \\ \cdot \\ \cdot \\ \delta_p \end{bmatrix}$$

We now show how we can obtain the coordinates of any element of $\ker \Phi(S)$ with respect to any basis when given those for Δ. For this purpose, we introduce the notion of multiplication by z in $\ker \Phi(S)$. In other words, we give the vector space $\ker \Phi(S)$ a z-module structure. The formula for z-multiplication is motivated by Kalman's structure theory for linear systems [17] in which the state space admits a z-module structure in a natural way. For $g(z) = n(z)/\Phi^r(z) \in \ker \Phi(S)$ we take

$$z \bullet g(z) \triangleq \frac{zn(z) \bmod \Phi(z)}{\Phi^r(z)}$$

and we may extend this definition by linearity to multiplication of $g(z)$ by any power series $f(z)$, at least formally.

Notice that we have a unique additive decomposition:

$$zg(z) = z \bullet g(z) + g^c(z)$$

where the first term is in $\ker \Phi(S)$ and $g^c(z) = k\ \Phi(z)/\Phi^r(z)$ for some constant k. It follows easily that $g^c(z) \in \ker S\,\Phi(S)$ and that $\Phi(S)[g^c(z)] = k$. This result holds more generally.

Proposition. For $g(z) \in \ker \Phi(S)$ and $f(z)$ a (formal) power series such that $f(z)g(z) \in \ker \Phi(S)$, then

$$f(z)g(z) = f(z) \bullet g(z) = \sum_{i=0}^{\infty} \frac{z^i\, n(z) \bmod \Phi(z)}{\Phi^r(z)} \tag{A1}$$

Proof. Define $g_i(z)$ and $g_i^c(z)$ so that

$$zg_{i-1}(z) = g_i(z) + g_i^c(z)$$

where $g_i(z) = zg_{i-1}(z)$, $g_0(z) = g(z)$, $g_i^c(z) = \varsigma_i\, \Phi(z)/\Phi^r(z) \in \ker S\,\Phi(S)$ and

$\Phi(S)[g_i^c(z)] = \varsigma_i$, a constant. Then

$$f(z)g(z) = f_0\, g(z) + \sum_{i=1}^{\infty} f_i [g_i(z) + \sum_{j=0}^{i-1} z^j g_{i-j}^c(z)]$$

Since $f(z)g(z)$ and all of the $\{g_i(z)\}$ are in ker $\Phi(S)$,

$$\Phi(S)[f(z)g(z)] = \sum_{i=1}^{\infty} f_i \sum_{j=0}^{i-1} \Phi(S)[z^j g_{i-j}^c(z)] = 0$$

Thus $\sum_{i=1}^{\infty} f_i \sum_{j=0}^{i-1} \varsigma_{i-j} z^j = 0$. Since $\varsigma_{i-j} z^j \Phi(z)/\Phi^r(z) = z^j g_{i-j}^c(z)$ we have also that

$$\sum_{i=1}^{\infty} f_i \sum_{j=0}^{i-1} z^j g_{i-j}^c(z) = 0$$

Thus

$$f(z)g(z) = \sum_{i=0}^{\infty} f_i\, g_i(z)$$

and the assertion is proved.

This is what we need to express any basis in terms of Δ. It is easily verified that if $g(z)$ has coordinate vector $\vec{g}$ in Δ, then $z \bullet g(z)$ has coordinate vector $Z_S \vec{g}$, where

$$Z_S = \begin{bmatrix} 0 & \cdot & 0 & -a_{p+1} \\ 1 & & & \cdot \\ & \cdot & & \cdot \\ & & 1 & -a_1 \end{bmatrix} \tag{A2}$$

Suppose we have the coordinates of an element of some basis from which all others may be obtained by multiplications by functions like $f(z)$ in (A1). We can calculate the coordinates of the elements of this basis with respect to Δ using (A2) and apply the appropriate coordinate transformation to A_Δ to obtain the matrix associated with the shift operator.

For example, consider the basis Θ introduced in the second section of this chapter, with $\beta_0 = 0$. With respect to Δ, the coordinate vector of $\theta_0 = 1$ is $\alpha_0 = (1, a_1, \cdots, a_p)'$, since $a_{p+1} = 0$, and the others satisfy

$$\alpha_i = \frac{\gamma_i}{\gamma_{i-1}} (Z_S - \overline{\beta}_{i-1} I)(I - \beta_i Z_S)^{-1} \alpha_{i-1}, \quad 1 \le i \le p$$

Thus

$$\Theta_p(z) = \begin{bmatrix} \theta_0 \\ \cdot \\ \cdot \\ \theta_p \end{bmatrix} = T\, \Delta_p(z) = T \begin{bmatrix} \delta_0 \\ \cdot \\ \cdot \\ \delta_p \end{bmatrix}$$

where T is the $(p+1)\times(p+1)$ matrix $[\alpha_0, \cdots, \alpha_p]'$. The matrix associated with the shift operator and basis Θ is $A = TA_0T^{-1}$, where

$$A_0 = \begin{bmatrix} -a_1 & \cdot & -a_p & 0 \\ 1 & & & \cdot \\ & \cdot & & \cdot \\ & & 1 & 0 \end{bmatrix}$$

satisfies $S[\Delta_p(z)] = A_0\Delta_p(z)$.

It turns out that T can be evaluated conveniently. Using the matrix inversion lemma, it can be shown that

$$(Z_S - \overline{\beta}_{i-1}I)(I - \beta_i Z_S)^{-1} = (Z - \overline{\beta}_{i-1}I)(I - \beta_i Z)^{-1} - \left[\frac{\beta_i(Z - \overline{\beta}_{i-1}I)(I - \beta_i Z)^{-1} + I}{1 - \beta_i b^* a}\right] ab^*$$

where $a = (0, -a_p, \cdots, -a_1)'$ and $b = (\beta_i^p, \cdots, \beta_i, 1)^*$.

Other bases are handled in the same way. Observe that β_0 need not be 0. For example if $\beta_0 \neq 0$, $\gamma_0/(1-\beta_0 z)$ has coordinate vector $\alpha_0 = (\alpha_{00}, \cdots, \alpha_{0p})'$ where α_{0i} is the coefficient of z^i in $\gamma_0 \prod_{i=1}^{p}(1-\beta_i z)$ since $\gamma_0/(1-\beta_0 z) = \gamma_0 \prod_{i=1}^{p}(1-\beta_i z)/\Phi^r(z)$. Note also that since $A = M_B(Z) = TA_0T^{-1}$, T is the similarity transformation taking the Möbius transform into controller form.

This approach can be used to find the matrix representation of any transformation given its representation with respect to Δ. An application is the computation of Z_n^Θ, and by extension P_n^Θ and Q_n^Θ for (21). Let $\Delta_{1n}(z) = (1/\Phi_{1n}^r(z), z/\Phi_{1n}^r(z), \cdots, z^n/\Phi_{1n}^r(z))'$ with $\Phi_{1n}^r(z) = z^n\overline{\Phi}_{1n}(\overline{z}^{-1}) = \sum_{i=0}^{n} a_{in} z^i$ and $a_{0n} = 1$. The components of $\Delta_{1n}(z)$ form a basis for ker $\Phi_{1n}(S)$. We know that for $g(z) \in$ ker $\Phi_{1n}(S)$ the operation $z \bullet g(z)$ can be done in terms of coordinates via $\vec{g}_z = Z_n^\Delta \vec{g}$, where $\vec{g}_z$ and $\vec{g}$ are, respectively, the coordinate vectors of $z\,g(z)$ and $g(z)$ with respect to Δ_{1n} and

$$Z_n^\Delta = \begin{bmatrix} 0 & \cdot & 0 & -a_{nn} \\ 1 & & & \cdot \\ & \cdot & & \cdot \\ & & 1 & -a_{1n} \end{bmatrix}$$

Since we know that $\phi_{n-1}(z)(z - \overline{\beta}_{n-1})/(1-\beta_n z)$ and $\phi_{n-1}^b(z)(z - \beta_{n-1}z)/(1-\beta_n z)$ (see (19)) are in ker $\Phi_{1n}(S)$, we have that $P_n^\Delta = (Z_n^\Delta - \overline{\beta}_{n-1}I)(I - \beta_n Z_n^\Delta)^{-1}$. P_n^Θ is obtained using the similarity transformation T_n defined by $T_n = [\alpha_1, \cdots, \alpha_n]$ where the α_i are vectors obtained from

$$\alpha_i = \frac{\gamma_i}{\gamma_{i-1}}(Z_n^\Delta - \overline{\beta}_{i-1}I)(I - \beta_i Z_n^\Delta)^{-1}\alpha_{i-1}, \quad 2 \le i \le n$$

with $\alpha_1' \Delta_{1n}(z) = \gamma_1/(1-\beta_1 z)$. Thus $P_n^{\Theta} = T_n P_n^{\Delta} T_n^{-1}$. A similar calculation gives the matrix Q_n^{Θ}.

Appendix 2

In this appendix we prove a result used in the second section.

Proposition. Consider the space H of formal power series in z and the shift operator S described by $S\, d(z) = z^{-1}(d(z) - d(0))$. Let M be an n-dimensional S-invariant subspace of H with $n < \infty$. Also, let $A = S|_M$ denote the restriction of S to M (so that $Am = Sm$ for all $m \in M$), and let $\Phi(z) = \det(zI - A)$ be its characteristic polynomial. Then $M = \ker \Phi(S)$.

Proof. The Cayley-Hamilton theorem implies that $\Phi(A) = 0$, so if $m \in M$ we have that $\Phi(S)\, m = \Phi(A)\, m = 0$. Hence $M \subseteq \ker \Phi(S)$.

Next we show that $\dim \ker \Phi(S) = \deg \Phi(z) = n = \dim M$, so that $M = \ker \Phi(S)$. Taking one factor of Φ at a time, first look at $\ker(S - \alpha I)$. Clearly, $(S - \alpha I)\, d(z) = z^{-1}(d(z) - d(0)) - \alpha\, d(z) = 0$ if and only if $d(z) = d(0)(1 - \alpha z)^{-1}$. We can now prove by induction that $(S - \alpha I)^k\, d(z) = 0$ if and only if $d(z)$ has the form

$$d(z) = \sum_{i=1}^{k} \frac{c_i z^{i-1}}{(1 - \alpha z)^i} \tag{A3}$$

This is clear for $k = 1$. Assuming it is true for $k-1$, $(S - \alpha I)^k\, d(z) = 0$ if and only if $(S - \alpha I)^{k-1}(S - \alpha I) d(z) = 0$. Now by the induction hypothesis

$$(S - \alpha I)\, d(z) = z^{-1}(d(z) - d(0)) - \alpha d(z) = \sum_{i=1}^{k-1} \frac{\gamma_i z^{i-1}}{(1 - \alpha z)^i}$$

Solving for $d(z)$ we obtain the desired form, (A3).

From the form of the elements of $\ker(S - \alpha I)^k$, it is clear that the dimension of this subspace is k. Since $\Phi(z)$ is a monic polynomial of degree n and $\Phi(S)\, m = 0$ for all $m \in M$, it follows from a standard result of linear algebra (e.g. Theorem 6, p. 213 of [22]) that M is the direct sum of the subspaces $\{U_i,\ 1 \le i \le r\}$, where $U_i = \ker(S - \alpha_i I)^{\nu_i}$, $\{\alpha_i\}$ being the distinct roots appearing in the factorization of $\Phi(z) = \prod_{i=1}^{r} (z - \alpha_i)^{\nu_i}$.

Hence the dimension of M equals the degree of $\Phi(z)$ and the proof of the result is complete. This result and the square summability of elements of l_2 give the following:

Corollary. Any S-invariant $(p+1)$-dimensional subspace of l_2 is the kernel of some $(p+1)^{th}$-degree monic polynomial $\Phi(S)$ whose zeros all have modulus less than 1.

References

1. G. E. P. Box and G. M. Jenkins, *Time Series Analysis: Forecasting and Control,* revised edition, Holden-Day, San Francisco, CA, 1976.
2. H. Akaike, in *System Identification: Advances and Case Studies,* (R. H. Mehra and D. G. Lainiotis, eds.), Academic Press, New York, NY, pp. 27-96, 1976.
3. P. Dewilde, A. C. Vieira, and T. Kailath, *IEEE Trans. Circuits and Systems,* **CAS-25** (9), pp. 663-675, 1978.
4. P. Dewilde and H. Dym, *IEEE Trans. Information Theory,* **IT-27,** (4), pp. 446-461, 1981.
5. N. I. Akheizer, *The Classical Moment Problem,* Oliver and Boyd, Edinburgh, 1965.
6. J. Makhoul, *Proc. IEEE,* **63,** (4), pp. 561-580, 1975.
7. N. Levinson, *J. Mathematical Physics,* **25,** (1), pp. 261-278, 1947.
8. P. Dewilde, *Proc. Colloq. on Fast Algorithms for Linear Dynamical Systems,* Aussois, France, pp. 6.1-6.67, 1981.
9. J. D. Klein, *Fast Parameter Estimation Algorithms for Stationary and Near-Stationary Processes,* Ph. D. Thesis, Princeton University, 1983.
10. W. Rudin, *Real and Complex Analysis,* McGraw-Hill, New York, NY, 1966.
11. V. Pták and N. J. Young, *Linear Algebra and its Applications,* **29,** pp. 357-392, 1980.
12. L. A. Harris, *Proc. on Infinite Dimensional Holomorphy,* Springer Lecture Notes 364, Springer-Verlag, Berlin, pp. 13-40, 1974.
13. J. D. Klein and B. W. Dickinson, *Proc. 18th Conf. on Information Sciences and Systems,* Princeton University, Princeton, NJ, pp. 11-15, 1984.
14. D. T. L. Lee, *Canonical Ladder Form Realizations and Fast Estimation Algorithms,* Ph. D. Thesis, Stanford University, 1980.
15. A. Bultheel and P. Dewilde, *Proc. Internat. Symp. on Mathematical Theory of Networks and Systems,* **3,** Delft, the Netherlands, pp. 207-212, 1979.
16. P. A. Fuhrmann, *Linear Systems and Operators in Hilbert Space,* McGraw-Hill, New York, NY, 1981.
17. R. E. Kalman, in *Topics in Mathematical System Theory,* (R. E. Kalman, P. L. Falb, and M. A. Arbib), McGraw-Hill, New York, NY, pp. 237-339, 1969.
18. C. A. Schwartz, B. W. Dickinson, and E. D. Sontag, *Stochastics,* **11,** pp. 159-172, 1984.
19. B. W. Dickinson and E. D. Sontag, *IEEE Trans. Information Theory,* **IT-31,** (5), 1985.
20. B. W. Dickinson, *J. Time Series Analysis,* **3,** pp. 165-168, 1983.
21. E. Deprettere and P. Dewilde, in *Digital Signal Processing,* (A. G. Constantinides, ed.), Academic Press, New York, NY, 1980.
22. S. Lang, *Linear Algebra,* Addison-Wesley, Reading, MA, 1966.

2

q-MARKOV COVARIANCE EQUIVALENT REALIZATIONS

ROBERT E. SKELTON[1] and B.D.O. ANDERSON[2]

ABSTRACT

This paper describes a class of reduced order models which match both the first q Markov Parameters and the first q Output Covariances for linear systems subject to white noise inputs. The class of reduced models contains the earlier models [2] as a special case.

1.0 INTRODUCTION

Let the time-invariant linear stable system

$$\dot{x}_1 = A_1 x_1 + D_1 w\ , \quad w \epsilon \underline{R}^{n_w}$$

$$y_1 = C_1 x_1 \qquad , \quad y_1 \epsilon \underline{R}^{n_y} \tag{1.1}$$

with white noise w(t) with unit intensity, be reduced to

$$\dot{x}_2 = A_2 x_2 + D_2 w$$

$$y_2 = C_2 x_2\ , \quad y_R \epsilon R^{n_y}\ , \tag{1.2}$$

where x_1 is of dimension n and x_2 is of dimension $r < n$.

Definition:

If the realization (A_2, D_2, C_2) *has these two properties:*

$$C_1 A_1^i D_1 = C_2 A_2^i D_2 \qquad , \quad i = 0, 1, \ldots, q-1 \tag{1.3a}$$

$$C_1 A_1^i X_1 C_1^* = C_2 A_2^i X_2 C_2^*\ , \quad i = 0, 1, \ldots, q-1 \tag{1.3b}$$

where

$$0 = X_1 A_1^* + A_1 X_1 + D_1 D_1^*\ , \quad 0 = X_2 A_2^* + A_2 X_2 + D_2 D_2^*$$

$$X_1 > 0,\ X_2 > 0\ , \tag{1.4}$$

the realization (A_2, D_2, C_2) *is called a "q-Markov COVariance Equivalent Realization"*

---

1 School of Aeronautics and Astronautics, Purdue University, West Lafayette, IN 47907.

2 Systems Engineering Dept. National Australian University, Canberra Australia, GPO Box 4, ACT 2600.

of (A_1, D_1, C_1), *or simply a "q-Markov COVER."*

Systems (1.1) and (1.2) have the same first q Markov parameters and the same output covariance and (q-1) derivatives of output covariance when w is white. When $q = \infty$ the q-Markov COVER is a stochastically equivalent realization in the sense of Anderson [1], since the entire autocorrelation is matched.

$$\underline{E}[y_1(t+\tau)y_1^*(\tau)] = \sum_{i=0}^{\infty} [C_1 A_1^i X_1 C_1^*] \frac{t^i}{i!} .$$

An algorithm is available [2] which constructs a q-Markov COVER, for any finite of and this algorithm has been applied to the controller reduction problem [3]. This procedure generates a specific member of a broader class of q-Markov COVERs which are described in this paper.

One motivation for studying the q-Markov COVER problem is made manifest by our first result.

Let the controller (with controller input z and output u)

$$u = G x_c + G_z z$$

$$\dot{x}_c = A_c x_c + F z \tag{1.5}$$

driving the plant (with external input u_c, feedback input u, output z, measurement z, process noise w and measurement noise v)

$$\dot{x} = A_p x + B_p(u+u_c) + D_p w$$

$$y = C_p x \tag{1.6}$$

$$z = M_p x + v$$

yield the closed loop relation (setting v and w to zero)

$$\begin{bmatrix} y(s) \\ u(s) \end{bmatrix} = [Z(s)]u_c , \quad Z(s) = \sum_{i=0}^{\infty} \frac{CA^iD}{s^{i+1}} \tag{1.7}$$

where

$$C = \begin{bmatrix} C_p & 0 \\ G_z M_p & G \end{bmatrix}, \quad A = \begin{bmatrix} A_{11} & A_{12} \\ A_{21} & A_{22} \end{bmatrix} = \begin{bmatrix} A_p + B_p G_z M_p & B_p G \\ F M_p & A_c \end{bmatrix}, \quad D = \begin{bmatrix} B_p \\ 0 \end{bmatrix} \tag{1.8}$$

Theorem 1:

Any reduction of any linear controller which preserves the first (q−1) *Markov parameters (of the controller) will also preserve the first q-Markov parameters of the closed loop system between external inputs* u_c *and feedback inputs* u *and plant outputs*

y.

Proof:

We must show that any two controllers with the same q-1 values of

$$GA_c^iF \ , \ i = 0, 1, ..., q-2 \tag{1.9}$$

and the same G_z, will yield exactly the same first q values of

$$CA^iD \ , \ i = 0, 1, ..., q-1 \ . \tag{1.10}$$

To see this, note that if the first q-1 values of (1.9) are preserved, then so are the first q-1 values of

$$A_{12} A_{22}^i A_{21} = B_pG A_c^i FM_p \ . \tag{1.11}$$

The final step is to observe that only the first n columns of A^i (denoted by $(A^i)_1$) are involved in the calculation

$$CA^iD = \begin{bmatrix} C_p & 0 \\ G_zM_p & G \end{bmatrix} A^i \begin{bmatrix} B_p \\ 0 \end{bmatrix} = \begin{bmatrix} C_p & 0 \\ G_zM_p & G \end{bmatrix} (A^i)_1 B_p \tag{1.12}$$

and that

$$\begin{bmatrix} C_p & 0 \\ G_zM_p & G \end{bmatrix} \begin{bmatrix} \begin{bmatrix} A_{11} & A_{12} \\ A_{12} & A_{22} \end{bmatrix}^i \end{bmatrix}_1 = f(A_{12}A_{22}^jA_{21}), \ j = 0,1,2, ..., i-1 \tag{1.13}$$

where the parameters (A_{12}, A_{22}, A_{21}) appear in $f(\cdot)$ *only* in this combination $A_{12}A_{22}^j$ A_{21}. Hence, the first q values of CA^iD, $i = 0, 1, 2, ...$ q-1 will be modified by a change in controller iff the controller changes the first q-1 values of $A_{12} A_{22}^j A_{21}$, or equivalently, of GA_c^iF, $i = 0, 1, ..., q-2$. #

This theorem can also be proven for discrete-time systems. Theorem 1 provides strong motivation for using q-Markov COVER methods for controller reduction since this method preserves a specified number of Markov parameters.

Note that (1.3a) is equivalent to requiring of a realization of order r with transfer matrix $G_R(s)$ that

$$\frac{d^i}{ds^i}[G_R(s^{-1})]_{s=0} = \frac{d^i}{ds^i}[G(s^{-1})]_{s=0}, \ i = 1, ..., q \tag{1.14}$$

$$G_R(s) = C_2(sI-A_2)^{-1}D_2 \tag{1.15}$$

$$G(s) = C_1(sI-A_1)^{-1}D_1 \tag{1.16}$$

In addition the reduced realization (A_2, C_2, D_2) must match the covariance derivatives

$$\frac{d^i}{dt^i}E[y(t+\tau)y^*(\tau)]_{t\to 0+} = \frac{d^i}{dt^i}E[y_R(t+\tau)y^*(\tau)]_{t\to 0+} \ , \ i = 0,1,\ldots,q-1 \ . \qquad (1.17)$$

Section 2.0 defines a new set of coordinates from which one may generate a class of solutions to the q-Markov COVER problem. Section 3.0 describes a special covariance *control problem* which arises out of the *model reduction* discussion of Section 2.0.

2.0 Structure of the q-Markov COVER Problem

We define (A_2, D_2, C_2) to be a *projection* of (A_1, D_1, C_1) iff

$$A_2 = \underline{L}A_1\underline{R} \ , \ D_2 = \underline{L}D_1 \ , \ C_2 = C_1\underline{R} \ , \ \underline{L}\underline{R} = I \ . \qquad (2.1)$$

We may construct an $\underline{L}$ and $\underline{R}$ satisfying $\underline{L}\underline{R} = I$ by

$$\underline{L} = P^{-1}[I \ 0] \ T_o^{-1} \ , \ \underline{R} = T \begin{bmatrix} I \\ G \end{bmatrix} P \qquad (2.2)$$

for some (P, T, G) where the nxn matrix T serves to place the system (A_1, C_1, D_1) in a convenient basis prior to the projection $\begin{bmatrix} I \\ G \end{bmatrix}[I \ 0]$, and the rxr matrix P plays the role of a similarity transformation of the *reduced system*.

Now define the matrices M_{iq}, Θ_{iq}, by

$$M_{iq} \triangleq \Theta_{iq}[D_i, X_iC_i^*], \ \Theta_{iq}^* = [C_i^*, A_i^*C_i^*, \ldots, A_i^{q-1*}C_i^*] (i=1,2,) \qquad (2.3)$$

Hence, from our first definition, (A_2, D_2, C_2) is a q-Markov COVER iff

$$M_{2q} = M_{1q} \qquad (2.4a)$$

$$M_{2q} \triangleq \Theta_{2p}[D_2, X_2, C_2^*] \ , \ \Theta_{2q} \triangleq [C_2^*, A_2^*C_2^*, \ldots, A_2^{q-1*}C_2^*] \ . \qquad (2.4b)$$

The matrix M_{1q} plays an important role in linear systems. Any realization (A_2, D_2, C_2) which preserves the first q-Markov parameters of (A_1, D_1, C_1) satisfies

$$M_{1q}\begin{bmatrix} I_{n_w} \\ 0 \end{bmatrix} = M_{2q}\begin{bmatrix} I_{n_w} \\ 0 \end{bmatrix}, \ I_{n_w} \triangleq \text{identity matrix } (n_w x n_w) \qquad (2.5)$$

and any realization which preserves the first q-output covariances satisfies

$$M_{1q}\begin{bmatrix} 0 \\ I_{n_y} \end{bmatrix} = M_{2q}\begin{bmatrix} 0 \\ I_{n_y} \end{bmatrix}. \qquad (2.6)$$

Hence, the stochastically equivalent realizations [1] satisfy (2.6) for $q = \infty$. Realizations preserving both q Markov parameters and q covariances satisfy (2.4). This implies that if the original system were described by state x, then

$$x = T\underline{x}\ ,\ \ \underline{x}^* \triangleq (\underline{x}_R^*, \underline{x}_T^*)\ ,\ \ x^* \triangleq (x_R^*\ x_T^*)\ ,$$

$$\begin{bmatrix} \dot{\underline{x}}_R \\ \dot{\underline{x}}_T \end{bmatrix} = \begin{bmatrix} A_R & A_{RT} \\ A_{TR} & A_T \end{bmatrix} \begin{bmatrix} \underline{x}_R \\ \underline{x}_T \end{bmatrix} + \begin{bmatrix} D_R \\ D_T \end{bmatrix} w \tag{2.7}$$

$$y = [C_R\ C_T] \begin{bmatrix} \underline{x}_R \\ \underline{x}_T \end{bmatrix}$$

and applying the projection (2.2) to the reduced model

$$\begin{aligned} A_2 &= \underline{L}A_1\underline{R} = P^{-1}[A_R + A_{RT}G]P = P^{-1}\hat{A}_2P \\ D_2 &= \underline{L}D_1 = P^{-1}D_R \qquad\quad = P^{-1}\hat{D}_2 \\ C_2 &= C_1\underline{R} = [C_R + C_TG]P \quad = \hat{C}_2P \end{aligned} \tag{2.8}$$

which has the transfer function

$$\begin{aligned} G_2(s) &= [C_R + C_TG][sI-(A_R + A_{RT}G)]^{-1}D_R \\ &= \hat{C}_2(sI-\hat{A}_2)^{-1}\,\hat{D}_2 = C_2(sI-A_2)^{-1}D_2\ , \end{aligned} \tag{2.9}$$

where (2.8) defines $(\hat{A}_2, \hat{D}_2, \hat{C}_2)$ to be the reduced model when $P = I$. The effect of P on the state covariance of the reduced model is clearly present,

$$X_2 = P^{-1}\hat{X}_2P^{-*}\ , \tag{2.10}$$

where $\hat{X}_2$ is the state covariance of $(\hat{A}_2, \hat{D}_2)$. Of course, P has *no* effect on the matrix M_{2q} and hence on our ability to match M_{1q}. To wit;

$$\begin{aligned} M_{2q} &= \Theta_{2q}[D_2, X_2C_2^*] = \hat{\Theta}_{2q}P[P^{-1}\hat{D}_2, P^{-1}\hat{X}_2P^{-*}P^*\hat{C}_2^*] \\ &= \hat{\Theta}_{2q}[\hat{D}_2, \hat{X}_2\hat{C}_2^*]\ . \end{aligned} \tag{2.11}$$

However, P plays a role in the choice of coordinates of the reduced model and hence we note its presence here. Coordinate transformations T on the higher dimensional model have no affect on M_{1q}. Hence,

$$M_{1q} = \Theta_{1q}[D_1, X_1C_1^*] = \hat{\Theta}_{1q}[\hat{D}_1, \hat{X}_1\hat{C}_1^*] = \hat{M}_{1q}$$

where

$$\hat{\Theta}_{1q} = \Theta_{1q}T,\ \hat{X}_1 = T^{-1}X_1T^{-*},\ \hat{D}_1 = T^{-1}D_1\ ,\ \hat{C}_1 = C_1T\ .$$

More specific results can be stated if we *begin* with the following special coordinates, which we define as q-normalized Hessenberg coordinates.

$$\hat{A}_1 = T^{-1}A_1T = \begin{bmatrix} A_R & A_{RT} \\ A_{TR} & A_T \end{bmatrix},$$

$$\hat{C}_1 = C_1T = [C_{11}\ 0\ \ 0] = [C_R\ C_T]\ ,\ |C_{11}| \neq 0 \qquad (2.12a)$$

$$\hat{D}_1 = T^{-1}D_1 = \begin{bmatrix} D_R \\ D_T \end{bmatrix}$$

$$A_R = \begin{bmatrix} A_{11} & A_{12} & 0 & 0 & 0 & \cdots & 0 \\ A_{21} & A_{22} & A_{23} & 0 & 0 & \cdots & 0 \\ A_{31} & A_{32} & A_{33} & A_{34} & 0 & \cdots & 0 \\ . & . & . & . & . & . & 0 \\ A_{q-1,1} & . & . & . & . & . & A_{q-1,q} \\ A_{q1} & \cdots & \cdots & \cdots & . & \cdots & A_{qq} \end{bmatrix},\ A_{RT} = \begin{bmatrix} 0 & \cdots & 0 \\ 0 & \cdots & 0 \\ 0 & \cdots & 0 \\ . & \cdots & . \\ . & \cdots & . \\ 0 & \cdots & 0 \\ A_{q,q+1} & \cdots & A_{q,n} \end{bmatrix}$$

A_{TR}, A_T have no structural constraints, save the property

$$\hat{X}_1 \triangleq T^{-1}X_1T^{-*} = I\ , \qquad (2.12b)$$

where

$$\text{rank } A_{i,i+1} = n_{i+1} \leq n_i\ ,\ \ i = 1, 2, \ldots, q-1\ . \qquad (2.12c)$$

$$\sum_{i=1}^{p} n_i = n\ ,\ n_1 = n_y\ ,\ A_{11}\epsilon \underline{R}^{n_1 x n_1},\ r = \sum_{i=1}^{q} n_i,\ A_R \epsilon \underline{R}^{r x r}\ .$$

To simplify the notation, we define

$$B_q \triangleq [A_{q,q+1}\ \ \cdots\ \ A_{q,n}]\ , \qquad (2.12d)$$

The form (2.12) may be constructed in steps (setting the row block of zeros after A_{12} in A on step 1 and the row blocks of zeros after A_{23} on step 2; etc.) to give the form (2.12). This form is slightly less restrictive than block Hessenberg (since rows of zeros are assigned after the blocks $A_{i,i+1}$ only up to i = q-1. The form (2.12) (*including* the $X_1 = I$ property) may be constructed by the algorithm in the Appendix of [3]. (However, we simply stop the block Hessenberg construction of [3] after step q-1, not wishing to require the entire matrix $\hat{A}_1 = T^{-1}A_1T$ to be block Hessenberg, only the first q-1 block of rows).

In these coordinates (2.12) note that if (A_R, D_R) are partitioned like so

$$A_R = \begin{bmatrix} A_{11} & \Gamma & 0 \\ \Psi & \Lambda & \Phi \\ A_{q1} & \Xi & A_{qq} \\ n_1 & n_2 + \cdots n_{q-1} & n_q \end{bmatrix} \quad \begin{matrix} n_1 \\ n_2 + \cdots + n_{q-1}, \\ n_q \end{matrix} \quad D_R = \begin{bmatrix} D_a \\ D_b \\ D_q \end{bmatrix} \tag{2.12e}$$

then the first q upper left blocks $n_1, n_2, \cdots n_q$ of

$$0 = \hat{A}_1\hat{X}_1 + \hat{X}_1\hat{A}_1^* + \hat{D}_1\hat{D}_1^*\ ,\ \hat{X}_1 = I \tag{2.12f}$$

satisfy

$$0 = A_{11} + A_{11}^* + D_aD_a^* \tag{2.12g}$$

$$0 = \Gamma + \Psi^* + D_aD_b^*\ ,\quad \Gamma \triangleq [A_{12}\ 0] \tag{2.12h}$$

$$0 = A_{q1}^* + D_aD_q^* \tag{2.12i}$$

$$0 = \Lambda + \Lambda^* + D_bD_b^* \tag{2.12j}$$

$$0 = \Phi + \Xi^* + D_bD_q^* \quad \Phi^* \triangleq [0\ A_{q-1,q}^*] \tag{2.12k}$$

$$0 = A_{qq} + A_{qq}^* + D_qD_q^* \tag{2.12l}$$

These expressions will be needed later.

Define the projection (2.8)

$$\underline{\hat{L}}\hat{A}_1\underline{\hat{R}} = A_2 \triangleq P^{-1}[A_R + BG]P = P^{-1}\hat{A}_2P\ ,\quad B \triangleq A_{RT} = \begin{bmatrix} 0 \\ B_q \end{bmatrix}$$

$$\underline{\hat{L}}\hat{D}_1 = D_2 \triangleq P^{-1}D_R = P^{-1}\hat{D}_2 \tag{2.13}$$

$$\hat{C}_1\underline{\hat{R}} = C_2 \triangleq [C_R + C_TG]P = \hat{C}_2P\ ,\ \hat{C}_2 = C_R + C_TG = [C_{11}\ 0]\ .$$

Lemma 2:

The coordinates (2.12) and the projection (2.13) have these properties:

$$\text{(i)}\quad \hat{\Theta}_{1q} = [\hat{\Theta}_{2q}\ 0]\ ,\quad \hat{\Theta}_{2q}^* \triangleq [\hat{C}_2^*, \hat{A}_2^*\hat{C}_2^*, \ldots, \hat{A}_2^{q-1*}\hat{C}_2^*] \tag{2.14}$$

$$\text{(ii)}\quad C_RA_R^iB = 0,\ i = 0, 1, \ldots, q-1 \tag{2.15}$$

(iii) $\hat{\Theta}_{2q}$ is independent of the choice G

Proof:

(i) By construction

$$\hat{\Theta}_{1q} = \begin{bmatrix} C \\ \cdot \\ \cdot \\ \cdot \\ CA^{q-1} \end{bmatrix} = \begin{bmatrix} d_1 & 0 & 0 & \cdots & 0 & 0 \\ x & d_2 & 0 & \cdots & 0 & 0 \\ x & x & d_3 & \cdots & 0 & 0 \\ \cdot & \cdot & \cdot & \cdots & \cdot & \cdot \\ \cdot & \cdot & \cdot & \cdots & \cdot & \cdot \\ \cdot & \cdot & \cdot & \cdots & \cdot & \cdot \\ x & \cdot & \cdot & \cdots & d_q & 0 \end{bmatrix} = [\hat{\Theta}_{2q}, 0]$$

where the matrices

$$d_2 \triangleq C_{11}A_{12} \;, \qquad d_1 = C_{11}$$

$$d_3 \triangleq C_{11}A_{12}A_{23}$$

$$\vdots$$

$$d_q \triangleq C_{11}A_{12}A_{23}A_{34} \;\cdots\; A_{q-1,q}$$

all have *full column rank* by property (2.12c). Hence the rank of $\hat{\Theta}_{1q}$ is equal to the rank of $\hat{\Theta}_{2q}$ which is $(n_1+n_2+\cdots n_q) \triangleq r$, the order of the reduced model. (ii) and (iii) both follow from

$$\hat{\Theta}_{2q} = \begin{bmatrix} [C_{11}\ 0] \\ [C_{11}\ 0][A_R+BG] \\ \cdot \\ \cdot \\ \cdot \\ [C_{11}\ 0][A_R+BG]^{q-1} \end{bmatrix} = \begin{bmatrix} [C_{11}\ 0] \\ [C_{11}\ 0][A_R] \\ \cdot \\ \cdot \\ \cdot \\ [C_{11}\ 0]\ A_R^{q-1} \end{bmatrix}$$

where the essential property is

$$C_R A_R^i B = [C_{11}\ 0\ 0\ 0] \begin{bmatrix} A_{11} & A_{12} & 0 & \cdots & 0 \\ A_{21} & A_{22} & A_{23} & \cdots & 0 \\ A_{31} & A_{32} & A_{33} & \cdots & . \\ . & . & . & \cdots & . \\ . & . & . & \cdots & A_{q-1,q} \\ A_{q1} & A_{q2} & A_{q3} & \cdots & A_{qq} \end{bmatrix}^i \begin{bmatrix} 0 \\ 0 \\ 0 \\ . \\ . \\ . \\ B_q \end{bmatrix} = 0 \quad ,$$

$i = 0, 1, \ldots, q\text{-}1$.

Lemma 3:

The projection defined by (2.13), (2.12) matches the first q Markov parameters for *any* choice of $\underline{G}$.

Proof:

From (2.7) we require

$$\hat{M}_{1q} \begin{bmatrix} I_{n_w} \\ 0 \end{bmatrix} = \hat{\Theta}_{1q}\hat{D}_1 = [\hat{\Theta}_{2q}, 0] \begin{bmatrix} D_R \\ D_T \end{bmatrix} = \hat{\Theta}_{2q} D_R = \hat{\Theta}_{2q}\hat{D}_2 = \hat{M}_{2q} \begin{bmatrix} I_{m_w} \\ 0 \end{bmatrix} . \tag{2.16}$$

where property (i) of lemma 2 is used. #

Thus, the matching of q Markov parameters is automatic (for all G) by the *structure* of (2.12). This now leaves only the task of matching covariances *by choice* of G.

Problem restatement: Find G such that $(A_R + BG, D_R)$ has state covariance $\hat{X}_2$ satisfying

$$\hat{\Theta}_{1q}\hat{X}_1\hat{C}_1^* = \hat{\Theta}_{2q}\hat{X}_2\hat{C}_2^* . \tag{2.17}$$

But from (2.14), (2.12) the left hand side becomes

$$\hat{\Theta}_{1q}\hat{X}_1\hat{C}_1^* = [\hat{\Theta}_{2q}, 0] \begin{bmatrix} X_a & X_b \\ X_c & X_d \end{bmatrix} \begin{bmatrix} C_{11}^* \\ 0 \end{bmatrix} = \hat{\Theta}_{2q} X_a C_{11}^* \tag{2.18}$$

where X_a is the upper left $r x n_y$ block of matrix $\hat{X}_1$. Now since the columns of $\hat{\Theta}_{2q}$ are linearly independent and the rows of C_{11}^* are linearly independent, we have $\hat{\Theta}_{1q}\hat{X}_1\hat{C}_1^* = \hat{\Theta}_{2q}\hat{X}_2\hat{C}_2^*$ or

$$\hat{\Theta}_{2q} X_a C_{11}^* = \hat{\Theta}_{2q}\hat{X}_2\hat{C}_2^* = \hat{\Theta}_{2q}[\hat{X}_{R1}, \hat{X}_{R2}] \begin{bmatrix} C_{11}^* \\ 0 \end{bmatrix} = \hat{\Theta}_{2q}\hat{X}_{R1} C_{11}^* \tag{2.19}$$

iff

$$\hat{X}_{R1} = X_a \,. \tag{2.20}$$

This proves the following.

Lemma 4:

The projection (2.13) from coordinates (2.12) yields a q-Markov COVER of

$(\hat{A}_1, \hat{D}_1, \hat{C}_1)$ iff there exists a G such that $(A_R + BG, D_R)$ has state covariance $\hat{X}_2$ which takes on the preassigned value for the first n_y columns (and hence also rows, by symmetry),

$$\hat{X}_{R1} = X_a \,.$$

3.0 A Covariance Assignment Problem

We now seek to solve the problem suggested by Lemma 4. This is recognized as a mathematical problem in covariance assignment using state feedback control. Such problems were introduced in [9, 10] for *control* design purposes, and we shall need one of those results here for our *model reduction* problem.

Define E_B such that

$$E_B^* B B^+ E_B = \begin{bmatrix} I_{r_b} & 0 \\ 0 & 0 \end{bmatrix}, \quad E_B^* E_B = I_r \,, \ \text{rank } B = r_b \tag{3.1}$$

where B^+ is the Moore-Penrose inverse of B, and I_r denotes the identity matrix of size rxr. Now the needed result from [9] is

Theorem 2: [9]

Let $(E_B^* \, Q \, E_B)_{22}$ *denote the lower right* $(r-r_b)x(r-r_b)$ *block of matrix* $E_B^* \, Q \, E_B$, *and define*

$$Q \triangleq A_R \overline{X}_2 + \overline{X}_2 A_R^* + D_R D_R^* \,. \tag{3.2}$$

There exists a G *such that* $(A_R + BG, D_R)$ *has a specified state covariance*

$\hat{X}_2 = \overline{X}_2 > 0$ *iff* $(A_R + BG, D_R)$ *is controllable and*

$$(E_B^* \, Q \, E_B)_{22} = 0 \,. \tag{3.3}$$

The class of all nontrivial solutions (define trivial ones such that BG = 0) which exist are given by

$$G = -\frac{1}{2}B^+(Q+E_BSE_B^*)\overline{X}_2^{-1}\ ,\ S = \begin{bmatrix} S_{11} & S_{12} \\ -S_{12}^* & 0 \\ r_b & r-r_b \end{bmatrix} \tag{3.4}$$

subject to $(A_R + BG, D_R)$ *controllable, where*

$$S_{12} = -(E_B^*QE_B)_{12}\ ,\ E_B^*QE_B = \begin{bmatrix} ()_{11} & ()_{12} \\ ()_{12}^* & ()_{22} \\ r_b & r-r_b \end{bmatrix}, \tag{3.5}$$

and where S_{11}, *is an arbitrary skew symmetric matrix. Note that* (A_R+BG) *will be asymptotically stable if* (A_R+BG, D_R) *is controllable since* $\overline{X}_2 > 0$.

Theorem 2 holds for any (A_R, B, D_R), but we wish now to specialize Theorem 2 for our problem. This is accomplished by the use of Lemma 4, which *requires* the choice of $\overline{X}_2$ in our use of Theorem 2 to be

$$\overline{X}_2 = \begin{bmatrix} I_{n_1} & 0 \\ 0 & X_f \end{bmatrix},\ X_f = X_f^* > 0 \tag{3.6}$$

where X_f is arbitrary. In other words, in the special coordinates (2.12), X_a in (2.20) is

$$X_a = \begin{bmatrix} I_{n_1} \\ 0 \end{bmatrix}. \tag{3.7}$$

Next, to simplify our presentation assume rank $B = r_b = n_q$. (This assumption is not necessary for application of the concepts). Then

$$BB^+ = \begin{bmatrix} 0 & 0 \\ 0 & B_qB_q^+ \end{bmatrix} = \begin{bmatrix} 0 & 0 \\ 0 & I_{n_q} \end{bmatrix} \tag{3.8}$$

$$E_B = \begin{bmatrix} 0 & I_{r-n_q} \\ I_{n_q} & 0 \end{bmatrix} \begin{matrix} n_1 + \cdots + n_{q-1} = r-n_q \\ n_q \end{matrix} \tag{3.9}$$

and (3.3) becomes

$$Q_{11} \triangleq (A_R\overline{X}_2 + \overline{X}_2A_R^* + D_RD_R^*)_{11} = 0\ . \tag{3.10a}$$

where $Q_{11} \in \mathbb{R}^{(r-n_q)x(r-n_q)}$. Now in order to exploit the *freedom* in the choice of X_f, we proceed to use the partitioned notation of (2.12e) to write

$$A_R\overline{X}_2 = \begin{bmatrix} A_{11} & \Gamma & 0 \\ \Psi & \Lambda & \Phi \\ A_{q1} & \Xi & A_{qq} \end{bmatrix} \begin{bmatrix} I_{n_1} & 0 & 0 \\ 0 & X_{f11} & X_{f12} \\ 0 & X_{f12}^* & X_{f22} \end{bmatrix} \begin{matrix} r-n_q \\ n_q \end{matrix} \tag{3.11}$$

where $\Phi^* = [0\ A_{q-1,q}^*]$.

Hence, (3.10) becomes these three partitioned eqs.

$$0 = A_{11} + A_{11}^* + D_a D_a^* \tag{3.10b}$$

$$0 = \Gamma\, X_{f11} + \Psi^* + D_a D_b^* \tag{3.10c}$$

$$0 = \Lambda X_{f11} + X_{f11}\Lambda^* + \Phi X_{f12}^* + X_{f12}\Phi^* + D_b D_b^* \tag{3.10d}$$

But from (2.12g), (3.10b) is automatically satisfied and (2.12h) reduces (3.10c) to

$$0 = \Gamma\, [X_{f11} - I] \tag{3.12a}$$

and (2.12j) reduces (3.10d) to

$$0 = \Lambda[X_{f11} - I] + [X_{f11} - I]\Lambda^* + \Phi X_{f12}^* + X_{f12}\Phi^*\ . \tag{3.12b}$$

Hence (3.10a) holds iff (3.12) holds. Using (3.9) and (3.11), (3.5) becomes

$$S_{12} = -Q_{12}^*\ . \tag{3.13 }$$

Finally, (3.4) reduces to (using (3.9), (3.11))

$$G = -\frac{1}{2} B_q^+[Q_{22} + S_{11}][X_{f22} - X_{f12}^* X_{f11}^{-1} X_{f12}]^{-1}[O_{n_1}, -X_{f12}^* X_{11}^{-1}, I_{n_q}] \tag{3.14a}$$

$$Q_{22} = \Xi X_{f12} + X_{f12}^*\Xi^* + A_{qq}[X_{f22} - I] + [X_{f22} - I]A_{qq}^*\ . \tag{3.14b}$$

These results may now be summarized. In the following theorem we consider the system $(\hat{A}_1, \hat{D}_1, \hat{C}_1, \hat{X}_1)$ described by (2.12).

Definition:

An "admissible" state covariance of a q-Markov COVER is defined by Eqs. (3.6) *and* (3.12) *described in terms of coordinates* (2.12).

Theorem 3:

A q-Markov COVER $(\hat{A}_2, \hat{D}_2, \hat{C}_2, \hat{X}_2)$ *of* $(\hat{A}_1, \hat{D}_1, \hat{C}_1, \hat{X}_1)$ *exists if* $\hat{X}_2$ *is "admissible". If* $\hat{X}_2 = \overline{X}_2$ *is admissible then many q-Markov COVERs may exist satisfying*

$$\hat{A}_2 = A_R + BG \ , \hat{D}_2 = D_R \ , \ \hat{C}_2 = C_R \ , \tag{3.15a}$$

$$\hat{X}_2 = \overline{X}_2 = \begin{bmatrix} I_{n_1} & 0 \\ 0 & X_f \end{bmatrix}, X_f = \begin{bmatrix} X_{f11} & X_{f12} \\ X_{f12}^* & X_{f22} \end{bmatrix} \begin{matrix} r-n_1-n_q \\ n_q \end{matrix} \tag{3.15b}$$

where G *is given by* (3.14), *for arbitrary values of the skew-symmetric matrix* $S_{11} = -S_{11}^*$, *and for any admissible values of* $X_{f11}, X_{f12}, X_{f22}$.

Corollary 1:

The 1-Markov COVER is unique within a similarity transformation P, *and is given by* (3.15a) *with* $\hat{X}_2 = \overline{X}_2 = I_{n_1}$, $G = 0$, *and the reduced model is stable.*

Corollary 2:

For stable systems q-Markov COVERs always exist and the choice, $\hat{X}_2 = \overline{X}_2 = I_r$ *is always admissible.*

Corollary 3:

If $\lambda_i[A_{qq}] + \lambda_j[A_{qq}] \neq 0$ *for all* i,j *then q-Markov COVERs with the admissible choice* $\hat{X}_2 = \overline{X}_2 = I_r$, *are given by* (3.15a,b) *with*

$$G = -\frac{1}{2} B_q^+ S_{11}[0, 0, I_{n_q}] \tag{3.16}$$

and the reduced model is stable by choosing S_{11} *such that* $(A_R + BG, D_R)$ *is controllable. This is always possible.*

Proofs:

Corollary 1 follows by setting $r = n_1 = n_y$ in Theorem 4. Corollary 2 follows by noting that $X_f = I$ satisfies the admissible conditions (3.12). Corollary 3 is verified by substituting $X_f = I$ in (3.14). Since (A_R, D_R) is controllable by structure of (2.12) there is at least one choice $(S_{11} = 0)$ which makes $(A_R + BG, D_R)$ controllable. #

Theorem 3 provides the *explicit* structure of the q-Markov COVER problem and allows the designer *much* freedom in their construction. Specifically, the free parameters (S_{11}, X_f, P) are available to accomplish other objectives. We shall need this freedom to obtain a q-Markov COVER with desirable robustness properties. This is the subject of a future paper.

Example 1: Find all 1-Markov COVERs of

$$G(s) = \frac{\alpha s+1}{(s+\beta)(s+\gamma)}, \beta > 0 , \gamma > 0$$

Solution

Begin with the representation $(\hat{A}_1, \hat{D}_1, \hat{C}_1)$

$$\hat{A}_1 = \begin{bmatrix} -\frac{\alpha^2\beta\gamma(\beta+\gamma)}{1+\alpha^2\beta\gamma} & \frac{\sqrt{\beta\gamma}(1+\alpha(\alpha\beta\gamma-\beta-\gamma))}{1+\alpha^2\beta\gamma} \\ -\frac{\sqrt{\beta\gamma}(1+\alpha(\alpha\beta\gamma+\beta+\gamma))}{1+\alpha^2\beta\gamma} & \frac{-(\beta+\gamma)}{1+\alpha^2\beta\gamma} \end{bmatrix}$$

$$\hat{D}_1 = \begin{bmatrix} \alpha\sqrt{\frac{2\beta\gamma(\beta+\gamma)}{1+\alpha^2\beta\gamma}} \\ \sqrt{\frac{2(\beta+\gamma)}{1+\alpha^2\beta\gamma}} \end{bmatrix}$$

$$\hat{C}_1 = \begin{bmatrix} \sqrt{\frac{1+\alpha^2\beta\gamma}{2\beta\gamma(\beta+\gamma)}} & 0 \end{bmatrix}$$

since in this representation $\hat{X}_1 = I$ and (2.12) is satisfied. From Corollary 1, the 1-Markov COVER is unique to within a similarity transformation:

$$B = B_q = \frac{\sqrt{\beta\gamma}(1+\alpha^2\beta\gamma-\alpha(\beta+\gamma))}{1+\alpha^2\beta\gamma}, \quad G = 0, \text{ (Corollary 1)}$$

$$A_2 = \hat{A}_2 = -\frac{\alpha^2\beta\gamma(\beta+\gamma)}{1+\alpha^2\beta\gamma}, \quad D_2 = P^{-1}\hat{D}_2 = P^{-1}\alpha\sqrt{\frac{2\beta\gamma(\beta+\gamma)}{1+\alpha^2\beta\gamma}},$$

$$C_2 = \hat{C}_2P = P\sqrt{\frac{1+\alpha^2\beta\gamma}{2\beta\gamma(\beta+\gamma)}}$$

$$\hat{X}_2 = \overline{X}_2 = 1, \; X_2 = P^{-2} \text{ for any } P \neq 0 .$$

For existence we require (3.12), which is automatically satisfied because $\overline{X}_2 = I_{n_1}$ and Γ, Λ and X_f are all void in this problem. The transfer function of the 1-Markov COVER is

$$G_R(s) = \begin{bmatrix} \frac{\alpha}{s+\frac{\alpha^2\beta\gamma(\beta+\gamma)}{1+\alpha^2\beta\gamma}} \end{bmatrix}$$

Example 2:

Find all 1-Markov COVERs of stable third order systems.

Solution:

All stable third order systems may be transformed to coordinates (2.12) in which case the parameters are

$$\hat{A}_1 = \begin{bmatrix} -d_1^2 & -2d_1d_2-d_4 & 0 \\ d_4 & -d_2^2 & -2d_2d_3-d_5 \\ -2d_1d_3 & d_5 & -d_3^2 \end{bmatrix}, \quad \hat{D}_1 = \begin{bmatrix} d_1 \\ d_2 \\ d_3 \end{bmatrix} \sqrt{2}$$

$$\hat{C}_1 = [d_6 \; 0 \; 0] \; , \quad \hat{X}_1 = I \; .$$

From Corollary 1, all 1-Markov COVERs are similar to

$$\hat{A}_2 = -d_1^2 \; , \quad \hat{D}_2 = d_1 \sqrt{2} \; , \quad \hat{C}_2 = d_6, \quad \hat{X}_2 = 1$$

Example 3:

Find all 2-Markov COVERs of stable third order systems.

Solution:

Existence is guaranteed by (3.12), since X_{f22}, X_{f12}, Ξ and Φ are void and the admissible $\overline{X}_2 = \begin{bmatrix} 1 & 0 \\ 0 & X_{f11} \end{bmatrix}$, where $X_{f11} = 1$ from (3.12b). From (3.14),

$$Q_{22} = 2A_{qq}(X_{f11}-1) = 0 \; , \quad S_{11} = 0 \;\; \text{(since } S_{11} \text{ is scalar)}$$

Hence G = 0, and

$$\hat{A}_2 = A_R + BG = \begin{bmatrix} -d_1^2 & -2d_1d_2-d_4 \\ d_4 & -d_2^2 \end{bmatrix}$$

$$\hat{D}_2 = \begin{bmatrix} d_1 \\ d_2 \end{bmatrix} \sqrt{2} \; , \quad \hat{C}_2 = (d_6 \; 0) \; , \quad \hat{X}_2 = \begin{bmatrix} 1 & 0 \\ 0 & 1 \end{bmatrix}$$

The transfer function of the 2-Markov COVER is

$$G_R(s) = \hat{C}_2(sI-\hat{A}_2)^{-1}\hat{D}_2 = \frac{\sqrt{2}[(s+d_2^2)d_1 + d_2(2d_1d_2+d_4)]d_6}{(s+d_1^2)(s+d_2^2)+(2d_1d_2+d_4)d_4} \; .$$

The transfer function of the original system is $G(s) = N(s)/D(s)$, where

$$N(s) \triangleq \sqrt{2}\{[(s+d_2^2)(s+d_3^2) \; + \; d_5(2d_2d_3+d_5)]d_1 - (s+d_3^2)d_2(2d_1d_2+d_4)\}d_6 \; +$$

$$\sqrt{2}d_3(2d_1d_2+d_4)(2d_2d_3+d_5)\ d_6$$

$$D(s) \triangleq (s+d_1^2)(s+d_2^2)(s+d_3^2)+(s+d_3^2)(d_4+2d_1d_s)d_4+(s+d_1^2)(d_5+2d_2d_3)d_5+$$
$$2d_1d_3(sd_1d_2+d_4)(2d_2d_3+d_5)$$

Finally, in this Section we must show how the class of q-Markov COVERs defined in this paper relates to the specific q-Markov COVER derived earlier in [2].

Theorem 4:

The particular q-Markov COVER of $(\hat{A}_1, \hat{D}_1, \hat{C}_1)$ *defined by the projection*

$$\hat{A}_2 = \underline{L}\hat{A}_1\underline{R}\ ,\ \hat{D}_2 = \underline{L}\hat{D}_1\ ,\ \ \hat{C}_2 = \hat{C}_1\underline{R} \tag{3.17a}$$

$$\underline{L} \triangleq U_1^*\hat{\Theta}_{1q}\ ,\ \ \underline{R} \triangleq \hat{X}_1\ \underline{L}^*(\underline{L}\hat{X}_1\underline{L}^*)^{-1} \tag{3.17b}$$

$$\hat{\Theta}_{1q} = [U_1\ U_2]\begin{bmatrix}\Sigma & 0\\ 0 & 0\end{bmatrix}\begin{bmatrix}V_1^*\\ V_2^*\end{bmatrix} \tag{3.17c}$$

is related by a similarity transformation P *to that special q-Markov COVER in Theorem 3 which is associated with the values* $X_f = I$, $G = 0$. *The similarity transformation that relates them is* $P = [U_1^*\ \hat{\Theta}_{2q}]^{-1}$.

Proof:

It is proven in [2] that (3.17) is related by a similarity transformation P to a *truncation* of the cost decoupled coordinates (2.12). A *truncation* of (2.12) is equivalent to (2.13) with $G = 0$. To show the transformation write the singular value decomposition of $\hat{\Theta}_{1q}$. From Lemma 2

$$\hat{\Theta}_{1q} = [\hat{\Theta}_{2q}, 0] = [U_1\ U_2]\begin{bmatrix}\Sigma & 0\\ 0 & 0\end{bmatrix}\begin{bmatrix}V_1^*\\ V_2^*\end{bmatrix}$$

and hence by defining $P^{-1} \triangleq U_1^*\hat{\Theta}_{2q}$, (3.17b) becomes

$$\underline{L} = U_1^*\hat{\Theta}_{1q} = \Sigma V_1^* = U_1^*[\hat{\Theta}_{2q}, 0] = P^{-1}[I\ 0]$$

$$\underline{R} = \begin{bmatrix}I\\ 0\end{bmatrix}P^{-*}[P^{-1}[I\ 0]\begin{bmatrix}I\\ 0\end{bmatrix}P^{-*}]^{-1} = \begin{bmatrix}I\\ 0\end{bmatrix}P$$

which is precisely the projection (2.2) used to develop the results of Section 4 if $G = 0$ and $T_o = I$ (since we started with coordinates (2.12)). #

Conclusions

A class of reduced order models is described which preserve the first q-Markov parameters and the first q-Output Covariances. Earlier published results are shown to be special cases obtained by a similarity transformation of these more general results. The explicit formulas describing the freedom in the reduced models may be used to assign still other properties to the reduced order model, such as the treatment of non-white noise inputs and robustness with respect to parameter variations. These topics will follow in future work.

Acknowledgments:

Part of this research was sponsored by NSF INT-841244 under the U.S.-Australia Cooperative Program.

References

1. B.D.O. Anderson, The Inverse Problem of Stationary Covariance Generation, *J. Statistical Physics,* 1, 1969, 133-147.
2. A. Yousuff, R.E. Skelton, D.A. Wagie, "Linear System Approximation via Covariance Equivalent Realizations," *J. Math. Analysis and Applications,* Vol. 106, No. 1, Feb. 1985, pp. 91-115.
3. A. Yousuff and R.E. Skelton, "Controller Reduction by Component Cost Analysis," *IEEE Trans. Auto. Control,* Vol. AC-29, No. 6, June 1984, pp. 520-530.
4. G.A. Latham, B.D.O. Anderson, "Frequency Weighted Optimal Hankel-Norm Approximation of Stable Transfer Functions," *Systems and Control Letters,* 5 (1985), 229-236.
5. Dale Enns, "Model Reduction for Control System Design," Ph.D. Thesis, EE Dept. Stanford Univ., 1984.
6. B.D.O. Anderson, "Linear Controller Approximation: A Method with Bounds," *IFAC Workshop on Model Error Concepts and Compensation,* Boston, Pergammon Press, June 1985.
7. E. Tse, J. Medanic and W.R. Perkins, "Generalized Hessenberg Transformations for Reduced Order Modeling of Large Scale Systems," *Int. J. Control,* 27, 1978, 493-512.
8. S. Barnett and C. Storey, "Analysis and Synthesis of Stability Matrices," *J. Diff. Eq.,* 3 (1967) 414-422.
9. A. Hotz and R.E. Skelton, "A Covariance Control Theory," *IEEE CDC,* December 1985, Fort Lauderdale.
10. E. Collins and R.E. Skelton, "Covariance Control of Discrete Systems," *IEEE CDC, Dec. 1985,* Fort Lauderdale.

3

REDUCED-ORDER MODELLING OF STOCHASTIC PROCESSES WITH APPLICATIONS TO ESTIMATION

Uday B. Desai

Department of Electrical and Computer Engineering
Washington State University, Pullman, WA 99164, U.S.A.

ABSTRACT

The theory of canonical correlation analysis is used to develop a new algorithm for obtaining reduced-order continuous-time stochastic models. By establishing an explicit relationship between canonical correlation coefficient - canonical vector pair, and solutions to algebraic Riccati equations, a transformation approach for stochastic model reduction is developed. It is also established that the reduced-order model as well as its inverse are asymptotically stable. This result along with the relationship between canonical variables and the linear least-squares estimate of the state vector leads to the new algorithm for designing reduced-order Kalman-Bucy filters. Finally, the CCA approach is extended to the hierarchical estimation problem. Using oblique projections and the interaction between appropriately defined past and future residuals, a reduced-order hierarchical estimator is developed.

INTRODUCTION

Over the years a great deal of research has gone into modelling Gaussian stochastic processes (see for example Priestly [1]). Some of the most commonly used linear models are auto-regressive models, moving average models, a combination of these two--autoregressive moving average (ARMA) models, and state space models. All the four models have proved very useful in a variety of problems arising in estimation and control theory, image processing, geophysics, and various other control and signal processing problems (see for example Kailath [2], Willsky [3], Burg [4]).

A number of modelling algorithms often try to obtain models which match, say the covariance data exactly, and in the process obtain high dimensional models. Now it is well known that the computational load in the implementation of filtering or control algorithms is determined by the dimension of the model, and for high dimensional models this load could become unmanageable. Thus the need for good reduced-order models, which may not provide an exact match for the given covariance data, but they do provide an "acceptable" approximate match. It is this reduced-order modelling problem along with applications to estimation problems that is addressed in this chapter.

We shall consider only the reduced-order state space modelling problem. The main reason for focusing our attention on state space models is that having obtained the state space model (reduced or otherwise), its transfer function then gives the ARMA model. Naturally, if one is interested in obtaining only the reduced-order ARMA model, then this would be a rather round about approach. Thus, currently we are investigating the extension of the stochastic model reduction technique developed in this paper to ARMA model reduction.

The basic tool for our model reduction approach is the theory of canonical correlation analysis (CCA). The CCA approach was first used for the stochastic realization problem by Akaike [5]. Motivated by his work, model reduction techniques for *discrete-time* stochastic systems were developed by Fujishige et al [6] who used the principal component analysis of stochastic processes, and by Baram [7], Larimore [8], Desai et al. [9]-[11] who used the canonical correlation analysis of stochastic processes. Using the work of Rao [12] on principal component extraction of instrumental variables, Arun and Kung [13] also developed a model reduction scheme for discrete time stochastic processes. Vaccaro [14] developed a stochastic model reduction scheme essentially using the deterministic model reduction approach of Moore [15]. From a stochastic standpoint, the main difference between these algorithms lies in the predictor spaces and their basis. Skelton and his co-workers [18]-[19] have developed model reduction methods for stochastic systems using the concept of cost equivalent and covariance equivalent realizations. More recently Bernstein and Hayland [20] have developed reduced-order models using the optimal projection/maximum entropy approach to stochastic modelling.

Among the different model reduction methods mentioned above the one reported in [9] is such that it *immediately* gives the reduced-order Kalman filter. This is because the reduced-order model is essentially a "subsystem" of a particular innovations representation (referred to as the balanced stochastic realization) of y(•), and moreover the inverse of this reduced realization is asymptotically stable. Thus one can use the inverse as a reduced-order Kalman filter.

In this chapter the CCA approach of [9]-[11] is extended to continuous-time stochastic processes and used to obtain reduced-order: (i) stochastic models, (ii) Kalman-Bucy filters, and (iii) hierarchical estimators. In particular, one of the goals of this chapter is to obtain a *direct* method for stochastic model reduction; i.e. a transformation which will directly transform, say, a given high dimensional continuous-time stochastic model, to a special form such that an appropriate "subsystem" of this special form will immediately give the reduced-order model. Furthermore, the inverse of the reduced-order model should give the reduced-order Kalman-Bucy (K-B) filter. The direct approach has the added advantage that one is able to track the physical significance of the original state variables. As will be seen in the next two sections, the key steps in achieving the above objectives are the canonical decomposition of the Hankel operator (equation (6)), and the relationship between this decomposition and the solutions to two algebraic Riccati equations (Lemmas 1-3). It is to be noted that the results on continuous-time stochastic model reduction and reduced-order K-B filters were first presented in [16] and [17].

Next a hierarchical estimator based on oblique projections is considered. Using the residual canonical correlation analysis approach the reduced-order hierarchical estimator is developed. We refer to this approach as residual canonical correlation analysis because it is based on the interaction between appropriately defined past and future residuals.

Paper Outline

First the CCA for continuous-time stochastic processes is developed, using which the reduced-order (approximate) stochastic realization algorithm is obtained. Next, a direct approach to stochastic model reduction leading to reduced K-B filters is developed. Having laid the CCA framework for reduced-order centralized estimation, next its applications to reduced-order hierarchical estimation is presented. The chapter is concluded with some remarks on other applications of the CCA approach, use of optimal Hankel norm approximation for stochastic model reduction [21], and some future research directions on this subject.

CANONICAL CORRELATION ANALYSIS FOR CONTINUOUS-TIME STOCHASTIC PROCESSES AND THE REALIZATION ALGORITHM

Let y(t) be a continuous-time zero mean, stationary, purely non-deterministic stochastic process of dimension p x 1, generated by the state model

$$\dot{\bar{x}}(t) = \bar{F}\bar{x}(t) + \bar{J}\bar{v}(t) \quad , \quad y(t) = \bar{H}\bar{x}(t) + \bar{v}(t) \tag{1}$$

where $\bar{x}(\cdot)$ is an nx1 state vector process with $E\bar{x}(t)\bar{x}'(t) = \Pi$; $\bar{v}(\cdot)$ is a zero mean Gaussian vector process of dimension $n \geq p$ with $E\bar{v}(t)\bar{v}'(s) = I\delta(t-s)$, $E\bar{x}(s)\bar{v}'(t) = 0$ for $t > s$. Primes will be used to denote matrix transpose. The objective of this section is to approximate (1) by a reduced-order model, such that the covariance of the reduced model "adequately" approximates the covariance of y(t) which is

$$R(t) = E[y(t+\tau)y'(\tau)] = \overline{H}e^{\overline{F}t}\overline{G}1(t) + \overline{G}'e^{-\overline{F}'t}\overline{H}'1(-t) + I\delta(t) = R'(-t) \qquad (2)$$

where $\overline{G} = \Pi\overline{H}' + \overline{J}$, and $1(\cdot)$ is the unit step function. We assume that $(\overline{F},\overline{G},\overline{H})$ is a minimal triple.

In order to achieve this objective we first investigate the continuous-time stochastic realization problem. For the realization problem the only data we have is R(t) the covariance of $y(\cdot)$. Note that the factorization of R(t) as depicted in (2) (or in a semi-separable form) is not available. A part of the realization problem is to obtain this factorization.

Canonical Decomposition

Let H(y) be the Hilbert space of random variables generated by the elements $\{y(s), s\epsilon \mathbf{R}\}$. Let the past space $H_t^-(y)$ and the future space $H_t^+(y)$ be respectively the closed linear hulls of $\{y(s), s < t\}$ and $\{y(s), s \geq t\}$. Note that $H_t^-(y) \vee H_t^+(y) = H(y)$. It can be shown [24]-[28] that a finite dimensional realization for $y(\cdot)$ exists if and only if the subspace obtained by projecting $H_t^+(y)$ (or $H_t^-(y)$) onto $H_t^-(y)$ (or $H_t^+(y)$) is finite dimensional. In fact, the state variable realization is obtained by picking an appropriate basis in either of the above subspaces. For the order reduction problem we are interested in a particular basis obtained from the canonical variables corresponding to $\{y(s), s < t\}$ and $\{y(s), s \geq t\}$. Towards this end, following [23], we first find the canonical vectors by solving the CCA problem

$$\int_t^{\infty} E[y(s_1)y'(\lambda)]u_i(t_- -\lambda)d\lambda - \sigma_i \int_{-\infty}^{t_-} E[y(s_1)y'(\mu)]v_i(t_- -\mu)d\mu = 0$$

$$-\sigma_i \int_t^{\infty} E[y(s_2)y'(\lambda)]u_i(t_- -\lambda)d\lambda + \int_{-\infty}^{t_-} E[y(s_2)y'(\mu)]v_i(t_- -\mu)d\mu = 0 \qquad (3)$$

where $s_1 < t$, and $s_2 \geq t$. Using a simple change of variable, (3) can be expressed in a more convenient and an illustrative form as

$$\int_{-\infty}^{0_-} R(\mu-\tau)u_i(\mu)d\mu - \sigma_i \int_0^{\infty} R(\lambda-\tau)v_i(\lambda)d\lambda = 0$$

$$-\sigma_i \int_{-\infty}^{0_-} R(\mu-\xi)u_i(\mu)d\mu + \int_0^{\infty} R(\lambda-\xi)v_i(\lambda)d\lambda = 0 \qquad (4)$$

where $\xi < 0$, $\tau \geq 0$. In the above $u_i(\cdot)$, and $v_i(\cdot)$ have dimension px1 and are referred to as the ***canonical vectors*** for the Hankel operator $R(\tau-\xi)$, while σ_i is the ***canonical correlation coefficient***. The presentation of equations (3) and (4) is formal since we are using quantities like 0_- and t_-. One way to avoid this would be to work with abstract operators operating on $H_t^+(y)$ and $H_t^-(y)$. Let Pr_+ and Pr_- be respectively the operators of orthogonal projection onto $H_t^+(y)$, and $H_t^-(y)$, then σ_i's are the square roots of the eigenvalues of the operator $Pr_+ \, Pr_-$ (or Pr_-Pr_+) and the canonical variables (to be defined later (8) in terms of the canonical vectors) will be the eigenvectors of Pr_+Pr_- and Pr_-Pr_+ restricted to $H_t^+(y)$ and $H_t^-(y)$ respectively. Our objective is to get an explicit representation of the canonical variables and the canonical vectors, so that an explicit algorithm for stochastic model reduction can be obtained. As such, we work with equations of type (4), and also because the final result will not be affected by this formalism.

Note that in an obvious operator notation the CCA problem (4) can be expressed as

$$R'_{\pm}u_i - \sigma_i R_+ v_i = 0$$

$$-\sigma_i R_- u_i + R_{\pm} v_i = 0 \tag{5}$$

If R_+ and R_- are the identity operators, then (5) reduces to the usual singular value decomposition of the Hankel operator $R_{\pm}$.

The stochastic realization problem is that of obtaining a Markovian representation of type (1) for $y(\cdot)$ given $R(\cdot)$. Assume that a finite dimensional (n) realization exists for $y(\cdot)$, then σ_i will be non-zero only for $i = 1, \ldots, n$. Let $\Sigma = \text{diag}\,[\sigma_1, \ldots, \sigma_n]$, $1 \geq \sigma_1 \geq \cdots \geq \sigma_n > 0$, $U(\mu) = [u_1(\mu), \ldots, u_n(\mu)]$, $V(\lambda) = [v_1(\lambda), \ldots, v_n(\lambda)]$ then the canonical vectors and the canonical variables satisfy

$$\int_{-\infty}^{0_-}\int_{-\infty}^{0_-} U'(\xi)R(\mu-\xi)U(\mu)d\mu d\xi = I \tag{5a}$$

$$\int_0^{\infty}\int_0^{\infty} V'(\lambda)R(\tau-\lambda)V(\tau)d\tau d\lambda = I \tag{5b}$$

$$\int_0^{\infty}\int_{-\infty}^{0_-} U'(\xi)R(\tau-\xi)V(\tau)d\xi d\tau = \Sigma, \quad 0 < \sigma_i \leq 1 \tag{5c}$$

and the ***canonical decomposition*** for the Hankel operator $R(\tau-\xi)$ is given by

$$R(\tau-\xi) = \int_{-\infty}^{0_-} R(\mu-\xi)U(\mu)d\mu\Sigma\int_0^{\infty} V'(\lambda)R(\tau-\lambda)d\lambda = O(\xi)C(\tau) \tag{6}$$

where $\tau \geq 0$, $\xi < 0$, and $0(\cdot)$, $C(\cdot)$ are defined by

$$O(\xi) = \int_{-\infty}^{0_-} R(\mu-\xi)U(\mu)d\mu\Sigma^{1/2} \triangleq He^{-F\xi} \tag{7a}$$

$$C(\tau) = \Sigma^{1/2}\int_{0}^{\infty} V'(\lambda)R(\tau-\lambda)d\lambda \triangleq e^{F\tau}G \tag{7b}$$

where F, G, H are respectively n x n, n x p and p x n matrices, whose computation is described below equation (20). The pair (F,G) is completely controllable and the pair (F, H) is completely observable. Next the canonical variables $\alpha(t_-)$ and $\beta(t)$ are defined by

$$\alpha(t_-) = \int_{0}^{\infty} V'(\lambda)y(t_- -\lambda)d\lambda, \quad \beta(t) = \int_{-\infty}^{0_-} U'(\mu)y(t-\mu)d\mu \tag{8}$$

Using the identities (5) it is easily established that

$$E\alpha(t_-)\alpha'(t_-) = I = E\beta(t)\beta'(t), \text{ and } E\alpha(t_-)\beta'(t) = \Sigma$$

Moreover $\alpha(t_-)$ and $\beta(t)$ are the basis for $Pr_-[H_t^+(y)]$ and $Pr_+[H_t^-(y)]$ respectively. The mutual information between $H_t^-(y)$ and $H_t^+(y)$ is given by

$$I(\alpha,\beta) = -\frac{1}{2}\sum_{i=1}^{n} \log(1-\sigma_i^2) \tag{9}$$

State Vector and its Dynamics

The state vector $x_*(t_-)$ is defined using (8) as a weighted linear combination of $\alpha(t_-)$, namely

$$x_*(t_-) = \Sigma^{1/2}\alpha(t_-) = \Sigma^{1/2}\int_{0}^{\infty} V'(\lambda)y(t_- -\lambda)d\lambda \tag{10}$$

and from (8) it is easily seen that $Ex_*(t_-)\, x_*'(t_-) = \Sigma$. Note that the state vector is also a basis for $Pr_-[H_t^+(y)]$ and this leads to a forwards realization for $y(\cdot)$. On the other hand if the state vector is defined as a basis for $Pr_+[H_t^-(y)]$ then we will obtain a backwards realization. Subsequently in the section on the transformation approach we shall use this to develop the balanced stochastic realization.

The forwards innovations process $\nu(t)$ for $y(t)$

$$\nu(t) = y(t) - Pr_-[y(t)] = y(t) - Pr[y(t)/H(x_*(t_-))] \tag{11}$$

where $H(x_*(t_-)) = Pr_-(H_t^+(y))$ is the subspace of $H_t^-(y)$ spanned by $\{x_{*i}(t_-),\ i = 1, ..., n\}$ and $Pr[y(t)/H(x_*(t_-))]$ is the orthogonal projection of y(t) onto $H(x_*(t_-))$. Now using the projection theorem, (6)-(7), and (5b) we get

$$Pr[y(t)/H(x_*(t_-))] = \int_0^\infty R(\lambda-t_-+t)V(\lambda)d\Sigma^{-1/2}x_*(t_-)$$

$$= He^{-F0-}\Sigma^{1/2}\int_0^\infty\int_0^\infty V'(\tau)R(\lambda-\tau)V(\lambda)d\tau d\lambda\Sigma^{-1/2}x_*(t_-) = x_*(t_-) = Hx_*(t_-)$$

Hence the output equation is

$$y(t) = Hx_*(t_-) + \nu(t) \tag{12}$$

Let $s_- \leq t_-$ then

$$x_*(t_-) = Pr[x_*(t_-)/H(x_*(s_-))] + \{x_*(t_-) - Pr[x_*(t_-)/H(x_*(s_-))]\} \tag{13}$$

Now consider the first term in (13). Then successively using the projection theorem, (10), (6)-(7), (7b) and (5b), (13) can be expressed as

$$Pr[x_*(t_-)/H(x_*(s_-))] =$$

$$= \int_0^\infty[\Sigma^{1/2}\int_0^\infty V'(\lambda_1)R(t_--s_-+\lambda_2-\lambda_1)d\lambda_1]V(\lambda_2)d\lambda_2\Sigma^{-1/2}x_*(s_-)$$

$$= e^{F(t_--s_-)}\int_0^\infty e^{F\lambda_2}GV'(\lambda_2)d\lambda_2\Sigma^{-1/2}x_*(s_-)$$

$$= e^{F(t_--s_-)}\Sigma^{1/2}\int_0^\infty\int_0^\infty V(\lambda_1)R(\lambda_2-\lambda_1)V'(\lambda_2)d\lambda_2 d\lambda_1\Sigma^{-1/2}x_*(s_-)$$

$$= e^{F(t_--s_-)}x_*(s_-) \tag{14}$$

Next we express the second term in (13) in terms of the innovations process $\nu(\cdot)$. Towards this end using (14) we have

$$x_*(t_-) - \Pr[x_*(t_-)/H(x_*(s_-))] = e^{Ft_-}[\rho(t_-) - \rho(s_-)] \tag{15}$$

where $\rho(t_-) \triangleq e^{-Ft_-}x_*(t_-)$ is easily seen to be a forwards martingale w.r.t. $H(x_*(s_-))$, i.e.,

$$\Pr[\rho(t_-)/H(x_*(s_-))] = \rho(s_-), \quad s_- \leq t_- \tag{16}$$

Furthermore, it is also easily seen that $\Pr[(\rho(t_-) - \rho(s_-))/H_s^-(y)] = \Pr[(\rho(t_-) - \rho(s_-))/H(x_*(s_-))] = 0$. Thus $\rho(t_-) - \rho(s_-)$ is orthogonal to $H_s^-(y)$. Also from the definition of $\rho(t_-)$ and $x_*(t_-)$ one observes that $\Pr[\rho(t_-) - \rho(s_-)/H_t^-(y)] = \rho(t_-) - \rho(s_-)$. Now let $H_s^{t_-}(\nu)$ be a closed subspace of $H(y)$ spanned by $\{\nu(\tau), s \leq \tau < t\}$, then $H_t^-(y) = H_s^-(y) \oplus H_s^{t_-}(\nu)$ and $\rho(t_-) - \rho(s_-) = \Pr[(\rho(t_-) - \rho(s_-))/H_s^{t_-}(\nu)] = \Pr[\rho(t_-)/H_s^{t_-}(\nu)]$. Hence

$$\rho(t_-) - \rho(s_-) = \int_{s_-}^{t_-} E[\rho(t_-)\nu'(\tau)]\nu(\tau)d\tau \tag{17}$$

The variable t_- was introduced for explicitness and now we can interpret (12) as $y(t) = Hx_*(t) + \nu(t)$. Using this and the definition of $\rho(t_-)$ we get

$$E\rho(t_-)\nu'(\tau) = e^{-Ft_-}[Ex_*(t_-)y'(\tau) - Ex_*(t_-)x_*'(\tau)H'] \tag{18}$$

Once again following an approach similar to the one used for the derivation of (12) and (14) it can be shown that

$$Ex_*(t_-)y'(\tau) = e^{F(t_- - \tau)}G, \quad Ex_*(t_-)x_*'(\tau) = e^{F(t_- - \tau)}\Sigma \tag{19}$$

Combining everything from (14) through (19) we have the *new realization for* $y(\cdot)$

$$x_*(t_-) = e^{F(t_- - s_-)}x_*(s_-) + \int_{s}^{t_-} e^{F(t_- - \tau)}[G - \Sigma H']\nu(\tau)d\tau$$

which in differential form will be

$$\dot{x}_*(t) = Fx_*(t) + (G - \Sigma H')\nu(t)$$
$$y(t) = Hx_*(t) + \nu(t) \tag{20}$$

where G = C(0), H = 0(0), and using (7) it is easily shown that F can be computed as

$$F = \int_0^\infty \dot{C}(\tau)C^{+}(\tau)d\tau = \int_{-\infty}^{0_-} \int_0^\infty \Sigma^{-1/2}U'(\xi)[\frac{\partial}{\partial\tau} R(\tau-\xi)]V(\tau)\Sigma^{-1/2}d\tau d\xi$$

Also from (11) it is clear that $E\nu(t)\nu'(\tau) = I\delta(t-\tau)$, and a "forwards Markovian" property holds true for (20), namely $Ex_*(t)\nu'(\tau) = 0$, $\tau < t$.

REDUCED-ORDER MODEL

In the previous section the n x 1 state vector $x(t_-)$ represented a basis for $Pr_-[H_t^+(y)]$. Thus the reduced-order model is obtained by picking a reduced basis for $Pr_-[H_t^+(y)]$. This is readily achieved if $\sigma_r >> \sigma_{r+1}$; for then the mutual information between $H_t^-(y)$ and $H_t^+(y)$ is adequately approximated by

$$I(\alpha,\beta) \simeq -\frac{1}{2}\sum_{i=1}^{r} \log(1-\alpha_i^2) = I(\alpha^1,\beta^1)$$

where $\alpha^1(t_-)$ and $\beta^1(t)$ is a subvector consisting of the first r entries of $\alpha(t_-)$ and $\beta(t)$ respectively. Corresponding to the reduced canonical variable α^1 we define the reduced state vector as

$$x_*^1(t_-) = \hat{\Sigma}^{1/2}\int_0^\infty \hat{V}'(\lambda)y(t_- -\lambda)d\lambda = \hat{\Sigma}^{1/2}\,\alpha^1(t_-) \tag{21a}$$

which represents the first r x 1 entries in $x_*(t)$. In (21), $\hat{\Sigma} = \text{diag}[\sigma_1, ..., \sigma_r]$, and $\hat{V}'$ consists of the first r rows of V'. Now using (20), (21) can be expressed in differential form as

$$\dot{x}_*^1(t) = \hat{F}x_*^1(t) + F_{12}x_*^2(t) + (\hat{G} - \hat{\Sigma}\hat{H}')\nu(t)$$
$$y(t) = \hat{H}x_*^1(t) + \tilde{H}x_*^2(t) + \nu(t) \tag{21b}$$

where F_{12}, $\hat{G}$, $\hat{H}$ and $\tilde{H}$ are defined by the following partition of F, G, and H

$$F = \begin{bmatrix} \hat{F} & F_{12} \\ F_{21} & \tilde{F} \end{bmatrix}, \quad G = \begin{bmatrix} \hat{G} \\ \tilde{G} \end{bmatrix}, \quad H = [\hat{H} \ \tilde{H}]$$

Now let x_{*i} be the i-th entry of x_*, then since $Ex_*(t)x_*'(t) = \Sigma$, we have

$$Ex_{*i}^2(t) = ||x_{*i}||^2 = \sigma_i$$

Furthermore since we assumed $\sigma_r >> \sigma_{r+1}$, and the fact that canonical correlations are bounded by one and zero, we have for i = r+1, ..., n

$$||x_{*i}||^2 = \sigma_i << 1$$

implying $||x_{*i}||^2 \simeq 0$, which in turn implies x_{*i} is a zero random process. Thus (21b) gives the (approximate) *reduced-order model* for y(•) as

$$\dot{\hat{x}}_*(t) = \hat{F}\hat{x}_*(t) + (\hat{G} - \hat{\Sigma}\hat{H}')\hat{\nu}(t)$$

$$y(t) \simeq \hat{H}\hat{x}_*(t) + \hat{\nu}(t) \triangleq \hat{y}(t) \tag{22}$$

where $E\hat{\nu}(t)\,\hat{\nu}'(\tau) = I\delta(t-\tau)$, $E\hat{x}(t)\,\hat{\nu}'(\tau) = 0$ for $\tau > t$. From (22) it is seen that the covariance of y(•) is approximated by the covariance of $\hat{y}$(•), namely

$$\hat{R}(t) = I\delta(t) + \hat{H}e^{\hat{F}t}\hat{G}1(t) + \hat{G}'e^{-\hat{F}'t}\hat{H}'1(-t)$$

Also from the above discussion it is seen that

$$\eta = \sum_{i=1}^{r}\log(1 - \sigma_i^2)/\sum_{i=1}^{n}\log(1 - \sigma_i^2)$$

is approximately the fraction of the information about the future in the past retained by the approximate state vector $\hat{x}_*$(•). The performance of (22) can be judged either by η or by the relative error between R(•) and $\hat{R}$(•).

A TRANSFORMATION APPROACH

The reduced-order realization algorithm developed in the previous section is appropriate, if to start with, only the covariance function R(t) was available. Now suppose a realization of the form (1) is available. Then it would be a rather roundabout approach to first obtain R(•) from (1), and then use the realization approach to obtain the reduced-order model. Also in this approach the physical significance of the state vector $\bar{x}(\cdot)$ would be lost because one does not have any explicit relationship between $\bar{x}(\cdot)$ and $x_*(\cdot)$. Thus it seems more appropriate to develop a model reduction method that will directly exploit the parameters of $(\bar{F}, \bar{J}, \bar{H}, I)$ of (1) and at the same time track the physical significance of $\bar{x}(\cdot)$. This is where the transformation approach to stochastic model reduction comes in.

In order to develop the transformation approach we first consider the backwards stochastic realization obtained by using the canonical variables $\beta(\cdot)$ (namely a basis for $Pr_+[H_t^-(y)]$). Let the state vector be defined by

$$z_*(t) = \Sigma^{1/2} \int_{-\infty}^{0_-} U'(\mu)y(t-\mu)d\mu = \Sigma^{1/2} \beta(t)$$

Now, following an approach analogous to the development of (20) it can be shown that

$$\dot{z}_*(t) = -F' z_*(t) - (H' - \Sigma G)\nu_b(t)$$

$$y(t) = G' z_*(t) + \nu_b(t) \tag{23}$$

where $\nu_b(\cdot)$ is the backwards innovation process for $y(\cdot)$ with $E\nu_b(t)\nu_b'(s) = I\delta(t-s)$, and satisfying the "backwards Markovian" property $Ez_*(t)\nu_b'(\tau) = 0$, $\tau < t$. Also $Ez_*(t)z_*'(t) = \Sigma$. Next we observe two special properties of (20) and (23):

(i) $Pr(x_*(t_-)/H_t^-(y)] = x_*(t_-)$, and $Pr[z_*(t)/H_t^+(y)] = z_*(t)$

(ii) Both $x_*(\cdot)$ and $z_*(\cdot)$ have covariance Σ, and Σ satisfies the following algebraic Riccati equations

$$0 = F\Sigma + \Sigma F' + (G - \Sigma H')(G - \Sigma H')'$$

$$0 = F'\Sigma + \Sigma F + (H' - \Sigma G)(H' - \Sigma G)' \tag{24}$$

From property (i) it is seen that (20) is a forwards innovations representation (IR) of $y(\cdot)$, while (23) is a backwards IR of $y(\cdot)$. Property (ii) shows that both the IRs have the same state covariance Σ. These two properties lead to the following definition of balanced stochastic realizations.

Definition: A balanced stochastic realization (BSR) is an innovations representation (forwards or backwards) with the state covariance matrix equal to the canonical correlation coefficient matrix Σ of the output process.

With this definition we see that the reduced-order model (22) is essentially a subsystem of a BSR. (Note a backwards reduced-order model can be obtained by dropping those states from (23) whose variance σ_i is "very small"). Thus the transformation approach boils down to transforming (1) to a BSR.

The above discussion suggests that one way to obtain the BSR is as follows: (i) first transform (1) and its backwards model to their respective IRs, (ii) find a transformation that will simultaneously transform their state covariances to Σ, and (iii) apply this transformation to the forwards and backwards IR to obtain the forwards and backwards BSR. In order to implement these three steps, we first examine the backwards realization corresponding to (1) ([26], [29]-[30]), namely,

$$\dot{\bar{z}}_b(t) = -\bar{F}'\bar{z}_b(t) + \Pi^{-1}\bar{J}\bar{v}_b(t), \quad y(t) = \bar{G}'\bar{z}_b(t) + \bar{v}_b(t) \tag{25}$$

where $\bar{z}_b(\cdot) = \Pi^{-1}\bar{x}(\cdot)$, $\bar{v}_b(\cdot)$ is zero mean with $E\bar{v}_b(t)\bar{v}_b'(s) = I\delta(t-s)$, and $E\bar{v}_b(t)\bar{z}'(s) = 0$ for $t < s$. The IRs for (1) and (25) are

$$\dot{\bar{x}}_*(t) = \bar{F}\bar{x}_*(t) + (\bar{G} - P\bar{H}')\ \nu(t)$$

$$y(t) = \bar{H}\bar{x}_*(t) + \nu(t) \tag{26a}$$

$$\dot{\bar{z}}_*(t) = -\bar{F}'\bar{z}_*(t) - (\bar{H}' - N\bar{G})\nu_b(t)$$

$$y(t) = \bar{G}'\bar{z}_*(t) + \nu_b(t) \tag{26b}$$

where $\bar{x}_*(t) = Pr[\bar{x}(t)/H_t^-(y)]$, $\bar{z}_*(t) = Pr[\bar{z}(t)/H_t^+(y)] = Pr[\Pi^{-1}\bar{x}(t)/H_t^+(y)]$, and $P = E\bar{x}_*(t)\bar{x}_*'(t)$, $N = E\bar{z}_*(t)\bar{z}_*'(t)$ satisfy the algebraic Riccati equations

$$0 = \bar{F}P + P\bar{F} + (\bar{G} - P\bar{H}')(\bar{G} - P\bar{H}')'$$

$$0 = \bar{F}'N + N\bar{F} + (\bar{H}' - N\bar{G})(\bar{H}' - N\bar{G})' \tag{27}$$

The next four Lemmas show how (26a) and (26b) can be transformed to the BSR (20) and (23) respectively. To maintain the continuity of the presentation, the proofs of these lemmas are relegated to the Appendix.

Lemma 1: The eigenvalues of NP are equal to the square of the canonical correlation coefficients: $\{\sigma_i^2\}_{i=1}^n$. *Proof:* See the Appendix.

In fact $\{\sigma_i^2\}_{i=1}^n$ represents an ***invariant set*** for the process $y(\cdot)$.

Lemma 2: Let the eigenvalue-eigenvector decomposition of PN be $PN = T^{-1}\Sigma^2 T$ where $T = \Sigma^{1/2}U_n' \Lambda_p^{-1/2} U_p'$ with $U_p \Sigma_p U_p' = P$, and $U_n \Sigma^2 U_n' = \Lambda_p^{1/2} U_p' N U_p \Lambda_p^{1/2}$. Furthermore let

$$x_d = T\bar{x}_* , \quad z_d = (T^{-1})' \bar{z}_*$$

Then $Ex_d(t)x_d'(t) = \Sigma = Ez_d(t)z_d'(t)$, and with $F_d = T\bar{F}T^{-1}$, $G_d = T\bar{G}$, $H_d = \bar{H}T^{-1}$, (26) becomes

$$\dot{x}_d = F_d x_d + (G_d - \Sigma H_d')\nu, \quad y = H_d x_d + \nu \tag{28a}$$

$$\dot{z}_d = -F_d' z_d - (H_d' - \Sigma G_d)\nu_b , \quad y = G_d' z_d + \nu_b \tag{28b}$$

Proof: Direct verification (see also [31]).

The following Lemma establishes the equivalence between (20), (23) and (28a), (28b) respectively.

Lemma 3: Assume σ_i, i = 1, ..., n are distinct.
(i) In terms of the parameters of (1) and the canonical vectors defined by (4), the transformation T of Lemma 2 is

$$T = S\Sigma^{-1/2} \int_{-\infty}^{0_-} U'(\xi)\bar{H}e^{-\bar{F}\xi} d\xi \tag{29}$$

where S = diag[± 1].
(ii) The triple (F,G,H) for (20) and (23) is related to (F_d,G_d,H_d) of (28) by a similarity transformation S, namely

$$F_d = SFS , \quad G_d = SG , \quad H_d = HS$$

Moreover

$$x_d(t) = Sx_*(t) \text{ and } z_d(t) = Sz_*(t)$$

Proof: See the Appendix:

From Lemma 3 we see that either of the realizations in (28) is a BSR. Moreover when σ_i's are distinct, the forwards (or the backwards) BSR is unique modulo a sign matrix S. If σ_i's are not distinct, say $\sigma_1 > \sigma_2 \cdots > \sigma_{k+1} = \ldots = \sigma_{k+j} > \sigma_{k+j-1} \cdots > \sigma_n$, then Lemma 3 generalizes with S = block diag[$\pm I_k$, S_2, $\pm I_{n-k-j}$], where S_2 is any j x j orthogonal matrix.

Now, if $\sigma_r >> \sigma_{r+1}$, then the upper r x r "subsystem" of (28a) will give the same reduced-order model as (22) modulo a sign matrix. The steps for obtaining the reduced-order model using the transformation approach are as follows.

Step 1 Obtain the IR (26a); this will give $\bar{x}_*(\cdot)$.
Step 2 Solve (27) for P and N (one could use the Schur decomposition method of Laub [32] for solving the Riccati equations).
Step 3 Obtain T via Lemma 2.
Step 4 Apply the transformation T to the IR of Step 1; this will give

$$x_d = T\bar{x}_*, \; F_d = T\bar{F}T^{-1}, \; G_d = T\bar{G}, \; H_d = \bar{H}T^{-1}.$$

Step 5 Suppose $\sigma_r >> \sigma_{r+1}$ for some $r < n$, then analogous to (22) the reduced-order model is given by.

$$\dot{\hat{x}}_d(t) = \hat{F}_d\hat{x}_d(t) + (\hat{G}_d - \hat{\Sigma}\hat{H}_d')\hat{\nu}(t)$$

$$y(t) = \hat{H}_d\hat{x}_d(t) + \hat{\nu}(t) \tag{30}$$

where $\hat{F}_d$, $\hat{G}_d$ and $\hat{H}_d$ are respectively the upper r x r, r x p, and p x r blocks of F_d, G_d, and H_d.

REDUCED-ORDER KALMAN-BUCY FILTER

In this section we show that the inverse of the reduced-order realization (22) (with $\hat{y}(\cdot)$ replaced by $y(\cdot)$), gives a new design for reduced-order Kalman-Bucy filters. This is achieved by showing that both (22) ((30)) and its inverse are asymptotically stable. In turn, we will have established that (22) ((30)) is an IR for $\hat{y}(\cdot)$.

Lemma 4: The balanced stochastic realization (20) (or equivalently (28)) is asymptotically stable: $\mathrm{Re}[\lambda_i(F)] < 0$.

Proof: Let $K = G - \Sigma H'$; now since (F,G) is a completely controllable pair, (F,K) will also be a completely controllable pair. Furthermore, since Σ is a symmetric positive definite matrix satisfying $0 = F\Sigma + \Sigma F' + K'K$, we have $\mathrm{Re}[\lambda_i(F)] < 0$.

Lemma 5: The reduced-order realization (22) ((30)) is asymptotically stable if $\{\sigma_i\}_{i=1}^{r}$ and $\{\sigma_i\}_{i=r+1}^{n}$ have no common entries.

The proof is analogous to the one given in Pernebo and Silverman [38] for deterministic reduced-order realizations, obtained from deterministic balanced realizations [15], and as such is omitted (also see (22)).

Theorem 1: Let $\Sigma^{-1} - \Sigma > 0$, then
(i) the inverse of the balanced stochastic realization (20) ((28a)) is asymptotically stable, i.e. $\mathrm{Re}[\lambda_i(F-KH)] > 0$ for i = 1, ..., n, where $K = G - \Sigma H'$.
(ii) also assume that $\{\sigma_i\}_{i=1}^{r}$ and $\{\sigma_i\}_{i=r+1}^{n}$ have no common entries, then the inverse of the reduced-order system (22) ((30)) is asymptotically stable, i.e. $\mathrm{Re}[\lambda_i(\hat{F}-\hat{K}\hat{H})] < 0$, i = 1, ..., r, with $\hat{K} = \hat{G} - \hat{\Sigma}\hat{H}'$.

Proof: From (23) we have

$$0 = F\Sigma + \Sigma F' + (G - \Sigma H')(G - \Sigma H')$$

$$0 = F\Sigma^{-1} + \Sigma^{-1}F' + (\Sigma^{-1}H' - G)(\Sigma^{-1}H' - G)'$$

Combining the above two equations, one obtains

$$0 = (F - KH)(\Sigma^{-1} - \Sigma) + (\Sigma^{-1} - \Sigma)(F - KH)' + (\Sigma^{-1} - \Sigma)H'H(\Sigma^{-1} - \Sigma) \tag{31}$$

where $K = G - \Sigma H'$. Since $\Sigma^{-1} - \Sigma > 0$, the above equation simplifies to

$$0 = \Omega(F - KH) + (F - KH)'\Omega + H'H \tag{32}$$

where $\Omega = (\Sigma^{-1} - \Sigma)^{-1}$. Now, since (F,H) is an observable pair, it is a straightforward matter to show that ((F-KH), H) is also an observable pair. This fact, coupled with $\Omega > 0$ and (32), implies that $\mathrm{Re}[\lambda_i(F-KH)] < 0$ for i = 1, ..., n; and part (i) is proved.

To prove part (ii), consider the upper rxr block of (32), which since Ω is diagonal will be

$$0 = \hat{\Omega}(\hat{F}-\hat{K}\hat{H}) + (\hat{F}-\hat{K}\hat{H})'\hat{\Omega} + \hat{H}'\hat{H} \tag{33}$$

where $\hat{\Omega} = (\hat{\Sigma}^{-1} - \hat{\Sigma})^{-1}$, $\hat{K} = \hat{G} - \hat{\Sigma}\hat{H}'$, and $\hat{F},\hat{G},\hat{H},\hat{\Sigma}$ are as defined in (22). Now let $\hat{\lambda}$ be any eigenvalue of $(\hat{F}-\hat{K}\hat{H})$ and $\hat{q}$ the corresponding eigenvector, then $(\hat{F}-\hat{K}\hat{H})\hat{q} = \hat{\lambda}q$ and $\hat{q}^{\#}(\hat{F}-\hat{K}\hat{H})' = \hat{\lambda}^{\#}\hat{q}^{\#}$ (# denotes complex conjugate transpose). Pre and post multiplying (31) by $\hat{q}^{\#}$ and $\hat{q}$ respectively gives

$$2\mathrm{Re}(\hat{\lambda})\hat{q}^{\#}\hat{\Omega}\hat{q} = -\hat{q}^{\#}\hat{H}'\hat{H}\hat{q} \tag{34}$$

Since $\hat{\Omega} > 0$, $\hat{q}^{\#}\hat{\Omega}\hat{q} > 0$, and also $\hat{q}^{\#}\hat{H}'\hat{H}\hat{q} \geq 0$, we have $\mathrm{Re}(\hat{\lambda}) \leq 0$. Suppose $\mathrm{Re}(\hat{\lambda}) = 0$, then $\hat{q}^{\#}\hat{H}'\hat{H}\hat{q} = 0$ implying $\hat{H}\hat{q} = 0$, which in turn implies $\hat{F}\hat{q} = \hat{\lambda}\hat{q}$, i.e. $\hat{\lambda}$ is an eigenvalue of $\hat{F}$, s.t. $\mathrm{Re}(\hat{\lambda}) = 0$. But this contradicts Lemma 5. Hence, $\mathrm{Re}(\hat{\lambda}) \neq 0$ and consequently we have $\mathrm{Re}(\hat{\lambda}) < 0$, which proves part (ii).

Remark 1: Even though part (i) of the Theorem is well known, the proof is included for the sake of completeness.

Remark 2: Under the assumptions of Theorem 1, and following an analogous proof, it can be shown that (i) $\mathrm{Re}[\lambda_i(-F' + K_b'G')] > 0$, $i = 1, ..., n$, where $K_b = H - G'\Sigma$, and (ii) $\mathrm{Re}[\lambda_i(-\hat{F}' + \hat{K}_b'\hat{G}')] > 0$, $i = 1, ..., r$, where $\hat{K}_b' = \hat{H} - \hat{G}'\hat{\Sigma}$.

Remark 3: Lemma 5, along with Theorem 1 implies that the forwards reduced-order model (22) is a forwards innovations representation for the process $\hat{y}(\cdot)$. Moreover, Remark 2 implies that the backwards reduced-order realization obtained from (23), namely

$$\dot{\hat{z}}(t) = -\hat{F}'\hat{z}(t) - \hat{K}_b'\hat{\nu}_b(t), \quad y(t) \simeq \hat{G}'\hat{z}(t) + \hat{\nu}_b(t) = \hat{y}(t) \tag{35}$$

where $E\hat{\nu}_b(t)\hat{\nu}_b'(s) = I\delta(t-s)$, and $E\hat{z}(t)\hat{\nu}_b'(s) = 0$ for $s < t$, is a backwards innovations representation for $\hat{y}(\cdot)$. Thus ***both (22) and (35) are causal and causally invertible.***

Remark 4: From (24) it is seen that $\hat{\Sigma}$ obeys

$$0 = \hat{F}\hat{\Sigma} + \hat{\Sigma}\hat{F}' + \hat{K}\hat{K}'$$

$$0 = \hat{F}'\hat{\Sigma} + \hat{\Sigma}\hat{F} + \hat{K}_b'\hat{K}_b \tag{36}$$

and since (22) and (35) are respectively the forwards and backwards IR for $\hat{y}(\cdot)$, (22) and (35) are also BSRs.

Remark 5: The proof for Theorem 1 has an interesting stochastic interpretation. Consider the forwards realization corresponding to the backwards BSR (23), namely

$$\dot{x}^*(t) = Fx^*(t) + (G - \Sigma^{-1}H')\nu^*(t)$$

$$y(t) = Hx^*(t) + \nu^*(t) \tag{37}$$

where $E\nu^*(t)(\nu^*(\tau))' = I\delta(t-\tau)$ and $E\nu^*(\tau)(x^*(t))' = 0$ for $\tau > t$, and $Ex^*(t)(x^*(t))' = \Sigma^{-1}$. Now we also have that

$$x_*(t) = P[x^*(t)/H_t^-(y)] \tag{38}$$

Next using (20) and (37) one obtains

$$\dot{x}^*(t) - \dot{x}_*(t) = (F - KH)(x^*(t) - x_*(t)) - (\Sigma^{-1} - \Sigma)H'\nu^*(t) \tag{39}$$

Moreover, using (38) it is seen that

$$E[(x^*(t) - x_*(t))(x'(t) - x_*(t))'] = \Sigma^{-1} - \Sigma$$

and also that $E\nu^*(\tau)x_*'(t) = 0$ for $\tau > t$. Then stochastically we can interpret $\Sigma^{-1} - \Sigma$ as the error co-variance for the Kalman-Bucy filter corresponding to the model (37), and (31) as the (error) state co-variance equation for (39).

Remark 6: In the proof of Theorem 1, the analysis following (34) also shows that (F,H) is an observable pair. Also, the proof involved in establishing Remark 2 will show that $(\hat{F},\hat{G})$ is a controllable pair. This provides an alternate derivation for establishing the minimality of $(\hat{F},\hat{G},\hat{H})$ [22].

Now using Lemmas 2 and 3, Theorem 1, and the reduced-order model (30) we have the following design for the *approximate reduced-order Kalman-Bucy filter.*

<u>Step 1</u> Implement Steps 1 through 5 for the transformation approach to stochastic model reduction, and obtain the reduced-order model (30).

<u>Step 2</u> Obtain $\hat{x}_d(t)$ from the inverse of (30) which is asymptotically stable by Theorem 1

$$\dot{\hat{x}}_d(t) = (\hat{F}_d - \hat{K}_d\hat{H}_d)\hat{x}_d(t) + \hat{K}_d y(t)$$

$$\hat{K}_d = \hat{G}_d - \hat{\Sigma}\hat{H}_d \tag{40}$$

<u>Step 3</u> The linear least-square estimate $\overline{x}_*(t) = Pr[\overline{x}(t)/H_t^-(y)]$ is given by

$$\overline{x}_*(t) \simeq \hat{Q}\hat{x}_d(t) \tag{41}$$

where $\hat{Q}$ is an nxr matrix consisting of the first r columns of T^{-1}, and in terms of the factorization of P and N in Lemma 2, it is given by

$$\hat{Q} = U_p \Sigma_p^{1/2} U_n \begin{bmatrix} \hat{\Sigma}^{1/2} \\ 0 \end{bmatrix} \tag{42}$$

Remark 7: In the reduced-order K-B filter (40)-(41) we see that there is a reduction in the dynamic part of the filter. We have an r-th order filter as opposed to an n-th (n > r) order filter. But this reduction is not completely without cost, we now have an additional "static" computation step, namely (41).

Reduced-oder Spectral Factors

Since both the forwards and the backwards reduced-order models (22) and (35) are causal and causally invertible (Remark 3), our order reduction algorithm also provides a reduced-order approximation for $\Phi(s)$ the power spectrum of $y(\cdot)$ and its canonical spectral factors. Let $W_*(s)$ and $W_{b*}(s)$ be respectively the transfer functions of (20) or (26a), and (23) or (26b), then $\Phi(s)$ has the canonical spectral factorization

$$\Phi(s) = W_*(s)W_*'(-s) = W_{b*}(s)W_{b*}'(s)$$

Now since (22) and (35) are reduced-order approximation to (20) and (23) respectively, we have the reduced-order approximation to $\Phi(s)$ as

$$\Phi(s) \simeq \hat{\Phi}(s) = \hat{W}_*(s)\hat{W}_*'(-s) = \hat{W}_{b*}(s)\hat{W}_{b*}(-s)$$

where $\hat{W}_*(s)$ and $\hat{W}_{b*}(s)$ are the transfer functions of (22) and (35) respectively, namely

$$\hat{W}_*(s) = I + \hat{H}(sI - \hat{F})^{-1}(\hat{G} - \hat{\Sigma}\hat{H}')$$

$$\hat{W}_{b*}(s) = I + \hat{G}'(-sI - \hat{F}')^{-1}(\hat{H}' - \hat{\Sigma}\hat{G})$$

Also from Lemma 5 and Theorem 1 we have that $\hat{W}_*$ is a minimum phase (canonical) spectral factor of $\hat{\Phi}$.

REDUCED-ORDER HIERARCHICAL ESTIMATION

Consider a linear stochastic system

$$\dot{\bar{x}}(t) = \bar{F}\bar{x}(t) + w(t) \tag{43}$$

where $w(\cdot)$ is a zero mean white process with $Ew(t)w'(s) = \overline{W}\delta(t-s)$ and $E\overline{x}(t)w'(s) = 0$ for $s > t$. At each local station i we have a p_i x 1 measurement vector

$$y_i(t) = \overline{H}_i\overline{x}(t) + v_i(t), \quad i = 1, 2 \tag{44}$$

where $v_i(\cdot)$ is a zero mean white process with

$$Ev_i(t)v_i'(s) = I_{p_i}\delta(t-s), \quad Ev_1(t)v_2'(s) = 0 \quad \text{for all t, s, and}$$

$$Ev_i(t)w'(s) = J_i\delta(t-s), \quad Ev_i(t)\overline{x}'(s) = 0 \quad \text{for } t > s\,.$$

For simplicity we consider the case of only two local stations. Generalizations to more than two stations are obvious. We assume that $y_i(\cdot)$ is stationary and purely non-deterministic. Thus $Re[\lambda(F)] < 0$. Now let $\Pi = Ex(t)x'(t)$, then Π satisfies the Lyapunov equation

$$0 = \overline{F}\Pi + \Pi\overline{F}' + \overline{W}$$

Let $\overline{G}_i = \Pi\overline{H}_i' + \overline{J}_i$, $\overline{G} = [\overline{G}_1\ \overline{G}_2]$, and $\overline{H}' = [\overline{H}_1'\ \overline{H}_2']'$. We assume that the triples $(\overline{F},\overline{G},\overline{H})$ and $(\overline{F},\overline{G}_i,\overline{H}_i)$ are minimal. Next let the linear least-squares estimate of $\overline{x}(t)$ based on all the past observations be

$$\overline{x}_*(t) = Pr[\overline{x}(t) \mid \{y_1(s), y_2(s), \quad s < t\}]$$

Then we know that $\overline{x}_*(t)$ is given by the Kalman-Bucy filter

$$\dot{\overline{x}}_*(t) = (\overline{F} - \overline{K}\overline{H})\overline{x}_*(t) + \overline{K}_1 y_i(t) + \overline{K}_2 y_2(t) \tag{45a}$$

where

$$\overline{K} = (\overline{G} - P\overline{H}') = [\overline{K}_1, \overline{K}_2], \quad \overline{K}_i = \overline{G}_i - P\overline{H}_i' \tag{45b}$$

$P = E\overline{x}_*(t)\overline{x}_*(t)$ obeys the ARE

$$0 = \overline{F}P + P\overline{F}' + (\overline{G} - P\overline{H}')(\overline{G} - P\overline{H}')' \tag{46}$$

Now using the superposition principle in (45), $\overline{x}_*(t)$ can be expressed as

$$\bar{x}_*(t) = \bar{x}_1^0(t) + \bar{x}_2^0(t) \tag{47}$$

where $\bar{x}_i^0(t)$ satisfies the differential equation

$$\dot{\bar{x}}_i^0(t) = (\bar{F} - \bar{K}\bar{H})\bar{x}_i^0(t) + \bar{K}_i y_i(t) \tag{48}$$

As was shown in [37], $\bar{x}_i^0(\cdot)$ has an interesting geometric interpretation in terms of oblique projections. This geometry is depicted in Figure 1 and is made precise by the following definition of oblique projections.

Definition: Let M be a Hilbert space of random variables with inner product $<\alpha,\beta> = E\alpha\beta$ for α, $\beta \in M$. Also let $H_t^-(y_i)$ be the closed subspace of M spanned by the elements of $\{y_i(s),\ s < t\}$. Let $\bar{x}_{*i}$, $\bar{x}_{1i}^0$ and $\bar{x}_{2i}^0$ be the ith element of $\bar{x}_*$, $\bar{x}_1^0$, and $\bar{x}_2^0$ respectively. Then $\bar{x}_{*i}(t) = Pr[\bar{x}_i(t)/H_t^-(y)]$ and the oblique projection of $\bar{x}_{*i}(t)$ onto $H_t^-(y_1)$ along $H_t^-(y_2)$, denoted by $\bar{x}_{1i}^0(t) = Pr[\bar{x}_{*i}(t)/H_t^-(y_1)$ along $H_t^-(y_2)]$, is the unique element belonging to $H_t^-(y_1)$ and satisfying the orthogonality condition

$$<Pr[\bar{x}_{*i}/H_t^-(\tilde{y}_1)] - \bar{x}_{1i}^0,\ \tilde{\gamma}_1> = 0 \quad \text{for all } \tilde{\gamma}_1 \in H_t^-(\tilde{y}_1) \tag{49}$$

where (with a slight abuse of notation) $H_t^-(\tilde{y}_1)$ is defined as

$$H_t^-(\tilde{y}_1) = H_t^-(y_1) - Pr[H_t^-(y_1)/H_t^-(y_2)]$$

Pr is a projection operator (whether it is an orthogonal or an oblique projection operator will be clear from the contents between the square brackets). Moreover $\bar{x}_{*i}$ has the unique decomposition

$$\bar{x}_{*i} = \bar{x}_{1i}^0 + \bar{x}_{2i}^0 \tag{50}$$

which in vector form is indeed (47).

Note that (47)-(48) provide a hierarchical estimation scheme. At each local station the oblique estimate $\bar{x}_1^0$ is computed, this is then transmitted to a central processor where the oblique estimates are simply added together to produce the centralized orthogonal estimate $\bar{x}_*$. We would like to mention at this point that the idea of using oblique estimates for hierarchical estimation was suggested by H.L. Weinert [39]. Our objective here is to develop an order-reduction algorithm for the oblique estimator.

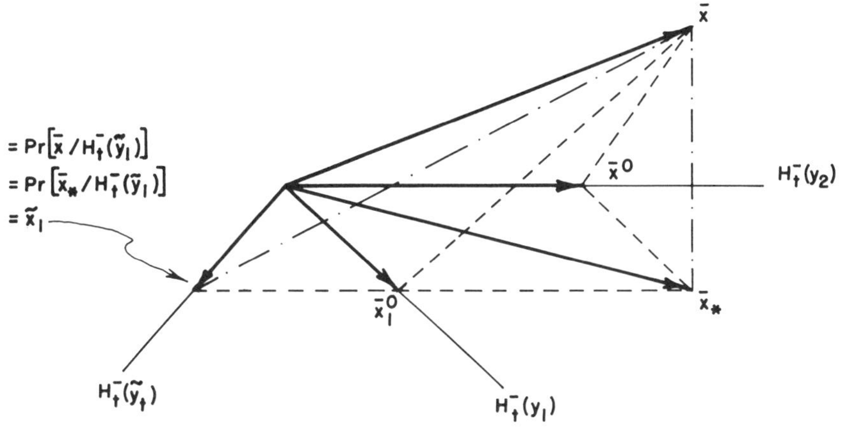

FIGURE 1.

Noninteracting Models

It is interesting to note that (48) provides a new model for $y_i(\cdot)$ in correspondence with $y_j(\cdot)$, namely

$$\dot{\bar{x}}_1^0(t) = (\bar{F} - \bar{K}_2\bar{H}_2)\bar{x}_1^0(t) + \bar{K}_1\nu_1^0(t) \qquad \dot{\bar{x}}_2^0(t) = (\bar{F} - \bar{K}_1\bar{H}_1)\bar{x}_2^0(t) + \bar{K}_2\nu_2^0(t)$$

$$y_1(t) = \bar{H}_1\bar{x}_1^0(t) + \nu_1^0(t) \qquad y_2(t) = \bar{H}_2\bar{x}_2^0(t) + \nu_2^0(t)$$

where $\nu_i^0(t) = y(t) - \Pr[y_i(t)/H_t^-(y_i)$ along $H_t^-(y_j)]$ will be called the "oblique innovations" of $y_i(\cdot)$. Note that $\nu_i^0(\cdot)$ is not a white process. The above models for $y_1(\cdot)$ and $y_2(\cdot)$ can be viewed as "noninteracting" from an estimation viewpoint; this is because the centralized estimate $\bar{x}_*$ is simply the sum of the state vectors for the two models, namely $\bar{x}_* = \bar{x}_1^0 + \bar{x}_2^0$. Perhaps such models may prove useful for studying interactions between systems or processes.

Performance Degradation

Suppose the processor for computing $\bar{x}_1^0$ or $\bar{x}_2^0$ fails, let's say processor one fails, and we are able to shut it off. Then $\bar{x}_*(t)$ is approximated by $\bar{x}_2^0(t)$ and we would like to know the performance degradation. Let

$$e(t) = \bar{x}(t) - \bar{x}_*(t)\,, \quad P_e = E\, e(t)\, e'(t))$$

Then P_e is the error covariance for the optimal situation. Now the error with the failure of processor one will be

$$e_1(t) = x(t) - \bar{x}_2^0(t) = e(t) + \bar{x}_*(t) - x_2^0(t) = e(t) + x_1^0(t)$$

Now, since $\bar{x}_1^0$ is orthogonal to e, taking covariances of both sides of the above equation we obtain the performance degradation measure

$$P_{e1} = P_e + P_1^0$$

Thus the performance degradation as compared to the optimal situation is by an additive matrix equal to the covariance of the oblique estimate.

Order Reduction

The reduced-order oblique estimator can be obtained by following an approach analogous to the stochastic realization approach for obtaining reduced-order K-B filters. This would then lead to the transformation approach for reducing the oblique estimators. But, in the present problem, the parameters of a realization are already available, namely, $(\bar{F},\bar{Q},\bar{H}_1,\bar{H}_2)$, hence it would be appropriate to use the transformation approach directly.

We would like to remark that the justification and the insight for using the appropriate transformation follows from the realization approach. For the discrete-time hierarchical estimation problem we went through this route and justified the transformation approach (see [37]). Since the essential process would be the same for the continuous-time case, we directly present the transformation approach for reducing the continuous-time hierarchical estimation algorithm (47)-(48).

In the problem for obtaining reduced-order K-B filters we examined the interaction (canonical correlations) between the past and the future of the observed process y(•). Unfortunately this approach does not work for reducing the oblique estimator (48). Instead, one needs to examine the interaction between the residuals of y_1 and y_2 with respect to y_2 and y_1, respectively. Let

$$H_t^-(\tilde{y}_i) = H_t^-(y_i) - Pr[H_t^-(y_i)/H_t^-(y_j)] \tag{51}$$

$$H_t^+(\tilde{y}_{bi}) = H_t^+(y_i) - Pr[H_t^+(y_i)/H_t^+(y_j)] \tag{52}$$

Next let

$$\tilde{x}_i(t) = Pr[\overline{x}(t)/H_t^-(\tilde{y}_i)], \quad \tilde{z}_i(t) = Pr[\Pi^{-1}\overline{x}(t)/H_t^+(\tilde{y}_{bi})] \tag{53}$$

and

$$\tilde{P}_i = E\tilde{x}_i(t)\,\tilde{x}_i'(t)\,, \quad \tilde{N}_i = E\tilde{z}_i(t)\tilde{z}_i'(t) \tag{54}$$

Using the various orthogonalities afforded by the geometry of oblique projections (Figure 1) it can be shown that if L is the operator which when operated on $H_t^-(y_i)$ yields $\overline{x}_1^0$, then the same operator L when operated on $H_t^-(\tilde{y}_i)$ yields $\tilde{x}_i$. Since the differential form for this L is (48), the differential equation for computing $\tilde{x}_i$ will be

$$\dot{\tilde{x}}_i(t) = (\overline{F} - \overline{K}\overline{H})\tilde{x}_i(t) + \overline{K}_i\tilde{y}_i(t) \tag{55}$$

Now with a little bit of work it is established that the innovations of $\tilde{y}_i$ is given by

$$\tilde{\nu}_i(t) = \tilde{y}_i(t) - Pr[\tilde{y}_i(t)/H_t^-(\tilde{y}_i)] = \tilde{y}_i(t) - \overline{H}_i\tilde{x}_i(t)$$

Using this in (55) we have the IR for $\tilde{y}_i$

$$\dot{\tilde{x}}_i(t) = (\overline{F} - \overline{K}_j\overline{H}_j)\tilde{x}_i(t) + \overline{K}_i\tilde{\nu}_i(t)$$
$$\tilde{y}_i(t) = \overline{H}_i\tilde{x}_i(t) + \tilde{\nu}_i(t) \tag{56}$$

Using the IR (56) it is easily shown that $\tilde{P}_i$ satisfies

$$0 = (\overline{F} - \overline{K}_j\overline{H}_j)\tilde{P}_i + \tilde{P}_i(\overline{F} - \overline{K}_j\overline{H}_j)' + \overline{K}_i\overline{K}_i' \tag{57}$$

Analogously it can be shown that the innovations representation for $\tilde{y}_{bi}$ is a backwards evolving equation and is given by

$$\dot{\tilde{z}}_i(t) = -(\overline{F}' - \overline{K}_{bj}'\overline{G}_j')\tilde{z}_i(t) - K_{bi}'\tilde{\nu}_{bi}(t)$$

$$\tilde{y}_{bi}(t) = \overline{G}_i\tilde{z}_i(t) + \tilde{\nu}_{bi}(t) \tag{58a}$$

where

$$\overline{K}_{bi} = \overline{H}_i - \overline{G}_i'N\ ,\ \ N = E\overline{z}_*(t)\overline{z}_*'(t)\ ,\ \ \overline{z}_*(t) = Pr[\Pi^{-1}\overline{x}(t)/H_t^+(y)] \tag{58b}$$

and N satisfies

$$0 = N\overline{F} + \overline{F}'N + (\overline{H} - \overline{G}'N)'(\overline{H} - \overline{G}'N) \tag{59}$$

From the backwards IR (58a) it follows that $\tilde{N}_i$ satisfies

$$0 = (\overline{F}' - \overline{K}_{bj}'\overline{G}_j')\tilde{N}_i + \tilde{N}_i(\overline{F}' - \overline{K}_{bj}'\overline{G}_j')' + K_{bi}'K_{bi} \tag{60}$$

Now the IR (56) is reduced by examining the interaction between the residuals $\tilde{y}_i$ and $\tilde{y}_{bi}$. This interaction can be quantified in terms of the eigenvalues of $\tilde{P}_i\tilde{N}_i$. In fact it can be shown that the square root of the eigenvalues of $\tilde{P}_i\tilde{N}_i$ will be the ***residual canonical correlations*** between the residual processes $\tilde{y}_i$ and $\tilde{y}_{bi}$; namely the cosine of the angles between the subspaces $H_t^-(\tilde{y}_i)$ and $H_t^+(\tilde{y}_{bi})$ (another term used for residual canonical correlations is bipartial canonical correlations [38]). Thus one can develop an approach analogous to the transformation approach of the previous section to reduce the IR (56). This in turn will yield a reduction scheme for (55) which gives a reduced-order representation for the operator L. Now, since the operators for obtaining $\tilde{x}_i$ from $\tilde{y}_i$, and $\overline{x}_i^0$ from y_i are the same, namely L, the reduced-order representation for L will yield a reduced-order differential equation for the oblique estimator (48).

Let the eigenvalue-eigenvector decomposition of $\tilde{P}_i\tilde{N}_i$ be

$$\tilde{P}_i\tilde{N}_i = T_i^{-1}\Sigma_i T_i \tag{61}$$

where $T_i = \Sigma_i^{1/2}U_{ni}'\Lambda_{pi}^{-1/2}U_{pi}'$. Matrices U_{pi}, U_{ni}, Λ_{ni}, and Σ_i are obtained from the following factorization of $\tilde{P}_i$ and $\tilde{N}_i$.

$$\tilde{P}_i = U_{pi}\Lambda_{pi}U_{pi}'\ ,\quad \Lambda_{pi}^{1/2}U_{pi}\tilde{N}_i\Lambda_{pi}^{1/2} = U_{ni}\Sigma_i^2U_{ni}'$$

$$\Sigma_i = [\sigma_{i1}, ..., \sigma_{in}]\ ,\quad 1 \geq \sigma_{i1} \cdots \geq \sigma_{in} > 0$$

Now consider

$$x_i^0(t) = T_i\bar{x}_i^0(t) \tag{62}$$

and suppose $\sigma_{ir_i} >> \sigma_{i(r_i+1)}$. Then the reduced-order oblique estimator is given by

$$\dot{\hat{x}}_i^0(t) = \hat{F}_i\hat{x}_i^0(t) + \hat{K}_iy_i(t) \tag{63}$$

where

$$\hat{F}_i = \text{upper principal } r_i \text{ x } r_i \text{ block of } T_i(\overline{F} - \overline{K}\overline{H})T_i^{-1} \tag{64a}$$

$$\hat{K}_i = \text{first } r_i \text{ rows of } T_i\overline{K}_i \tag{64b}$$

and $\bar{x}_i^0$ is approximately obtained by

$$\bar{x}_i^0(t) \simeq \hat{Q}_i\hat{x}_i^0(t) \tag{65}$$

where $\hat{Q}_i = U_{pi}\Lambda_{pi}^{-1/2}U_{ni}[\hat{\Sigma}_i^{1/2}\ 0]'$ = first r_i columns of T_i^{-1}, and $\hat{\Sigma}_i$ = diag $[\sigma_{i1}, ..., \sigma_{ir_i}]$. Analogous to Theorem 1 it can be shown that if $\hat{\Sigma}_i$ and $\tilde{\Sigma}_i$ = diag $[\sigma_{i(r_i+1)}, ..., \sigma_{in}]$ have no common entries then $Re[\lambda(\hat{F}_i)] < 0$, i.e. *(63) is asymptotically stable.* The reduced-order hierarchical estimation algorithm is depicted in Figure 2.

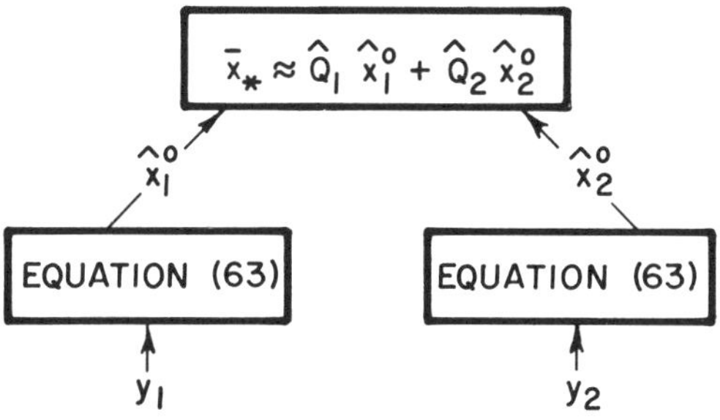

FIGURE 2.

Steps for obtaining the reduced-order hierarchical estimator:

Step 1 Compute the gains $\overline{K}_i$ using (45b) and (46), and $\overline{K}_{bi}$ using (58b) and (59).

Step 2 Solve the Lyapunov equations (57) and (60) for $\tilde{P}_i$, and $\tilde{N}_i$, respectively.

Step 3 Compute the transformation T_i as outlined below (61).

Step 4 Compute the parameters for (63) using (64).

Step 5 Compute $\hat{Q}_i$ as given below (65).
Step 6 Implement Figure 2.

Remark 8: When we compare the above order reduction algorithm with the order reduction algorithm of the previous section, one notices that for the hierarchical estimation problem the reduction is based on the covariances of $\tilde{x}_i$ and $\tilde{z}_i$ and not on the covariances of $\bar{x}_i^0$ and $\bar{z}_i^0 = \Pr[\Pi^{-1}\bar{x}(t)/H_t^+(y_i)$ along $H_t^+(y_j)]$. From our study of this problem from the realization perspective [37], we believe that indeed $\tilde{P}_i$ and $\tilde{N}_i$ are the relevant covariances. Nevertheless, since it does not agree with intuition, one way to check it would be through extensive simulations. This is currently being carried out. Also, other alternative approaches, for example CCA for more than two data sets [39] should be explored.

Remark 9: An alternate way to compute $\tilde{P}_i$ and $\tilde{N}_i$ is as follows. Let

$$\bar{x}_{i*}(t) = \Pr[\bar{x}(t)/H_t^-(y_i)], \quad \bar{z}_{i*}(t) = \Pr[\Pi^{-1}\bar{x}(t)/H_t^+(y_i)]$$

Then from the various orthogonality relations in Figure 1 we have

$$\bar{x}_* = \bar{x}_{j*} + \tilde{x}_i\,, \quad \bar{z}_* = \bar{z}_{j*} + \tilde{z}_i$$

Moreover $\bar{x}_{j*}$ is orthogonal to $\tilde{x}_i$, and $\bar{z}_{j*}$ is orthogonal to $\tilde{z}_i$. Thus taking covariances of both sides of the above equations we obtain

$$\tilde{P}_i = P - P_j\,, \quad \tilde{N}_i = N - N_j$$

CONCLUSION

In this paper we have presented the CCA approach for designing reduced-order Kalman-Bucy filters and reduced-order hierarchical estimators. There are many other applications of CCA; for example, it has been used to develop reduced-complexity linear quadratic regulators [39]-[40], and it has been used to characterize the interaction between subsystems of a large scale interconnected system [39], [41]. We believe that CCA is a useful tool whenever the problem solution depends on the interaction between sets of variables. For example, in the hierarchical estimation case the sets of variables are the past and the future residuals. Another example is that of identification of a linear system which essentially depends on the interaction between the input and output. Thus CCA should prove useful in identifying reduced-order systems.

A void in the present use of CCA for model reduction is the development of some bounds on the errors incurred by using approximate models. This is currently being pursued. In this context the error analysis developed by Glover [42] for deterministic balanced approximation and optimal Hankel norm approximation should prove useful. Also, another area of research is to develop optimal Hankel norm approximation methods for reducing stochastic systems. After all, central to our approach is the decomposition of the Hankel operator. Major steps in this direction have been taken by Jonckheere and Helton [21], Harshavardhana et al. [43], and Bacon Frazho [44].

APPENDIX

First we rewrite the covariance of $y(\cdot)$ (equation (2)) as

$$R(t-s) = I\delta(t-s) + K(t-s) \tag{A1}$$

where $K(t-s) = \overline{H}e^{\overline{F}(t-s)}\overline{G}1(t-s) + \overline{G}'e^{-\overline{F}'(t-s)}\overline{H}'1(s-t)$. In operator notation (A1) will be written as

$$\mathbf{R} = \mathbf{I} + \mathbf{K}$$

Now, because of our earlier assumptions for (1), operator $\mathbf{R}$ is invertible, and its inverse is given by (see for example Kailath [34])

$$\mathbf{R}^{-1} = \mathbf{I} - \mathbf{H}$$

where the kernel $H(t,s)$ of $\mathbf{H}$ is the Fredholm resolvent of $K(t\text{-}s)$. Geesey ([33], Lemma 3.6) showed that the solution to the Riccati equation can be expressed explicitly in terms of the Fredholm resolvent. A simple extension of Geesey's result gives the following expressions for the solutions to algebraic Riccati equations (27):

$$P = \int_0^\infty\int_0^\infty\int_0^\infty \overline{C}(\sigma+\tau_1)[I\delta(\tau_1)\delta(\tau_2) - H(\tau_1,\tau_2)]\overline{C}'(\sigma+\tau_2)d\tau_1 d\tau_2 d\sigma$$

$$N = \int_{-\infty}^{0_-}\int_{-\infty}^{0_-}\int_{-\infty}^{0_-} \overline{O}'(\sigma+\tau_1)[I\delta(\tau_1)\delta(\tau_2) - H(\tau_1,\tau_2)]\overline{O}(\sigma+\tau_2)d\tau_1 d\tau_2 d\sigma \tag{A2}$$

where $\overline{O}(t) = \overline{H}e^{-\overline{F}t}$ and $\overline{C}(t) = e^{\overline{F}t}\overline{G}$. Using an operator notation (A2) can be expressed as

$$P = \tilde{\mathbf{C}} \cdot \mathbf{R}^{-1} \cdot \overline{\mathbf{C}}', \quad N = \overline{\mathbf{O}}' \cdot \mathbf{R}^{-1} \cdot \overline{\mathbf{O}} \tag{A3}$$

The + and - in (A3) represent the positive and the negative interval of integration in (A2). Before proceeding to the proofs of Lemmas 1 and 3 we first prove the following Lemma, which establishes an explicit relationship between PN (the product of the solutions to Riccati equations (27)), and the canonical vectors $u_i(\cdot)$, $v_i(\cdot)$ and the canonical correlation coefficients σ_i defined by (4).

Lemma A1: The left and right eigenvectors of PN are respectively given by

$$\int_{-\infty}^{0_-} u_i'(s)\overline{O}(s)ds \triangleq \mathbf{u}_i' \cdot \overline{\mathbf{O}} , \quad \int_0^{\infty} \overline{C}(s)v_i(s)ds \triangleq \overline{\mathbf{C}} \cdot \mathbf{v}_i$$

Proof: From (A3) we have

$$\text{PN}\,\overline{\mathbf{C}} \cdot \mathbf{v}_i = \overline{\mathbf{C}} \cdot \mathbf{R}^{-1} \cdot \overline{\mathbf{C}}'\overline{\mathbf{O}}' \cdot \mathbf{R}^{-1} \cdot \overline{\mathbf{O}}\overline{\mathbf{C}} \cdot \mathbf{v}_i$$

Now using the canonical decomposition (6), namely $R = \overline{\mathbf{O}}\overline{\mathbf{C}}$ and also successively using (5a), (5b) we get

$$\text{PN}\,\overline{\mathbf{C}} \cdot \mathbf{v}_i = \sigma_i^2\overline{\mathbf{C}} \cdot \mathbf{v}_i$$

In an analogous fashion it can be shown that $\mathbf{u}_i' \cdot \overline{\mathbf{O}}$ is the left eigenvector of PN.

Proof of Lemma 1: From Lemma A1 it follows that $\{\sigma_i^2\}_{i=1}^n$ are the eigenvalues of PN.

Proof of Lemma 3:
Part (i)

Using Lemma A1, the eigenvalue-eigenvector decomposition of PN can be expressed as

$$\text{PN} = \overline{\mathbf{C}} \cdot \mathbf{V}\Sigma^2\mathbf{U}' \cdot \overline{\mathbf{O}} \tag{A4}$$

where $\mathbf{U}$ and $\mathbf{V}$ are operator representations for $U(\mu)$ and $V(\mu)$ defined above (5a). Next, since σ_i's are distinct, (A4) can be expressed as

$$\text{PN} = \overline{\mathbf{C}} \cdot \mathbf{V}\Sigma^{-1/2}S\Sigma^2S\Sigma^{-1/2}\mathbf{U}' \cdot \overline{\mathbf{O}} \tag{A5}$$

where $S = \text{diag}[+-1]$. Now comparing (A5) with the eigenvalue-eigenvector decomposition of PN in Lemma 2, we have

$$T = S\Sigma^{-1/2}\mathbf{U}' \cdot \overline{\mathbf{O}}$$

Part (ii)

First, from (A5) it follows that

$$T^{-1} = \overline{\mathbf{C}} \cdot \mathbf{V}\Sigma^{-1/2}S$$

From Lemma 2 we have

$$F_d = T\overline{F}T^{-1} = S\Sigma^{-1/2} \int_{-\infty}^{0_-} U'(\xi)\overline{H}e^{-\overline{F}\xi}d\xi\overline{F} \int_0^{\infty} e^{\overline{F}\tau}\overline{G}V(\tau)d\tau\Sigma^{-1/2}S$$

But for $\tau \geq 0$ and $\xi < 0$,

$$\overline{H}e^{-\overline{F}\xi}\overline{F}\,e^{\overline{F}\tau}\overline{G} = \frac{\partial}{\partial\tau} R(\tau-\xi)$$

The above two equations, together with the definition of F given below (20) imply

$$F_d = S\ F\ S$$

Now, once again using Lemmas 2 and 3, and noting that $R(t-s) = \overline{O}(t)\overline{C}(s) = \overline{H}e^{\overline{F}(t-s)}\overline{G}$, for $t > s$, G_d can be expressed as

$$G_d = T\overline{G} = S\ \Sigma^{-1/2} \int_{-\infty}^{0_-} U'(\xi)\overline{H}e^{-\overline{F}\xi}d\xi\overline{G} = S\ \Sigma^{-1/2} \int_{-\infty}^{0_-} U'(\xi)R(0-\xi)d\xi$$

Next, applying the identity (5a) to the canonical decomposition (6) and using the result in the above equation gives

$$G_d = S\ \Sigma^{-1/2}\ \Sigma \int_0^{\infty} V'(\lambda)R(0-\lambda)d\lambda = SC(0) = SG$$

Analogously it can be shown that

$$H_d = HS$$

REFERENCES

[1] Priestly, M.B., *Spectral Analysis and Time Series*, Academic Press, N.Y., 1981.

[2] Kailath, T., *IEEE Trans. Inf. Theory*, V. 20, pp. 146-181, 1974.

[3] Willsky, A.S., *Digital Signal Processing and Control and Estimation Theory,* MIT Press, 1979.

[4] Burg, J.P., *Maximum Entropy Spectral Analysis,* Ph.D. dissertation, Dept. of Geophysics, Stanford University, CA, May 1975.

[5] Akaike, H., *SIAM J. Control,* V. 13, pp. 162-173, 1975.

[6] Fujishige, S., Nagai, H. and Sawaaragi, Y., *Int. J. Control,* V. 22, pp. 807-819, 1975.

[7] Baram, Y., *IEEE Trans. on Auto. Contr.,* V. AC-26, No. 6, pp. 1225-1231, 1981.

[8] Larimore, W.E., *Proc. 1983 American Contr. Conf.,* San Fransicisco, CA, pp. 445-451, 1983.

[9] Desai, U.B. and Pal, D., *Proc. 21st IEEE Conf. on Decision and Control,* Orlando, pp. 1105-1112, 1982.

[10] Desai, U.B. and Pal, D., *IEEE Trans. on Auto. Control,* V. 29, pp. 1097-1100, 1984.

[11] Desai, U.B., Pal, D., Kirkpatrick, R.D., *Int. J. Control,* V. 42, pp. 821-838, 1985.

[12] Rao, C.R., *Sankhya* Series A, Vol. 26, 1964.

[13] Arun, K.S. and Kung, S.Y., *Proc. of 22nd IEEE Conf. on Decision and Control,* San Antonio, TX, pp. 1353-1355, 1983.

[14] Vaccaro, R.J., *IEEE Trans. on Auto. Control,* V. AC-30, pp. 921-923, 1985.

[15] Moore, B.C., *IEEE Trans. on Auto. Control,* V. 26, pp. 17-32, 1981.

[16] Desai, U.B., *Proc. 17th Annual Conf. on Info. Sci. and Sys.,* The Johns Hopkins University, pp. 629-624, 1983.

[17] Desai, U.B., *Proc. of 1983 Int. Symp. on Math. Thy. of Networks and Systems*, Springer-Verlag, N.Y., 1984.

[18] Skelton, R.E., *Int. J. Control,* V. 32, pp. 1031-1055, 1980.

[19] Yousuff, A., Wagie, D.A. and Skelton, R.E., *J. Math. Analysis and Applications*, 1984.

[20] Bernstein, D.S. and Hyland, D.C., *IFAC Workshop on Model Error Concepts,* Boston, MA, 1985 (proceedings to be published by Pergamon Press 1985).

[21] Jonckheere, E.A. and Helton, J.W., *IEEE Trans. on Auto. Control,* V. AC-3-, pp. 1192-1201, 1985.

[22] Harshaavardhana, P., Jonckheere, E.A. and Silverman, L.M., *Proc. 22nd IEEE Conf. on Decision and Control,* San Antonio, TX, pp. 1260-1265, 1983.

[23] Gelfand, I.M. and Yaglom, A.M., *American Math. Soc. Translations (2)*, pp. 199-246, 1959.

[24] Faurre, P., in *System Identificaiton: Advances and Case Studies,* Ed. R.K. Mehra and D.G. Lainiotis, Academic Press, N.Y., 1976.

[25] Picci, G., *Proc. IEEE,* V. 64, pp. 112-122, 1976.

[26] Lindquist, A. and Picci, G., *SIAM J. Control and Optimization,* V. 17, pp. 365-389, 1979.

[27] Lindquist, A., Picci, G. and Ruckebusch, G., *Math. Sys. Thy.,* V. 12. pp. 271-279, 1979.

[28] Lindquist, A. and Picci, G., in *Stochastic Systems: The Mathematics of Filtering and Identification and Applications,* Ed. M. Hazewinkel and J.C. Willems, Riedel Pub. Co., pp. 169-204, 1981.

[29] Sidhu, G.S. and Desai, U.B., *IEEE Trans. on Auto. Control,* V. 21, pp. 538-541, 1976.

[30] Ljung, L. and Kailath, T., *IEEE Trans. on Info Theory,* V. 22, pp. 488-491, 1976.

[31] Desai, U.B., Pal, D. and Hsu, C.S., *Proc. 21st Conf. on Decision and Control,* Orlando, FL, pp. 1238-1288, 1982; *IEEE Trans. on Auto. Control,* V. 29, pp. 269-271, 1984.

[32] Laub, A.J., *IEEE Trans. on Auto. Control,* V. 24, pp. 913-921, 1979.

[33] Geesey, R., *Canonical Representation of Second-Order Processes with Applications,* Ph.D. dissertation, Dept. of Elect. Engrg., Stanford University, CA, 1968.

[34] Kailath, T. *IEEE Trans. on Infor. Theory,* V. 15, pp. 665-672, 1969.

[35] Pernebo, L. and Silverman, L.M., *IEEE Trans. on Auto. Control,* V. 27, pp. 382-287, 1982.

[36] Weinert, H.L., Private Communication, 1983.

[37] Desai, U.B. and Kiaei, S., *Proc. of the 24th IEEE Conf. on Decision and Control,* Fort Lauderdale, FL, pp. 416-421, 1985.

[38] Timm, N.H. and Carlson, J.E., *Psychometrika,* 41, pp. 159-176, 1976.

[39] Desai, U.B., *IFAC Workshop on Model Error Concepts,* Boston, MA 1985 (proceedings to be published by Pergamon Press 1986).

[40] Desai, U.B. and Banerjee, S., *Proc. of the American Control Conf.,* San Diego, CA, pp. 160-165, 1984; *Automatica,* 1986.

[41] Desai, U.B., Benerjee, S. and Kiaei, S., *Proc. of the 23rd IEEE Conf. on Decision and Control,* Las Vegas, NV, pp. 1523-1528, 1984.

[42] Bacon, B.J. and Frazho, A.E., *IEEE Trans. on Auto. Control,* V. AC-30, pp. 1138-1140, 1985.

4

GENERALIZED PRINCIPAL COMPONENTS ANALYSIS AND ITS APPLICATION IN APPROXIMATE STOCHASTIC REALIZATION

by

K. S. Arun
Univ. of Illinois at Urbana-Champaign
Coordinated Science Lab.
Urbana, IL 61801

S. Y. Kung
Univ. of Southern California
Signal and Image Processing Inst.
Los Angeles, CA 90089-0272

ABSTRACT

The state of a linear system is an information interface betwen the past and the future, and approximate realization is essentially a problem of approximating an apparently high-dimensional interface by a low-order partial state. In this chapter, we generalize the ideas of principal components to the problem of approximating the information interface between two random vectors. Two such generalizations exist in the statistical literature [1, 2]. Applications of these generalizations to the partial-state selection problem lead to three approximate stochastic realization methods. We discuss these methods and the partial-state selection criterion that each optimizes. We also study their relation to determinstic identification and balanced model reduction.

INTRODUCTION

The problem of modeling a stochastic process as the output of a linear, rational, discrete-time system driven by white noise, arises in many applications. In almost all such applications, there *is* no low-order model that generates the stochastic process. In addition, the given covariance lags (that are either estimated from a finite record of the process, or are directly measured,) are perturbed, and no low order model will exactly match the given covariance sequence. Consequently, the problem reduces to that of fitting a low-order approximate model to the given covariances.

Some applications

In signal processing applications, approximate stochastic realization is an integral part of modern methods for high-resolution, power spectrum estimation. Some other applications where approximate stochastic realization comes into play are: Speech analysis for recognition and for coding; digital control (first, the plant has to be modelled); econometrics for forecasting and production control; meteorology for weather forecasting; and image analysis for coding and for classification purposes. Three other applications are outlined below.

Seismic deconvolution: An interesting application arises in blind seismic deconvolution, where we are given a seismic trace, and do not know either the source-wavelet or the earth's reflectivity sequence. Assuming that the reflectivity sequence is a Bernoulli-Gaussian white-noise process, one can treat the wavelet-estimation problem as a stochastic-system identification problem. The source-wavelet is interpreted as the unknown impulse-response of the hypothetical system, the reflectivity-sequence is the white input, and the seismic trace (which is the convolution of the source-wavelet with the reflectivity sequence) is the output process. We thus have a stochastic realization problem on our hands.

Interference spectrometry: The interference spectrometry problem is an application where the stochastic process $y(t)$ is not available, while its covariances are directly measured. The problem here is that of estimating the power (intensity) distribution $P(\sigma)$ of an electro-magnetic source (say an absorption spectrum or an emission spectrum) as a function of wave-number σ [3, 4]. In the classical, Michelson two-beam interferometer, the radiation is broken into two beams, and a variable path difference x is introduced in one. The two beams are recombined and the resultant total intensity (integrated over all frequencies/wave-numbers) $I(x)$ is measured for discrete values of the path difference x. This total intensity varies with the path difference x as

$$I(x) = \int_{\sigma=0}^{\infty} P(\sigma)\,d\sigma + \int_{\sigma=0}^{\infty} P(\sigma)\cos(2\pi\sigma x)\,d\sigma .$$

The variable part of this intensity is the so-called interference function, whose Fourier transform is the intensity spectrum $P(\sigma)$. Thus, the interfence function plays the role of the covariance sequence $r(m)$, and the problem is to estimate the power spectrum $P(\sigma)$ from measurements of $r(m)$ over a finite range of values for m.

This is an application where covariances are directly measured and the stochastic process $y(t)$ itself is not available. In other applications, the stochastic process may be given and covariances are estimated by time-averaging.

High-resolution power spectrum estimation: In the high-resolution spectrum estimation problem, we are given the estimates or measurements of the first few covariance lags $r(m)$, $m=0,1,2,\ldots,M$, of an underlying stochastic process, and we wish to estimate its power spectrum

$$P(\omega) = \sum_{m=-\infty}^{+\infty} r(m)e^{-j\omega m},$$

with high resolution in frequency. The covariance lags may be estimated using time-averages on a single sample sequence of the stochastic process. If we assume the unknown lags $r(m)$, $|m|>M$, are zero and estimate $P(\omega)$ as a finite Fourier sum, then the resultant spectral estimate is unable to resolve closely-spaced spectral peaks and instead introduces other spurious oscillations in the spectrum. These spurious peaks may be reduced by premultiplying the given covariance lags with a window function to provide a smooth transition to zero in the covariance sequence. However the resolution in the spectral estimate is still limited to aproximately the length ($2M$) of the given covariance segment.

The modern approach to the problem of obtaining high-resolution spectral estimates, is to extrapolate the covariance sequence outside the observation window, with the help of some prior information about the stochastic process. The currently popular approach uses linear, rational models for the covariance extrapolation. The process is modelled as the output of a linear, rational model driven by white noise. The method of maximum entropy [5], for instance, uses an all-pole model of order (M) for the purpose. The maximum entropy method owes its popularity to the computational simplification provided by the Levinson algorithm [6, 7] that generates estimates of the Mth order all-pole model's parameters in $O(M^2)$ time from M covariance lags. The maximum entropy method assumes the M covariance lags are exactly known, and as a consequence, is sensitive to covariance estimation errors and other forms of perturbations. It provides resolution far superior to Fourier transform methods at the expense of introducing many spurious peaks. The most recent trend in model-based spectral estimation is to use the more general, linear, rational model with arbitrary zero-locations inside the unit circle, in place of all-pole models which have all zeros at the origin. More significant has been the trend towards covariance approximation instead of covariance matching. The latter is a concession to the fact that the given covariance lags are themselves estimates or measurements, and are invariably tainted by errors.

Recent methods use low-order pole-zero models that approximate the given covariance lags and provide a non-zero extension outside the observation window. This is essentially an approximate stochastic realization problem. Once the model has been realized, the power spectrum estimate is very simply,

$$P(\omega) = \rho \left| H(e^{j\omega}) \right|^2$$

where ρ is the variance of the input white noise driving the model, and $H(z)$ is the model transfer function. Such an approach gives good spectral estimates only if the model is appropriate. For speech data, for instance, 4th–5th order rational models provide excellent spectral estimates. Heuristically, we need one pole per formant, and a few zeros to model nasal sounds. Another very popular application area is the estimation of frequencies of sinusoids that are embedded in additive white noise. In such cases, a rational model cannot reproduce the line-spectra accurately, but the model's poles provide good estimates of the locations of these spectral lines. On the other hand, a Fourier transform based spectral estimate cannot resolve closely-spaced spectral lines, and a model-based approximate realization approach is the better choice.

The approach

In this chapter, we will assume that we are directly given covariance measurements/estimates, and are concerned solely with model estimation from the covariance data. Thus, the methods described here, are applicable to problems like interference spectrometry, where only covariance is available, and the process itself is unavailable. It is important to note however, that only a *finite* segment of the covariance sequence is given, and even that is *perturbed* by estimation or measurement errors.

In this chapter, we will formulate the covariance approximation problem as one of selecting a partial-state vector that best condenses the (linear) past-future interdependence in the process. The state of a system is an information interface between the past and the future, and its size is equal to the system order. When the system has to be approximated by a low-order model, it is thus a question of constructing a "partial state" from the significant components of this information interface. We will develop a partial-state selection criterion that measures the partial state's efficiency in summarizing the past for predicting the future. The principal components approach is then derived as an optimal solution that maximizes this efficiency amongst all possible models of a given order.

PROBLEM FORMULATION

The model

In state-space notation, a pth order linear, rational, discrete-time model for $y(t)$, is

$$\begin{aligned} \mathbf{x}(t+1) &= \mathbf{F}\mathbf{x}(t) + \mathbf{T}v(t) \\ y(t) &= \mathbf{h}\mathbf{x}(t) + v(t) \end{aligned} \qquad \text{(a)}$$

where $v(t)$ is a white input sequence of variance ρ, $\mathbf{x}(t)$ is a px1 state vector process and $\mathbf{F}$, $\mathbf{T}$, and $\mathbf{h}$ are constant matrices of sizes pxp, px1 and 1xp respectively. Henceforth, bold letters and upper case Greek letters will be used to denote matrices and vectors, and the transpose operator will be denoted by a superscript t.

In terms of the state-space parameters, the transfer function of the system is given by

$$H(z) = \mathbf{h}(z\mathbf{I}-\mathbf{F})^{-1}\mathbf{T} + 1,$$

the poles of the model are the eigenvalues of $\mathbf{F}$, while the zeros are the eigenvalues of the matrix $(\mathbf{F}-\mathbf{Th})$. It can be shown that the relationship between the impulse response of the model and the state-space parameters is

$$i(k) = \mathbf{h}\mathbf{F}^{k-1}\mathbf{T}, \qquad k>0.$$

For a given transfer function, the parameter-triple $(\mathbf{F},\mathbf{T},\mathbf{h})$ is unique (if the order p is minimal) modulo a similarity (coordinate) transformation. For any invertible pxp matrix $\mathbf{Q}$, the transformed triple $(\mathbf{Q}^{-1}\mathbf{F}\mathbf{Q}, \mathbf{Q}^{-1}\mathbf{T}, \mathbf{h}\mathbf{Q})$ is also a valid choice for the state-space parameters, and it corresponds to a transformed coordinate system for the state. The new state is $\mathbf{Q}^{-1}\mathbf{x}$ instead of $\mathbf{x}$, but they both span the same space, and the input-output relationship and system transfer function is left unchanged.

An interesting set of coordinates that we will be using in this chapter is the so-called internally-balanced coordinate system [8]. The balanced realization is a special case of the principal-axis realization introduced in [9]. The principal axis realization is characterized by both the observability grammian **W** and the controllability grammian **K** being diagonal. In general, these grammians are the solutions of the following two Lyapunov equations [10]:

$$\mathbf{K} = \mathbf{FKF}^t + \mathbf{TT}^t$$

$$\mathbf{W} = \mathbf{F}^t\mathbf{WF} + \mathbf{h}^t\mathbf{h}$$

and are also explicitly given by

$$\mathbf{K} = \begin{bmatrix} \mathbf{T} & \mathbf{FT} & \mathbf{F}^2\mathbf{T} & \mathbf{F}^3\mathbf{T} & \cdots \end{bmatrix} \begin{bmatrix} \mathbf{T}^t \\ \mathbf{T}^t\mathbf{F}^t \\ \mathbf{T}^t\mathbf{F}^{t^2} \\ \mathbf{T}^t\mathbf{F}^{t^3} \\ \cdot \\ \cdot \\ \cdot \end{bmatrix} = \mathbb{C}\mathbb{C}^t$$

and

$$\mathbf{W} = \begin{bmatrix} \mathbf{h}^t & \mathbf{F}^t\mathbf{h}^t & \mathbf{F}^{t^2}\mathbf{h}^t & \mathbf{F}^{t^3}\mathbf{h}^t & \cdots \end{bmatrix} \begin{bmatrix} \mathbf{h} \\ \mathbf{hF} \\ \mathbf{hF}^2 \\ \mathbf{hF}^3 \\ \cdot \\ \cdot \\ \cdot \end{bmatrix} = \mathbb{O}^t\mathbb{O}$$

In linear systems terminology, the matrices $\mathbb{O}$ and $\mathbb{C}$ are known as the observability matrix and controllability matrix respectively. Note that these matrices and the two grammians are not unique for a given transfer-function, and change with the state coordinates. A transformation of the state from $\mathbf{x}$ to $\mathbf{Q}^{-1}\mathbf{x}$ changes the observability matrix to $\mathbb{O}\mathbf{Q}$ and the controllability matrix to $\mathbf{Q}^{-1}\mathbb{C}$, while changing the grammians to $\mathbf{Q}^{-1}\mathbf{K}\mathbf{Q}^{t^{-1}}$ and $\mathbf{Q}^t\mathbf{W}\mathbf{Q}$. A transformation $\mathbf{Q}$ that simultaneously diagonalizes both the grammians can always be found, and a principal-axis realization always exists [9]. In fact, for any given transfer-function, many such principal-axis realizations exist, and the balanced realizaton is one of them.

A realization is said to be internally-balanced [8] if the grammians **K** and **W** are not only diagonal, but also equal to each other:

$$\mathbf{K} = \mathbf{W} = \Sigma, \qquad \text{where} \quad \Sigma = \mathrm{diag}\,(\sigma_1, \sigma_2, \ldots, \sigma_p).$$

An important property of $\mathcal{O}$ that follows from the shift-invariance of the model, is that it satisfies

$$\begin{bmatrix} \mathbf{h} \\ \mathbf{hF} \\ \mathbf{hF}^2 \\ \mathbf{hF}^3 \\ \cdot \\ \cdot \\ \cdot \end{bmatrix} \mathbf{F} = \begin{bmatrix} \mathbf{hF} \\ \mathbf{hF}^2 \\ \mathbf{hF}^3 \\ \mathbf{hF}^4 \\ \cdot \\ \cdot \\ \cdot \end{bmatrix}$$

Hence, if we denote the matrix formed by deleting the last row of $\mathcal{O}$ by $\mathcal{O}_1$, and the matrix formed by deleting the first row of $\mathcal{O}$ by $\mathcal{O}_2$, then the above equation may be rewritten as

$$\mathcal{O}_1 \bullet \mathbf{F} = \mathcal{O}_2. \tag{b}$$

and $\mathbf{F}$ may be obtained from $\mathcal{O}$ as $\mathbf{F} = \mathcal{O}^{-1}\mathcal{O}_2$.

The notion of state

Before we formulate the approximate stochastic modeling problem as one of partial-state selection, we need to first develop a stochastic definition for the state of a system, along the lines of [11], when the input to the system is a white random process of variance ρ. Here, by white, we only mean second-order white, i.e. a sequence of zero-mean, uncorrelated random variables, all with the same variance:

$$\mathbf{E}\left[v(t)v(t+m)\right] = \begin{cases} \rho & m=0 \\ 0 & m\neq 0 \end{cases}$$

where $\mathbf{E}[\bullet]$ denotes the expectation operator.

For a zero-mean $n\mathrm{x}1$ random vector $\mathbf{Y} = (y_1, y_2, \ldots, y_n)^t$, Span($\mathbf{Y}$) will denote the Hilbert space of all random variables that are linear combinations of $\{ y_1, y_2, \ldots, y_n \}$. The inner product on this space of zero-mean random variables is the cross covariance, and its dimension (upper bounded by n) is the largest number of mutually uncorrelated random variables in the space.

We will use the notation $\mathbf{x}\backslash\mathbf{Y}$ to denote the linear, minimum-variance estimator of zero-mean random vector $\mathbf{x}$ from the zero-mean random vector $\mathbf{Y}$. It is also the orthogonal projection of $\mathbf{x}$ onto the subspace Span($\mathbf{Y}$). From elementary estimation theory [12], we know that

$$x\backslash Y = E\left[xY^t\right]\left(E\left[YY^t\right]\right)^{-1} Y. \qquad (c)$$

Here, E[•] denotes the expectation operator.

When the input is a white-noise process, the past and future inputs are uncorrelated, and as a result, the following two components of the future output vector $\mathbf{Y}^+$:

$$\begin{pmatrix} y(t) \\ y(t+1) \\ y(t+2) \\ \cdot \\ \cdot \\ \cdot \end{pmatrix} = \begin{pmatrix} i(1) & i(2) & i(3) & \cdots \\ i(2) & i(3) & i(4) & \cdots \\ i(3) & i(4) & i(5) & \cdots \\ \cdot & \cdot & \cdot & \cdots \\ \cdot & \cdot & \cdot & \cdots \\ \cdot & \cdot & \cdot & \cdots \end{pmatrix} \begin{pmatrix} v(t-1) \\ v(t-2) \\ v(t-3) \\ \cdot \\ \cdot \\ \cdot \end{pmatrix}$$

$$+ \begin{pmatrix} i(0) & 0 & 0 & \cdots \\ i(1) & i(0) & 0 & \cdots \\ i(2) & i(1) & i(0) & \cdots \\ \cdot & \cdot & \cdot & \cdots \\ \cdot & \cdot & \cdot & \cdots \\ \cdot & \cdot & \cdot & \cdots \end{pmatrix} \begin{pmatrix} v(t) \\ v(t+1) \\ v(t+2) \\ \cdot \\ \cdot \\ \cdot \end{pmatrix}$$

$$\mathbf{Y}^+ = \boldsymbol{\mathcal{H}}\,\mathbf{V}^- + \mathbf{L}\mathbf{V}^+$$

are orthogonal. Consequently, the orthogonal projection of $\mathbf{Y}^+$ on Span($\mathbf{V}^-$) must be $\boldsymbol{\mathcal{H}}\,\mathbf{V}^-$ itself

$$\text{i.e. } \mathbf{Y}^+\backslash\mathbf{V}^- = \boldsymbol{\mathcal{H}}\,\mathbf{V}^-.$$

However, it can be shown that this information is completely summarized in the state. Recall that the impulse response i(k) is related to the parameters as

$$i(k) = \mathbf{h}\mathbf{F}^{k-1}\mathbf{T} \qquad \forall \ t > 0.$$

This indicates that the infinite Hankel matrix $\boldsymbol{\mathcal{H}}$ factors into the product of the matrices $\mathbb{O}$ and $\mathbb{C}$.

$$\boldsymbol{\mathcal{H}} = \mathbb{O} \cdot \mathbb{C}.$$

Moreover, the state-transition equation $\mathbf{x}(t+1) = \mathbf{F}\mathbf{x}(t) + \mathbf{T}v(t)$ indicates that the state is a linear combination of past inputs:

$$\mathbf{x}(t) = \begin{bmatrix} \mathbf{T} & \mathbf{FT} & \mathbf{F}^2\mathbf{T} & \cdots \end{bmatrix} \begin{bmatrix} v(t-1) \\ v(t-2) \\ v(t-3) \\ \cdot \\ \cdot \\ \cdot \end{bmatrix} = \mathbb{C}\,\mathbf{V}^-.$$

Therefore, $\mathbf{Y}^+ \backslash \mathbf{V}^- = \mathbb{H}\,\mathbf{V}^- = \mathbb{O}\,\mathbf{x}$, which is a mathematical restatement of the fact that the state condenses all the information in the past input that is sufficient for predicting the future output.

Minimum-phase model (Innovations representation): In addition, when the model is minimum phase, we can show that the state can be also interpreted as a summary of the past *output* history (instead of past input history) for prediction of the future output.

Since the minimum phase model has all zeros within the unit circle, it has a stable inverse. The inverse filter is obtained by simply rearranging Eq. (a) as:

$$\mathbf{x}(t+1) = (\mathbf{F}-\mathbf{Th})\mathbf{x}(t) + \mathbf{T}y(t)$$

$$v(t) = -\mathbf{hx}(t) + y(t).$$

The minimum-phase property ensures that the zeros of the model, which are the eigenvalues of $(\mathbf{F}-\mathbf{Th})$ lie within the unit circle. But these eigenvalues are precisely the poles of the inverse filter, hence the inverse filter must be stable. Thus the state-process $\mathbf{x}(t)$ as well as the input process $v(t)$ can be obtained causally from the output $y(t)$ using the above filter. The state transition equation of the inverse filter indicates that

$$\mathbf{x}(t) = \begin{bmatrix} \mathbf{T} & (\mathbf{F}-\mathbf{Th})\mathbf{T} & (\mathbf{F}-\mathbf{Th})^2\mathbf{T} & (\mathbf{F}-\mathbf{Th})^3\mathbf{T} & \cdots \end{bmatrix} \bullet \begin{bmatrix} y(t) \\ y(t+1) \\ y(t+2) \\ y(t+3) \\ \cdot \\ \cdot \\ \cdot \end{bmatrix} = \mathbf{\Psi Y}^-(t)$$

Thus, for the minimum phase model, we can write

$$\mathbf{Y}^+ = \mathbb{O}\,\mathbf{\Psi Y}^- + \mathbf{LV}^+$$

When the input is white, the two terms are orthogonal, since the future input $\mathbf{V}^+$ is uncorrelated with the past output $\mathbf{Y}^-$. Hence,

$$Y^+ \backslash Y^- = \mathcal{O}\Psi Y^- = \mathcal{O}x.$$

Thus, for the minimum phase model, we have

$$x = \Psi Y^-, \quad \text{and} \quad Y^+ \backslash Y^- = \mathcal{O}x \qquad \text{(d)}$$

which means that the state of the minimum phase model summarizes the past *output* history for predicting the future output.

As a footnote, Eq. (d) indicates that the projection of $y(t)$ on the past space Span(Y^-) is nothing but

$$y(t) \backslash Y^- = hx(t).$$

Therefore, the part of $y(t)$ that cannot be predicted from the past Y^- is simply

$$y(t) - hx(t) = v(t).$$

Thus, the input white noise to a minimum phase model is the innovations process [13] for the output, and consequently, the minimum phase model is also called the innovations representation (IR) of the output process [14, 15].

Partial state selection

Using Eq. (c), one can verify that

$$Y^+ \backslash Y^- = HR^{-1}Y^-$$

where $R = E(Y^- Y^{-'})$ and $H = E(Y^+ Y^{-'})$ are the Toeplitz and Hankel matrices respectively, formed from the covariance lags of the output process. Combining this equation with Eq. (d):

$$x = \Psi Y^- \quad \text{and} \quad Y^+ \backslash Y^- = \mathcal{O}x,$$

leads to the following observations.

(1) HR^{-1} equals $\mathcal{O}\Psi$. Consequently HR^{-1} must have rank equal to the size of the state vector (i.e. equal to the model order p).

(2) Moreover, $HR^{-1}Y^- = \mathcal{O}x$, which means there are only p independent components in the space spanned by the infinite random variables of the vector $HR^{-1}Y^-$. In other words, the dimension of Span($HR^{-1}Y^-$) is equal to the model order p, and the state x is any basis for this space. Since the elements of $Y^+ \backslash Y^-$ all lie in this space, this space is the required summary of all the information in Y^- regarding Y^+. Consequently, Span($HR^{-1}Y^-$) is the state-space of the model, and any basis for this space is a valid state for the model.

Thus, the stochastic realization problem is, simply stated, the problem of picking a basis for Span($\mathbf{HR}^{-1}\mathbf{Y}^-$) [16].

However, when the covariance lags are estimated from a finite record of the stochastic process or are directly measured, then the perturbations in the lags will distort the rank structure of $\mathbf{HR}^{-1}$. It will have full rank, making the apparent state size much larger than the true model order. Then the problem is one of constructing a partial state from those components in Span($\mathbf{Y}^-$) that contain the most information regarding $\mathbf{Y}^+$. This partial-state must "effectively" summarize the information interface between $\mathbf{Y}^+$ and $\mathbf{Y}^-$.

Note that the problem is one of compressing $\mathbf{Y}^-$ while retaining maximal information not about $\mathbf{Y}^-$, but about $\mathbf{Y}^+$. Hence, the principal components analysis of $\mathbf{Y}^-$ will not suffice [17, 18], for the partial-state selection problem.[1] The compression of $\mathbf{Y}^-$ into its principal components is not appropriate because it is based on the selection of components containing the maximum information about $\mathbf{Y}^-$ itself, whereas only specific information about $\mathbf{Y}^+$ is of interest in the partial-state selection problem.

However, there exist in the statistical literature, generalizations of the concept of principal components (of a random vector) to the problem of compressing the information interface between two random vectors (which will henceforth be referred to as the 2-vector problem for the sake of brevity). We will present three approaches to approximate stochastic realization as applications of three such generalizations.

THREE APPROXIMATE REALIZATION METHODS

For a zero-mean nx1 random vector $\mathbf{Y}$, the p principal components of $\mathbf{Y}$ -

- are maximally correlated with $\mathbf{Y}$,
- have maximum self-information in the Gaussian case,
- and retain the maximum reconstruction (prediction) efficiency for $\mathbf{Y}$.

Generalizing these three properties to the 2-vector problem leads to the three methods of this section.

[1] Note that the covariance matrix $\mathbf{R}$ is not expected to have rank equal to the model order, even when the lags are exact. Hence, in the perturbed situation, a principal components approximation of $\mathbf{R}$ is not justified.

The principal components of H

Taking a hint from the correlation-maximizing property of the principal components of a single random vector, we could look for a partial-state in Span($\mathbf{Y}^-$) that maximizes some measure of its correlation with $\mathbf{Y}^+$. For instance, we could pick[2]

$$\text{px1 sized } \mathbf{x}_{\text{partial}} = \Psi\mathbf{Y}^- \text{ to } \underset{\text{constraint:}\Psi\Psi^t=\mathbf{I}_p}{\text{maximize}} \left\| \mathbf{E}\left[\mathbf{Y}^+\mathbf{x}^t_{\text{partial}}\right] \right\|_F \qquad \text{(e)}$$

The solution to this is constructed from the principal components of the covariance Hankel matrix $\mathbf{H}$. The rows of Ψ must be the orthonormal singular vectors of $\mathbf{H}$ corresponging to the p largest singular values. If the singular value decomposition (SVD) of $\mathbf{H}$ is

$$\mathbf{H} = \mathbf{U}\Sigma\mathbf{V}^t = \mathbf{U}_1\Sigma_1\mathbf{V}_1^t + \mathbf{U}_2\Sigma_2\mathbf{V}_2^t.$$

(where the subscript 1 stands for the dominant components corresponding to the p largest singular values,) then the solution to the minimization problem of Eq. (e) is $\Psi = \mathbf{V}_1^t$. This justifies the principal components approximation of $\mathbf{H}$, that has been extensively used for aproximate stochastic modeling [19, 20, 21, 22, 23]. We will henceforth refer to this approximation as the PC-H approximation.

Discussion of the PC-H approximation: The PC-H method picks p orthonormal combinations of the past ($x_k=\Psi_k\mathbf{Y}^-$, $\Psi_k\Psi_j^t$ = Kronecker δ_{ij}), that are maximally correlated with the future. But there is always the distinct possibility that these components x_k have large values of $\| \mathbf{E}[\mathbf{Y}^+x_k] \|$, only because they have large variances $\mathbf{E}|x_k|^2$ and not because they contain the most information about $\mathbf{Y}^+$. In such a case, the partial-state-components picked by the PC-H method, need not be the ones that best predict the future. Moreover, the components ($x_k=\Psi_k\mathbf{Y}^-$) are not mutually orthogonal ($\Psi_k\Psi_j^t=0$ does not imply that $\mathbf{E}[x_kx_j]=0$), and so there is redundancy of information between the different components. Consequently, $\sum_{\text{all } k} \| \mathbf{E}[\mathbf{Y}^+x_k] \|_2^2$ is not a good measure of the information in the partial-state regarding the future.

It appears that the PC-H criterion of Eq. (e) is not appropriate for the partial-state selection problem. The underlying reason is that the components are

---

[2]For any matrix $\mathbf{A}$, $\|\mathbf{A}\|_F$ denotes the Euclidean norm, Trace ($\mathbf{A}^t\mathbf{A}$).

not required to be normalized and mutually orthogonal. Hence, if we remove the $\Psi\Psi^t = I_p$ constraint in the PC-H criterion and instead constrain the partial-state components to be orthonormal, then we will get a better criterion for the approximation of the state-space.

The canonical correlations criterion

One such criterion was first proposed in statistics by Hotelling [1], and later used for the partial-state selection problem by Akaike [16, 11]. Here, any orthonormal basis $\mathbf{Z}^+$ is found for Span($\mathbf{Y}^+$), and the p partial-state components are selected as p orthonormal random variables from Span($\mathbf{Y}^-$) that have the maximum correlation with $\mathbf{Z}^+$. The constraint that the partial-state components be orthonormal translates into the constraint:

$$E\left[\mathbf{x}_{\text{partial}}\mathbf{x}_{\text{partial}}^t\right] = \Psi\mathbf{R}\Psi^t = \mathbf{I}_p.$$

If $\mathbf{R}^{1/2}$ is any square root of $\mathbf{R}$ (i.e. $\mathbf{R} = \mathbf{R}^{1/2}\mathbf{R}^{1/2^t}$), and $\mathbf{R}^{-1/2}$ is its inverse, then one choice for $\mathbf{Z}^+$ is $\mathbf{R}^{-1/2}\mathbf{Y}^+$, and so our problem is to

$$\underset{\text{constraint:}\Psi\mathbf{R}\Psi^t=\mathbf{I}_p}{\text{Maximize}} \left\| E\left[\mathbf{R}^{-1/2}\mathbf{Y}^+\mathbf{x}_{\text{partial}}^t\right] \right\|_F = \left\| \mathbf{R}^{-1/2}\mathbf{H}\Psi^t \right\|_F$$

The solution to this constrained optimization problem is constructed from the principal singular vectors of $\mathbf{R}^{-1/2}\mathbf{H}\mathbf{R}^{-1/2^t}$.

$$\mathbf{R}^{-1/2}\mathbf{H}\mathbf{R}^{-1/2^t} = \mathbf{U}\Sigma\mathbf{V}^t = \mathbf{U}_1\Sigma_1\mathbf{V}_1^t + \mathbf{U}_2\Sigma_2\mathbf{V}_2^t.$$

where as before, the subscript 1 denotes the principal components in the SVD, and

$$\Psi = \mathbf{V}_1^t\mathbf{R}^{-1/2}.$$

Though the square root of $\mathbf{R}$ is not unique, different choices of $\mathbf{R}^{1/2}$ will not change the singular values Σ of $\mathbf{R}^{-1/2}\mathbf{H}\mathbf{R}^{-1/2^t}$. Although the singular vectors $\mathbf{U}$ and $\mathbf{V}$ depend on the choice of $\mathbf{R}^{-1/2}$, the composition $\Psi = \mathbf{V}_1^t\mathbf{R}^{-1/2}$ is the same for all choices of the square root.

The singular values of $\mathbf{R}^{-1/2}\mathbf{H}\mathbf{R}^{-1/2^t}$ are the canonical correlation (c.c.) coefficients between the past $\mathbf{Y}^-$ and the future $\mathbf{Y}^+$ [1]. It was shown by [24, 25], that for the Gaussian case, the c.c. coefficients between $\mathbf{Y}^+$ and $\mathbf{Y}^-$ provide a measure of the *mutual information* between $\mathbf{Y}^-$ and $\mathbf{Y}^+$. A heuristic derivation of the formula for the mutual information between $\mathbf{Y}^+$ and $\mathbf{Y}^-$ follows.

It can be shown that the information in a Gausian random vector is the logarithm of the determinant of its covariance matrix (c.f. Appendix B in [7]). For instance, the information (or uncertainity) in $\mathbf{Y}^+$ is $0.5\log(\det \mathbf{R})$. Similarly, the uncertainity in $\mathbf{Y}^+$ that is not explained by $\mathbf{Y}^-$ is obtained from the conditional variance of $\mathbf{Y}^+$ conditioned on $\mathbf{Y}^-$ being given, and is equal to $0.5\log[\det(\mathbf{R} - \mathbf{H}\mathbf{R}^{-1}\mathbf{H}^t)]$. The mutual information between $\mathbf{Y}^+$ and $\mathbf{Y}^-$ is by definition, the uncertainity in $\mathbf{Y}^+$ that is resolved by knowledge of $\mathbf{Y}^-$ (or vice-versa). It is therefore,

$$0.5\log(\det\mathbf{R}) - 0.5\log\left|\det(\mathbf{R} - \mathbf{H}\mathbf{R}^{-1}\mathbf{H}^t)\right|$$

$$= -0.5\log\left|\det(\mathbf{I} - \mathbf{R}^{-1/2}\mathbf{H}\mathbf{R}^{-1}\mathbf{H}^t\mathbf{R}^{-1/2^t})\right|$$

$$= -0.5\sum_{\text{all } k}\log(1 - \sigma_k^2)$$

where the σ_k-s are the c.c. coefficients.

The canonical components of $\mathbf{Y}^-$ (with respect to $\mathbf{Y}^+$) are $x_k = \mathbf{v}_k^t\mathbf{R}^{-1/2}\mathbf{Y}^-$ and the mutual information between each x_k and $\mathbf{Y}^+$ is $-0.5\log(1-\sigma_k^2)$. Thus, the p components from Span($\mathbf{Y}^-$) that maximize the mutual information with $\mathbf{Y}^+$ are the p canonical components $x_1, x_2, \ldots, x_p$ with the p largest c.c. coefficients. Just as the principal components of a random vector maximize the self-information content, the canonical-components approximation maximizes the mutual information in the 2-vector problem.

Thus, it seems that a natural choice for the components of the partial state are the canonical components of $\mathbf{Y}^-$ that have the largest c.c. coefficients, and consequently, the maximum mutual information (with respect to $\mathbf{Y}^+$). Akaike first suggested the use of c.c. analysis for partial-state selection, and subsequently, many approximate modeling algorithms have been proposed [26, 27, 28] , that use such an approximation.

Relation to stochastic balancing: Desai and Pal [27, 29] have made an important contribution to stochastic realization by pointing out that the non-zero c.c. coefficients between $\mathbf{Y}^-$ and $\mathbf{Y}^+$ can be obtained from the full-order model as the eigenvalues of $\mathbf{P}^{*-1}\mathbf{P}_*$, where $\mathbf{P}_*$ is the state-variance of the minimum-phase model, and $\mathbf{P}^*$ is the state-variance of the corresponding maximum-phase model (with the same $\mathbf{F}$ and $\mathbf{h}$ matrices). For every minimum-phase model $(\mathbf{F},\mathbf{T},\mathbf{h},\rho)$ that generates a process $y(t)$, there exists a maximum-phase model $(\mathbf{F},\mathbf{T}^*,\mathbf{h},\rho^*)$ driven by a different white-noise sequence, that generates the same output process $y(t)$. The two models can be put in coordinate systems where their $\mathbf{F}$ and $\mathbf{h}$ matrices are the same, and they differ only in their input parameters: $\mathbf{T}$ and ρ. Then, $\mathbf{P}_*$ is the solution of the Lyanupov equation

$$\mathbf{P}_* \;=\; \mathbf{F}\mathbf{P}_*\mathbf{F}^t + \rho\mathbf{T}\mathbf{T}^t$$

and $\mathbf{P}^*$ is the solution of the other Lyapunov equation

$$\mathbf{P}^* \;=\; \mathbf{F}\mathbf{P}^*\mathbf{F}^t + \rho^*\mathbf{T}^*\mathbf{T}^{*t}.$$

Desai and Pal [27, 29] show that the product $\mathbf{P}^{*-1}\mathbf{P}_*$ is invariant to coordinate transformations of the state, and that they are in fact, the squares of the non-zero c.c. coefficients between the past and the future of the process. There exists a certain coordinate-system, the so-called stochastically-balanced coordinates, where $\mathbf{P}_*=\mathbf{P}^{*-1}$ and they are both diagonal. In the stochastically-balanced coordinates, since $\mathbf{P}^{*-1}=\mathbf{P}_*$, are both diagonal, the entries of $\mathbf{P}_*$ are basically the c.c. coefficients themselves. Therefore, the partial-state picked by the c.c. approach is composed from those elements of the stochastically balanced, full-order, state of the minimum-phase model, that have the largest variances, since the c.c. coefficients are proportional to these variances.

Discussion of mutual information: The mutual-information measure may be better understood by looking at the orthogonal decomposition of $\mathbf{Y}^+$ into

$$\mathbf{Y}^+ = \mathbf{H}\mathbf{R}^{-1}\mathbf{Y}^- + \mathbf{L}\mathbf{V}^+ = \mathbf{W} + \Delta$$

Recollect that $\mathbf{Y}^-$ contains no information about $\mathbf{L}\mathbf{V}^+$ (they are orthogonal to each other), and one is not predictable from the other. Hence, we might expect $\mathbf{Y}^-$ to contain equal amount of information regarding $\mathbf{Y}^+$ and $\mathbf{W}$. However, the mutual information between $\mathbf{Y}^-$ and $\mathbf{W}$ is larger than the mutual information between $\mathbf{Y}^-$ and $\mathbf{Y}^+$.

That is the case, because the mutual information between $\mathbf{Y}^-$ and $\mathbf{Y}^+$ is a measure of the information in $\mathbf{Y}^-$ regarding $\mathbf{Y}^+$ relative to the total self-information in $\mathbf{Y}^+$. We have

$$\text{Mutual–Info.}\,(\mathbf{Y}^-,\mathbf{Y}^+) = \text{Info.}\,(\mathbf{Y}^+) - \text{Info.}\,(\Delta) \neq \text{Info.}\,(\mathbf{W})$$

Hence, the value of our measure depends very much on Δ even though Δ is orthogonal to $\mathbf{Y}^-$. With regards to the approximation problem, this property means that the canonical components (of $\mathbf{Y}^-$) with the largest mutual information (with respect to $\mathbf{Y}^+$) can be dramatically changed simply by changing Δ. This fact is illustrated in [30]

Moreover, it is our contention that

Claim: The mutual information between a canonical component x_k and the future $\mathbf{Y}^+$ does not necessarily measure the contribution of x_k towards the prediction of $\mathbf{Y}^+$.

The argument: The best linear predictor of the future $\mathbf{Y}^+$ in terms of the past $\mathbf{Y}^-$ is

$$\mathbf{Y}^+\backslash\mathbf{Y}^- = \mathbf{H}\mathbf{R}^{-1}\mathbf{Y}^-$$

and it can always be decomposed into the sum of the individual contributions from the different canonical components.

$$\mathbf{Y}^+\backslash\mathbf{Y}^- = \mathbf{H}\mathbf{R}^{-1}\mathbf{Y}^- = \mathbf{H}\mathbf{R}^{-\frac{1}{2}t}\mathbf{V}(\mathbf{V}^t\mathbf{R}^{-\frac{1}{2}}\mathbf{Y}^-) = \mathbf{H}\mathbf{R}^{-\frac{1}{2}}\sum_{\text{all } k} \mathbf{v}_k x_k$$

This is a discrete-time version of Eq. (48) in [17], where it is proposed that the canonical components with the largest c.c. coefficients be used for the extrapolation of a stationary random process (for basically, the prediction of $\mathbf{Y}^+$ given $\mathbf{Y}^-$). However, in this decomposition, the different terms are *not* uncorrelated, and the variance ($\mathbf{v}_k^t\mathbf{R}^{-\frac{1}{2}}\mathbf{H}^t\mathbf{H}\mathbf{R}^{-\frac{1}{2}t}\mathbf{v}_k$) of each term is *not* equal to the c.c. coefficient σ_k. Hence, the c.c. coefficient does *not* measure the contribution of the canonical component x_k to the prediction of $\mathbf{Y}^+$ and consequently, a small c.c coefficient does not neccessarily mean that the canonical component x_k can be dropped from the partial-state.

□

Thus the partial-state with the largest mutual information (with respect to $\mathbf{Y}^+$) does not neccessarily make the largest contribution towards the prediction of $\mathbf{Y}^+$. Thus, mutual information is not the requisite information measure for the partial-state selection problem (or for the fundamentally identical problem of extrapolation in [17]), and the decomposition of $\mathbf{Y}^+$ and $\mathbf{Y}^-$ into canonical components is not appropriate. Instead, the contribution of $\mathbf{x}_{\text{partial}}$ to the prediction of $\mathbf{Y}^+$ may be better gauged by the amount of information in the predictor $\mathbf{Y}^+\backslash\mathbf{x}_{\text{partial}}$. Such a scheme will be developed next. But first, we will study mutual information and the c.c. approach a bit further.

The effect of normalization: The difference betweeen the c.c. criterion and the PC-H criterion is in the normalization of $\mathbf{Y}^+$ and $\mathbf{x}_{\text{partial}}$. While the normalization of $\mathbf{x}_{\text{partial}}$ is desirable, the normalization of $\mathbf{Y}^+$ to $\mathbf{R}^{-1/2}\mathbf{Y}^+$ is not as easily justifiable. In fact, it is this unneccessary normalization of $\mathbf{Y}^+$ that is the source of most of the problems of the c.c. approach.

We have already seen that the partial-state picked by c.c. analysis does not make the largest contribution towards the prediction of the future $\mathbf{Y}^+$. That this can lead to strange approximations of systems, is analytically illustrated in the counter-example that follows.

Assume, we are given the exact covariances of a 4th order model whose impulse-response is

$$i(t) = (1-\epsilon)\gamma^t\cos(\omega_1 t) + \epsilon\gamma^t\cos(\omega_2 t); \qquad \epsilon<<1, \ \gamma<1$$

so that one mode is much stronger than the other. We wish to reduce the model order from 4 to 2. Intuitively, it seems we should pick the stronger mode, but it turns out that when the poles are sufficiently close to the unit circle ($\gamma \to 1$), the 4 nonzero c.c. coefficients will be all equal, and will not display the difference in the amplitudes $1-\epsilon$ and ϵ, so that the c.c. approximation may not pick the stronger mode.

In the following analysis, we have to reduce the input white noise variance ρ to zero (as the pole-radius γ is increased to unity), in order to keep the variance $r(0)$ of the output process constant. When γ is sufficiently close to one, $\rho=2(1-\gamma^2)$ will ensure that $r(0)=1$.

Claim: As $\gamma \to 1$, and $\rho = 2(1-\gamma^2)$, the c.c. coefficients converge to $(1,1,1,1,0,0,\cdots)$.

Proof: It has been shown in [27, 29] that the non-zero c.c. coefficients between $\mathbf{Y}^-$ and $\mathbf{Y}^+$ are the singular values of the product $\mathbf{P}^{*-1}\mathbf{P}_*$ where $\mathbf{P}^*$ and $\mathbf{P}_*$ are the state-variances of the maximum-phase model and the minimum-phase model (respectively) for the process $y(t)$, when (the models are in any coordinate-system, as long as) they have the same $\mathbf{F}$ and $\mathbf{h}$ matrices.

Let the input parameters for the minimum-phase model be $\mathbf{T}$ and ρ, and those for the maximum-phase model be $\mathbf{T}^*$ and ρ^*. Then, the Lyapunov equations satisfied by $\mathbf{P}_*$ and $\mathbf{P}^*$ are

$$P_* = FP_*F^t + \rho TT^t.$$

and

$$P^* = FP^*F^t + \rho^* T^* T^{*t}$$

which means the difference $\Sigma = P^* - P_*$ which is always non-negative (since P_* and P^* are the smallest and largest solutions respectively [31] of the algebraic Riccati equation) must satisfy the equation:

$$\Sigma = F\Sigma F^t + \rho^* T^* T^{*t} - \rho TT^t$$

Moreover, the two models generate the same output process and the same output covariance (though they are driven by different white noise sequences of in fact, different variances). Therefore,

$$hP_*h^t + \rho = r(0) = hP^*h^t + \rho^*$$

$$\Rightarrow \quad \rho = h\Sigma h^t + \rho^*$$

Since both $h\Sigma h^t$ and ρ^* are non-negative, $h\Sigma h^t$ must be smaller than ρ.

Now we have the machinery to see what happens as $\gamma \longrightarrow 1$. Since $\rho = 2(1-\gamma^2)$, we can see that ρ must converge to zero, which implies that $h\Sigma h^t$ must converge to zero, since it is smaller than ρ.

However,

$$h\Sigma h^t = hF\Sigma F^t h^t + \rho^*(hT^*)^2 - \rho(hT)^2.$$

in which all terms are non-negative. Therefore,

$$hF\Sigma F^t h^t \leqslant h\Sigma h^t + \rho(hT)^2.$$

In the right-hand-side of the above inequality, the inner product hT is always equal to the first impulse-response coefficient $i(1)$, and this remains finite, even as ρ approaches zero. Hence, both terms in the right-hand-side converge to zero, as $\gamma \longrightarrow 1$. Therefore, we have

$$hF\Sigma F^t h^t \longrightarrow 0.$$

Continuing in the same line of argument, we can show that

$$hF^k \Sigma F^{t^k} h^t \longrightarrow 0 \quad \forall \quad k=0,1,2,\ldots,p-1.$$

which implies that the diagonal elements of $\mathcal{O}_p \Sigma \mathcal{O}_p^t$ converge to zero. Here, $\mathcal{O}_p$ is the pxp principal submatrix of the observability matrix $\mathcal{O}$.

Since $\mathcal{O}_p \Sigma \mathcal{O}_p^t$ is positive-definite, its non-diagonal terms are smaller than the diagonal terms. Consequently, the entire matrix $\mathcal{O}_p \Sigma \mathcal{O}_p^t$ converges to zero as $\gamma \longrightarrow 1$. Moreover, $\mathcal{O}_p$ is an invertible matrix (since the model is minimal-order), and so Σ itself must converge to zero. Therefore, we have

$$\mathbf{P}^* - \mathbf{P}_* \; \longrightarrow \; 0$$

$$=> \quad \mathbf{P}^{*-1}\mathbf{P}_* \; \longrightarrow \; \mathbf{I}_4 \qquad \text{as } \gamma \longrightarrow 1.$$

It is well known in numerical analysis that the eigenvalues of a Hermitian matrix are perfectly conditioned [32, 33], i.e.

$$| \lambda_j(\mathbf{I} + \Delta) - \lambda_j(\mathbf{I}) | \leqslant \| \Delta \|_s$$

Therefore, as the poles approach the unit circle, $\mathbf{P}^{*-1}\mathbf{P}_*$ approaches the identity matrix, and its eigenvalues, which are the 4 non-zero c.c. coefficients between $\mathbf{Y}^-$ and $\mathbf{Y}^+$ approach unity. Hence the result.

□

The above counter-example illustrates the effect of the inherent normalization in c.c. analysis: the c.c. coefficients are normalized correlations that do not contain any information regarding the strength of the component modes, and hence do not effectively measure the component's significance to the model. In the limiting case, ($\gamma \approx 1$,) c.c. analysis could very well pick the weaker mode. This lack of strength information in the c.c. coefficients makes them very sensitive to small perturbations. When the impulse-response consists of only one damped sinusoid, only two c.c. coefficients are non-zero, and the rest are zero. When we add a small perturbation of the form $\epsilon\gamma^t \cos\omega_2 t$ (possibly 60 Hz. leakage from the power lines) to the impulse-response, the third and fourth c.c. coefficients are dramatically changed from 0 to nearly 1, irrespective of how small ϵ is. This makes the approximate model obtained by the c.c. analysis sensitive to perturbations.

The predictive efficiency criterion

The previous two approaches to approximate stochastic modeling (the principal components of $\mathbf{H}$ and the canonical correlations method) were derived by generalizing the correlation-maximizing property and the information-maximizing property of the principal components (of a random vector) to the 2-vector problem. Recall however that the function of the partial state is to predict the future output well. Hence, instead of maximizing its correlation with $\mathbf{Y}^+$ or its mutual information with respect to $\mathbf{Y}^+$, it seems more appropriate to use the reconstruction-efficiency property of the principal components, and generalize it to the 2-vector problem.

The principal-components approximation of a random vector provides an optimal data compression that maximizes its ability to reconstruct the full sized vector. In the partial-state selection problem for approximate modeling, we come across a similar problem: that of compressing $\mathbf{Y}^-$ into a partial state that can best predict $\mathbf{Y}^+$. Taking a hint from the reconstruction-efficiency property of the principal components of a random vector, we might wish to compress $\mathbf{Y}^-$ into a partial state that has the smallest error in predicting $\mathbf{Y}^+$. Our partial-state selection problem is then to pick a partial state $\mathbf{x}_{\text{partial}} = \Psi \mathbf{Y}^-$ to

$$\text{minimize} \quad E\left[\left\| \mathbf{Y}^+ - \mathbf{Y}^+ \backslash \mathbf{x}_{\text{partial}} \right\|^2 \right].$$

The inherent constraint here is that Ψ should have only p rows.[3] Such a criterion was first used by C. R. Rao in multivariate statistics for the 2-vector problem [2]. Since $\mathbf{x}_{\text{partial}} = \Psi \mathbf{Y}^-$, it can be shown using Eq. (c) that

$$\mathbf{Y}^+ \backslash \mathbf{x} = \mathbf{H}\Psi^t \left[\Psi \mathbf{R} \Psi^t \right]^{-1} \mathbf{x} \tag{f}$$

and the prediction error to be minimized is $\text{Trace}\left[\mathbf{R} - \mathbf{H}\Psi^t(\Psi\mathbf{R}\Psi^t)^{-1}\Psi\mathbf{H}^t\right]$. Equivalently, we have to choose a $p\times\infty$ matrix Ψ that maximizes $\text{Trace}\left[(\Psi\mathbf{H}^t\mathbf{H}\Psi^t)(\Psi\mathbf{R}\Psi^t)^{-1}\right]$. The solution to this optimization problem is:

> The p rows of Ψ must be a basis for the space spanned by the p generalized eigenvectors of the matrix pencil $(\mathbf{H}^t\mathbf{H}, \mathbf{R})$, corresponding to the p largest generalized eigenvalues.

If $\mathbf{R}$ is invertible, as is the case when the model is strictly stable, we can obtain Ψ from the eigenvectors of $\mathbf{H}\mathbf{R}^{-1}\mathbf{H}^t$ instead.[4] Let the eigendecomposition (or SVD) of $\mathbf{H}\mathbf{R}^{-1}\mathbf{H}^t$ be

$$\mathbf{H}\mathbf{R}^{-1}\mathbf{H}^t = \mathbf{U}\Sigma^2\mathbf{U}^t = \mathbf{U}_1\Sigma_1^2\mathbf{U}_1^t + \mathbf{U}_2\Sigma_2^2\mathbf{U}_2^t$$

and let subscript '1' denote the principal components, as before. Then, the predictive-efficiency criterion is optimized when

[3]Without such a constraint, no size compression is required, and the entire past $\mathbf{Y}^-$ can be used as the state.

[4]When $\mathbf{R}$ is singular, the process is purely sinusoidal, and this solution is the same as the Toeplitz approximation method of [34, 35].

$$\Psi = \mathbf{A}\mathbf{U}_1^t\mathbf{H}\mathbf{R}^{-1}$$

where $\mathbf{A}$ is any pxp invertible matrix.

Note that this solution is different from Akaike's solution because under perturbations, $\mathbf{H}$ will be full rank and the principal components of $\mathbf{H}\mathbf{R}^{-1}\mathbf{H}^t$, $\mathbf{R}^{-1/2}\mathbf{H}\mathbf{R}^{-1/2^t}$, and $\mathbf{H}$ will not be the same. C.R. Rao himself states that his generalized principal components analysis for studying the association between two random vectors is different from Hotelling's canonical correlations analysis.

The Unweighted Principal Components (UPC) Algorithm: After picking the partial-state components using the predictive efficiency criterion, we still have to obtain the corresponding parameter estimates. The parameter-estimation step is taken from the deterministic-modeling algorithm of Kung [36]. It is assumed here that the model order p is estimated (or given) prior to the model parameter estimation. From that point on, the rest of the Unweighted Principal Components (UPC) algorithm is:

Step 1: Perform an eigendecomposition of

$$\mathbf{H}\mathbf{R}^{-1}\mathbf{H}^t = \mathbf{U}\Sigma^2\mathbf{U}^t = \mathbf{U}\Sigma_1^2\mathbf{U}_1^t + \mathbf{U}_2\Sigma_2^2\mathbf{U}_2^t$$

and retain only the principal components (denoted by subscript 1). Now Ψ can be any basis from the row span of $\mathbf{U}_1^t\mathbf{H}\mathbf{R}^{-1}$, i.e

$$\Psi = \mathbf{A}\mathbf{U}_1^t\mathbf{H}\mathbf{R}^{-1} \quad \text{for any invertible pxp matrix } \mathbf{A}.$$

Different choices of $\mathbf{A}$ will correspond to different coordinate transformations of the partial-state. The partial-state will be in balanced coordinates, if we choose

$$\Psi = \Sigma_1^{-1/2}\mathbf{U}_1^t\mathbf{H}\mathbf{R}^{-1}$$

Then, Eq. (f) indicates that

$$\mathbf{Y}^+\backslash\mathbf{x}_{\text{partial}} = \mathbf{H}\Psi^t(\Psi\mathbf{R}\Psi^t)^{-1}\mathbf{x}_{\text{partial}}$$

which implies that the observability matrix is

$$\mathcal{O} = \mathbf{H}\mathbf{R}^{-1}\mathbf{H}^t\mathbf{U}_1\Sigma_1^{-1/2}\left[\Sigma_1^{-1/2}\mathbf{U}_1^t\mathbf{H}\mathbf{R}^{-1}\mathbf{H}^t\mathbf{U}_1\Sigma_1^{-1/2}\right]^{-1} = \mathbf{U}_1\Sigma_1^{-1/2}$$

The state variance is $\Psi\mathbf{R}\Psi^t = \Sigma_1$ and the observability grammian $\mathcal{O}^t\mathcal{O}$ is also equal to Σ_1. Hence the partial-state is in balanced coordinates, except for a possible scale factor.

Step 2: But, the partial-state is not a "true state" of a linear time-invariant system, and the $\mathcal{O}$ and Ψ matrices don't have the required structure. Hence, as in the deterministic algorithm of Kung [36], we resort to a second approximation, and $\mathbf{F}$ is obtained as the least-squares solution of Eq. (b):

$$\mathcal{O}_1\mathbf{F} = \mathcal{O}_2.$$

where $\mathcal{O}_1$ and $\mathcal{O}_2$ are as defined earlier, in the context of Eq. (b). Moreover, $\mathbf{h}$ and $\mathbf{T}$ are the first row and column of $\mathcal{O}$ and Ψ respectively. Therefore, the parameter estimates are:

$$\mathbf{h} = 1^{st} \text{ row of } \mathcal{O}$$

$$\mathbf{T} = 1^{st} \text{ column of } \Psi$$

$$\mathbf{F} = \mathcal{O}_1^{\dagger}\mathcal{O}_2$$

where the superscript † stands for the pseudoinverse.

Alternate interpretation of c.c. analysis

The predictive-efficiency criterion sheds new light on the canonical correlations criterion, and clearly illustrates the inherent normalization in the c.c. analysis. It can be seen that the partial state components selected by the canonical correlations approach, maximize the predictive-efficiency of the partial state for a *normalized* future vector. The normalized problem is to

$$\text{pick } \mathbf{x}_{\text{partial}} = \Psi\mathbf{Y}^- \text{ to}$$

$$\text{minimize } E\left[\| \mathbf{R}^{-1/2}\mathbf{Y}^+ - \mathbf{R}^{-1/2}\mathbf{Y}^+ \backslash \mathbf{x} \|^2 \right] = \text{Trace}\left(\mathbf{I} - \mathbf{R}^{-1/2}\mathbf{H}\Psi^t(\Psi\mathbf{R}\Psi^t)^{-1}\Psi\mathbf{H}^t\mathbf{R}^{-1/2^t} \right)$$

and the solution is obtained from the principal generalized eigenvectors of the matrix pair $(\mathbf{H}\mathbf{R}^{-1}\mathbf{H}^t, \mathbf{R})$ or equivalently from the principal components of the matrix $\mathbf{R}^{-1/2}\mathbf{H}\mathbf{R}^{-1/2^t}$. Thus, Akaike's method maximizes the partial-state's predictive-efficiency for the normalized future vector $\mathbf{Z}^+ = \mathbf{R}^{-1/2}\mathbf{Y}^+$. Because the partial-state components picked by the unweighted predictive-efficiency criterion are also generalized principal components (generalized to the 2-vector problem), they are called Unweighted Principal Components.

The following simulation illustrates the counter example of the last section, and illustrates the differences bewtween the two methods.

Simulation of the counter example: The first 100 covariance lags of a 4th order ARMA model were estimated by time-averaging the impulse-response over 6000 samples. The impulse-response of the model considered was

$$i(t) = (0.95)^t\cos(0.45\pi t) + (0.1)(0.95)^t\cos(0.25\pi t).$$

The c.c. coefficients were computed as the singular values of the 50x50 matrix $\mathbf{R}^{-1/2}\mathbf{H}\mathbf{R}^{-1/2^t}$ and the 5 largest c.c coefficients were found to be:

$$0.9242,\ 0.9040,\ 0.6224,\ 0.5583,\ 9.06\times10^{-5}.$$

Observe that the difference in the amplitudes does not show up well. On the other hand, the square roots of the 5 largest singular values of the 50x50 matrix $HR^{-1}H^t$ constructed from the same 100 covariance estimates were found to be:

$$2.1330,\ 2.1045,\ 0.210,\ 0.186,\ 1.97\times10^{-4}.$$

Here, the difference in amplitudes clearly shows up in the singular values.

The relative strengths of the modes are ignored in canonical correlations analysis, because of the implicit normalization in the approximation criterion. In statistical applications, where c.c. analysis was first introduced [1] , the random vectors under consideration are usually vectors composed of measurements of different features from two populations. In studying the correlation between the populations, it is necessary to remove the effect of the units of measurement and the correlation between the features. So the vectors are ortho-normalized. But in the partial-state selection problem, the vectors under study (Y^+ and Y^-) are composed of succesive measurements of the same process, and the units of measurement are uniform. Then the normalization of Y^+ into $Z^+ = R^{-1/2}Y^+$ may be uncalled for, and in fact, leads to the problems illustrated by the counter example.

Alternate interpretations of UPC analysis

Karhunen Loeve decomposition: Ideally, we expect $W = Y^+\backslash Y^-$ to equal $\mathcal{O}\Psi Y^-$ so that Span(W) is p–dimensional and W can be compressed into a px1 state vector without any loss of information. But perturbations make the apparent dimension of Span(W) much larger than the true model order p, and W cannot be exactly summarized in a px1 vector. Then, we have to approximately summarize W in a px1 partial-state $x_{partial}$ so that it retains most of the information in W.

We know that the principal components in the Karhunen-Loeve (KL) expansion of W provides such an optimal data compression of W [37, 38, 2]. Here, $W = Y^+\backslash Y^- = HR^{-1}Y^-$ so that $E[WW^t] = HR^{-1}H^t$. Therefore, the principal components approximation of W is

$$x = U_1^tW = U_1^tHR^{-1}Y^-$$

which it turns out is exactly the partial-state picked by the UPC approach. Thus, the UPC method picks its partial-state by optimally compressing the oversized vector $W=Y^+\backslash Y^-$ [39]. We know that the principal components compression is optimal in terms of both predictive-efficiency and self-information [38]. In other words, $\mathbf{x}_{partial} = \mathbf{U}_1^t\mathbf{W}$ minimizes the reconstruction error $E \| \mathbf{W} - \mathbf{W}\backslash\mathbf{x} \|$ and maximizes the self-information in the partial-state: $0.5\log\left(\det E[\mathbf{x}\mathbf{x}^t]\right)$.

Maximizing Self-Information: The partial-state picked by the UPC method, not only maximizes its own self-information, it also maximizes the "self-information" in the predictor vector: $\mathbf{Y}^+\backslash\mathbf{x}_{partial}$ amongst all $p\times 1$ sized partial-states of the form $\mathbf{x}_{partial} = \Psi\mathbf{Y}^-$. The term self-information is used here within quotation marks, because the infinite-sized random vector $\mathbf{Y}^+\backslash\mathbf{x}_{partial}$ has only p independent random variables, and its self-information in the clasical sense is negative infinity. The covariance matrix of $\mathbf{Y}^+\backslash\mathbf{x}_{partial} = \mathbf{H}\Psi^t(\Psi\mathbf{R}\Psi^t)^{-1}\mathbf{x}_{partial}$ is $\mathbf{H}\Psi^t(\Psi\mathbf{R}\Psi^t)^{-1}\Psi\mathbf{H}^t$, which has only p positive eigenvalues, the rest being zero. Hence the self-information (which is half the logarithm of the determinant of the covariance matrix) is negative infinity. However, we define a new information measure based on only the nonzero eigenvalues:

$$0.5\log\prod_{k=1}^{p}\lambda_k = 0.5\sum_{k=1}^{p}\log\lambda_k$$

where λ_k are the eigenvalues of $\mathbf{H}\Psi^t(\Psi\mathbf{R}\Psi^t)^{-1}\Psi\mathbf{H}^t$, in descending order of magnitude. This new information measure is maximized when $\mathbf{x}_{partial}$ is the UPC partial-state.

In contrast to the mutual information (with respect to $\mathbf{Y}^+$) in the partial state, which is

$$\text{Info.}(\mathbf{Y}^+) - \text{Info.}(\mathbf{Y}^+ - \mathbf{Y}^+\backslash\mathbf{x}_{partial}) = 0.5\log\left(\det\mathbf{R}\right) - 0.5\log\left(\det\left[\mathbf{R} - \mathbf{H}\Psi^t(\Psi\mathbf{R}\Psi^t)^{-1}\Psi\mathbf{H}^t\right]\right)$$

the information criterion maximized by the UPC method is the "modified self-information" in the predictor $\mathbf{Y}^+\backslash\mathbf{x}_{partial}$. The two measures are different since information is not preserved in subtraction, i.e.

$$\text{Info.}(\mathbf{Y}^+\backslash\mathbf{x}_{partial}) \neq \text{Info.}(\mathbf{Y}^+) - \text{Info.}(\mathbf{Y}^+ - \mathbf{Y}^+\backslash\mathbf{x}_{partial}).$$

Maximizing correlations: Alternately, the unweighted principal components are those orthonormal random variables in Span($\mathbf{Y}^-$) that are maximally correlated with $\mathbf{Y}^+$. The solution to the problem of

$$\underset{\text{constraint:}\Psi\mathbf{R}\Psi^t=\mathbf{I}_p}{\text{Maximize}} \left\| E\left[\mathbf{Y}^+\mathbf{x}^t\right] \right\|_F = \text{Trace}\left(\Psi^t\mathbf{H}^t\mathbf{H}\Psi^t\right)$$

is also $\Psi = \mathbf{U}_1^t\mathbf{H}\mathbf{R}^{-1}$. Compare this criterion to the maximum-correlation criterion of the PC-H method:

$$\underset{\text{constraint}:\Psi\Psi^t=\mathbf{I}_p}{\text{Maximize}} \; \left\| E\left[\mathbf{Y}^+\mathbf{x}^t\right] \right\|_F$$

and the canonical correlations criterion:

$$\underset{\text{constraint}:\Psi\mathbf{R}\Psi^t=\mathbf{I}_p}{\text{Maximize}} \; \left\| E\left[\mathbf{R}^{-1/2}\mathbf{Y}^+\mathbf{x}^t\right] \right\|_F$$

and observe that the UPC criterion ensures that $\mathbf{x}$ is orthonormal without normalizing $\mathbf{Y}^+$, while the c.c. criterion does both normalizations, and the PC-H method does neither.

Relation to deterministic identification: The connections between the UPC algorithm and the PHC algorithm of [36] run much deeper than apparent at first sight. Not only do the methods share a similar second step, but also the first step approximations (the partial-state selection step) are closely related. Recall that for the minimum phase model, $\mathbf{Y}^+\backslash\mathbf{Y}^- = \mathcal{O}\,\mathbf{x}$ which in turn is equal to $\mathcal{H}\,\mathbf{V}^-$ because $\mathbf{x} = \mathcal{C}\,\mathbf{V}^-$. Moreover, using Eq. (c), we saw that $\mathbf{Y}^+\backslash\mathbf{Y}^-=\mathbf{H}\mathbf{R}^{-1}\mathbf{Y}^-$. Combining the two, we get

$$\mathbf{H}\mathbf{R}^{-1}\mathbf{Y}^- = \mathcal{H}\,\mathbf{V}^-$$

Therefore, the covariance matrices of the two vectors must also be the same. And thus, we come to the rather surprising result:

$$\mathbf{H}\mathbf{R}^{-1}\mathbf{H}^t = \rho\mathcal{H}\mathcal{H}^t$$

where ρ is the variance of the input white noise process.

Thus, an eigendecomposition of $\mathbf{H}\mathbf{R}^{-1}\mathbf{H}^t$ is equivalent to the SVD of the impulse-response Hankel $\mathcal{H}$. Hence, the UPC algorithm is a stochastic generalization of the PHC algorithm, and allows us to obtain the PHC model estimates directly from covariance data instead of impulse-response measurements.

At the same time, the above equation establishes a stochastic justification for the PHC algorithm in terms of maximizing the predictive-efficiency of the partial-state. It can be shown that the singular values of $\mathcal{H}$ are also the diagonal entries of $\mathbf{K} = \mathbf{W} = \Sigma$, when the state is in balanced coordinates, and that the PHC scheme is equivalent to balanced model reduction. The above equation shows that the UPC method for approximate stochastic modeling is also a balanced model reduction method, except that instead of starting from the full order model, we start from the covariances corresponding to a large order model.

The predictive-efficiency criterion thus provides a stochastic justification for balanced model reduction as well. In fact, a model reduction algorithm completely equivalent to balanced model reduction was derived independently along the above lines by Fujishige and co-researchers [40] even before Moore [8] ; however the term 'balanced' and the concept of balancing was introduced by Moore. Fujishige et al. used a least-squares prediction criterion to pick their partial state from the space spanned by the full-order state of the model. Just like in the UPC algorithm, their partial state was also picked to maximize its predictive efficiency for the future output. The major difference from the UPC method, is that in their model-reduction formulation, the full-order model is completely known, while the UPC method works with only output covariances. The Fujishige model reduction algorithm obtains its partial state from the principal components in the KL expansion of $\boldsymbol{\mathcal{H}}\,\mathbf{V}^-$ instead of $\mathbf{HR}^{-1}\mathbf{Y}^-$. Fujishige's method does not require the full-order model to be minimum-phase. On the other hand, in the UPC problem formulation, the phase of the model is unknown, and $\boldsymbol{\mathcal{H}}\,\mathbf{V}^-$ is not uniquely known. However, $\mathbf{HR}^{-1}\mathbf{Y}^-$ equals $\boldsymbol{\mathcal{H}}\,\mathbf{V}^-$ for the full-order, *minimum-phase* model corresponding to the given covariance sequence. In effect, the UPC method performs Fujishige model reduction or balanced model reduction on the full-order, minimum-phase model corresponding to the given covariances.

CONCLUDING REMARKS

In the previous section, it was shown that the matrix approximated by its dominant singular vectors in the UPC method, is $\boldsymbol{\mathcal{H}\mathcal{H}}^t$. It is shown in [30] that for a particular choice of square root $\mathbf{R}^{-1/2}$, the matrix used in c.c. analysis is equal to the Hankel matrix built from the impulse response of the all-pass system $\Phi(z) = \dfrac{H(z)}{H(z^{-1})}$. Since the c.c. approximation does depend on the choice of the square root, we can say that the matrix used in c.c. analysis is equal to the above-mentioned Hankel matrix. If we use the notation $\Gamma\left[\,\cdot\,\right]$ to denote the Hankel matrix constructed from the causal part of the inverse z-transform of the function "$\cdot$" within the square brackets, then we have

$$\boldsymbol{\mathcal{H}} = \Gamma\left[H(z)\right].$$

$$\mathbf{HR}^{-1}\mathbf{H}^{t} = \Gamma\left[H(z)\right] \bullet \Gamma\left[H(z)\right]^{t},$$

$$\mathbf{H} = \Gamma\left[S(z)\right], \quad \text{and}$$

$$\mathbf{R}^{-1/2}\mathbf{HR}^{-1/2^{t}} = \Gamma\left[\Phi(z)\right]$$

$$\text{where } S(z) = \rho H(z)H(z^{-1}) \text{ and } \quad \Phi(z) = \frac{H(z)}{H(z^{-1})}.$$

Thus the PC-H method works with the magnitude factor of the full-order system, the c.c. method works with the phase factor, and the UPC method works with the entire transfer function. Representative simulations of the three methods can be found in [30].

In conclusion, we have presented three different criteria and three methods for approximate stochastic realization, as generalizations of the principal components ideas from multivariate statistics. The notion of principal-components approximation as introduced in [37], applies to the problem of compressing a single random vector. However, the problem encountered in approximate stochastic realization involves compressing the information interface between two random vectors. For such problems as well, the ideas of principal-components approximation have been generalized in different contexts by Hotelling [1] and C. R. Rao [2]. Different generalizations lead to different criteria and to the three approximate stochastic realization methods of this chapter. The different criteria can only be compared in a subjective manner, and the ultimate choice between them depends on the nature of the intended application.

REFERENCES

1. Hotelling, H., "Relations between two sets of Variates," *Biometrika* **28** pp. 321-372 (1936).
2. Rao, C.R., "The use and Interpretation of Principal Component Analysis in Applied Research," *Sankhya Series A* **26** pp. 329-358 (1964).
3. Chamberlain, J.E., *The principles of interferometric spectroscopy*, Wiley (1979).
4. Bell, R.J., *Introductory Fourier transform spectroscopy*, Academic (1972).
5. Burg, J. P., "Maximum Entropy Spectral Analysis," Ph.D. Dissertation, Stanford University, Stanford California (1975).
6. Levinson, N., "The Wiener RMS (Root-Mean-Square) Error Criterion in Filter Design and Prediction," *J. Math. Phys.* **25** pp. 261-278 (Jan 1947).
7. Haykin, S. and Kesler, S., "Prediction-Error Filtering and Maximum-Entropy Spectral Estimation," pp. 8-72 in *Nonlinear methods of spectral analysis*, ed. Haykin S.,Springer Verlag (1979).
8. Moore, B. C., "Principal Component Analysis in Linear Systems: Controllability Observability and Model Reduction," *IEEE Transactions on Automatic Control* **AC-26**(1) pp. 17-31 (February 1981).
9. Mullis, C. T. and Roberts, R. A., "Synthesis of Minimum Round-off Noise Fixed Point Digital Filters," *IEEE Transactions on Circuits and Systems* **CAS-23** pp. 551-562 (1976).
10. Kailath, T., *Linear Systems*, Prentice Hall New York (1980).
11. Akaike, H., "Markovian Representation of Stochastic Processes by Canonical Variables," *SIAM Journal on Control* **13**(1) pp. 162-173 (January 1975).
12. Sorenson, H., *Parameter Estimation: Principles and Problems*, Marcel Dekker Inc. (1980).
13. Kailath, T., "The Innovations Approach to Detection and Estimation Theory," *Proceedings of the IEEE* **58** pp. 680-695 (May 1970).
14. Anderson, B. D. O. and Kailath, T., "The Choice of Signal Process Models," *Journal of Math. Analysis and Applications* **35**(3) pp. 659-668 (September 1971).
15. Gevers, M. and Kailath, T., "An Innovations Approach to Least Squares Estimation Part VI: Discrete-Time Innovations-Representations and Recursive Estimation," *IEEE Transactions on Automatic Control* **AC-18** pp. 588-600 (December 1973).
16. Akaike, H., "Markovian Representation of Stochastic Processes and its Application to the Analysis of Autoregressive Moving Average Processes," *Annals of the Institute of Statistical Mathematics* **26** pp. 363-387 (1974).
17. Yaglom, A. M., "Outline of some topics in linear extrapolation of stationary random processes," pp. 259-278 in *Proceedings of the 5th Berkeley Symposium on Mathematical Statistics and Probability*, University of California Press, Berkeley CA (1965).
18. Kailath, T., "A View of Three Decades of Linear Filtering Theory," *IEEE Trans. Inform. Theory* **IT-20**(2) pp. 145-181 (Mar 1974).
19. J.A.Cadzow,, "Spectral Estimation: An Overdetermined Rational Model Equation Appraoach," *Proceedings of the IEEE* **70**(9) pp. 907-939 (September 1982).

20. Prevosto, M., Benveniste, A., and Barnouin, B., "Identification of Vibrating Structures Subject to Nonstationary Excitation: A Nonstationary Stochastic Realization Problem," pp. 252-255 in *Proceedings of the Intl. Conf. on Acoustics Speech and Signal Processing*, IEEE, Paris France (May 1982).

21. Benveniste, A. and Fuchs, J. J., "Single sample modal identification of a nonstationary stochastic process," *IEEE Transactions on Automatic Control* **AC-30**(1) pp. 66-75 (January 1985).

22. Maciejowski, J. M., "The Use of Principal Components for Approximate Linearisation Stochastic Realisation and Spectral Factorisation," Note Presented at the IEEE Colloquium on 'Principal Components: Model Reduction and Control' at City University London January 7 1983. (1983).

23. Kung, S.Y. and Arun, K.S., "A Novel Hankel Approximation Method for ARMA Pole Zero Estimation from Noisy Covariance Data," pp. WA-19 in *Technical Digest of the Topical Meeting on Signal Recovery and Synthesis with Incomplete Information and Partial Constraints*, Optical Society of America, Incline Village Nevada (January 1983).

24. Gelfand, I.M. and Yaglom, A.M., "Calculation of the Amount of Information about a Random Function contained in other such Function," *American Mathematical Society Translations Series 2* **12** pp. 199-246 (1959).

25. Kullback, S. and Leibler, R.A., "On Information and Sufficiency," *Annals of Mathematical Statistics* **22** pp. 79-86 (1951).

26. Baram, Y., "Realization and Reduction of Markovian Models from Nonstationary Data," *IEEE Transactions Automatic Control* **AC-26**(6) pp. 1225-1231 (December 1981).

27. Desai, U.B. and Pal, D., "A Realization Approach to Stochastic Model Reduction and Balanced Stochastic Realization," pp. 613-620 in *Proc. of the 16th Annual Conference on Information Sciences and Systems*, Princeton University, New Jersey (March 1982). Also to appear in Intl. Journal of Control, 1985.

28. White, J., "Stochastic State Space Models from Emperical Data," pp. 243-246 in *International Conference on Acoustics Speech and Signal Processing*, IEEE, Boston MA (April 1983).

29. Desai, U. B. and Pal, D., "A transformation approach to stochastic model reduction," *IEEE Transactions on Automatic Control* **AC-29** pp. 1097-1099 (Dec. 1984).

30. Karalamangala, A. S., "A Principal Components Approach to Approximate Modeling and ARMA Spectral Estimation," Ph.D. Dissertation, Dept. of Electr. Engg., Univ. of So. California, Los Angeles (1984).

31. Faurre, P., "Stochastic Realization Algorithms," in *System Identification: Advances and Case Studies*, ed. Mehra R.K. and Lainiotis D.G.,Academic Press (1976).

32. Parlett, B. N., *The Symmetric Eigenvalue Problem*, Prentice-Hall Inc., Englewood Cliffs NJ (1980).

33. Wilkinson, J. H., *The algebraic eigenvalue problem*, Oxford University Press (1965).

34. Kung, S.Y., "A Toeplitz Approximation Method and some Applications," pp. 262-266 in *Proc. of the International Symposium on the Mathematical Theory of Networks and Systems*, , Santa Monica CA (August 5-7 1981).

35. Kung, S. Y., Arun, K. S., and BhaskarRao, D. V., "State-space and singular value decomposition based methods for harmonic retrieval problem," *Journal of the Optical Society of America* **73** pp. 1799-1811 (December 1983).

36. Kung, S.Y., "A New Identification and Model Reduction Algorithm via Singular Value Decomposition," pp. 705-714 in *Proceedings of the 12th Asilomar Conference on Circuits Systems and Computers*, IEEE, Pacific Grove CA (November 1978).

37. Hotelling, H., "Analysis of a Complex of Variables into Principal Components," *Journal of Educational Psychology* **24** pp. 417-441 and 498-520 (1933).

38. Brown, Jr. J.L., "Mean Square Truncation Error in Series Expansions of Random Functions," *Journal of SIAM* **8**(1) pp. 28-32 (November 1960).

39. Kung, S.Y. and Arun, K.S., "Approximate Realization Methods for ARMA Spectral Estimation," pp. 105-109 in *Proc. of the International Symposium on Circuits and Systems*, IEEE, Newport Beach CA (May 1983).

40. Fujishige, S., Nagai, H., and Sawaragi, Y., "System-theoretical Approach to Model Reduction and System-order Determination," *International Journal of Control* **22**(6) pp. 807-819 (1975).

5

FINITE-DATA ALGORITHMS FOR APPROXIMATE STOCHASTIC REALIZATION

Richard J. Vaccaro
Department of Electrical Engineering, University of Rhode Island
Kingston, RI, 02881, USA

ABSTRACT

This chapter deals with the problem of constructing a state-space model for a stochastic process from a finite number of estimated covariance lags. The approach is to first obtain a high-order model which exactly matches the estimated covariance sequence, and then use balanced model reduction techniques to obtain a lower order model which approximates the given sequence. It is shown that balanced models can be obtained from a realization algorithm which uses an infinite covariance sequence. Scaling ideas are then introduced so that balanced realizations can be obtained from finite covariance sequences.

INTRODUCTION

Mathematical realization formulas which specify a Markovian model for a stochastic process (that is, a state-space model driven by white noise) from its covariance sequence have been known for some time [1]. The model is constructed so that its output covariance sequence equals the given sequence, and so the resulting model may be called an *exact realization* of the given covariance sequence. The practical utility of exact realizations is limited, however, by the fact that most covariance sequences encountered in practice are not exact - they are usually estimated from a finite number of observations of a stochastic process. The estimated covariances differ from the "true" process covariances due to estimation errors, additive noise, etc. In addition, the order of the Markovian model which is identified from an estimated covariance sequence is usually quite high - one half the number of covariance lags with probability one. For these reasons, it is more desirable to obtain a low-order Markovian model whose output covariance sequence *approximates* the estimated sequence, an *approximate stochastic realization.*

The approach taken in this chapter for obtaining such a realization is to first obtain a high-order exact realization, and then use model reduction techniques to obtain a lower order approximate realization. We begin by reviewing two existing algorithms for obtaining exact realizations. The first works on finite covariance sequences, but gives no indication of how the exact realization can be reduced to obtain an approximate realization. The second algorithm was in fact derived from a model reduction point of view, but requires knowledge of an infinite covariance sequence; it is shown that this algorithm is equivalent to a well known deterministic model reduction algorithm. After gathering insights from the previous algorithms, two new algorithms for approximate stochastic realization are derived, both of which use scaling ideas to allow infinite-data algorithms to be used with finite data. These new algorithms are meant to be representative of a class of algorithms based on two different approaches to scaling [2].

STOCHASTIC REALIZATION FROM EXACT COVARIANCE DATA

Let $\{y_k\}$ be a stationary, scalar stochastic process which is zero mean and Gaussian. Assume that $\{y_k\}$ is a finite dimensional process of order p so that it can be realized by a Markovian model of the following form:

$$x_{k+1} = Fx_k + gu_k \ , \qquad E\,[x_k\, x_{k+j}{}^T] = P\,\delta_{k-j} \tag{1}$$
$$y_k \;= hx_k + u_k \ , \qquad \mathrm{var}(u_k) = \rho$$

where F is $p \times p$, and δ is the Kronecker delta function. If the zeros of the system in (1) lie inside the unit circle, (1) is referred to as an *innovations model* for the process $\{y_k\}$ [3-4]. The output covariance sequence of (1) is

$$r_j = E\,[y_k\, y_{k+j}] \ , \quad j = 0\,, 1\,, \cdots, L \tag{2}$$

In this section, two existing algorithms are reviewed which obtain exact realizations from infinite covariance data, i.e. from the sequence in (2) as N goes to infinity. The first algorithm is classical, and in fact gives exact realizations for any $L \geq 2p$; however, this algorithm gives no indication of how a good reduced-order model can be obtained in case an approximate stochastic realization is desired. This algorithm is characterized by the fact that it requires the solution to an algebraic Riccati equation. The second algorithm is representative of a class of algorithms which do not require a Riccati solution, but only a matrix factorization to obtain the stochastic realization. By using a particular factorization, an algorithm which is good for model reduction, and hence for approximate stochastic realization, is obtained. However, the factorization approach requires knowledge of the infinite covariance sequence. To be practical, a realization algorithm must be able to work with finite covariance data. In the next section, it is shown that by combining the two algorithms in an appropriate way, an algorithm which gives exact realizations from finite covariance data and which is appropriate for model reduction can be obtained.

Riccati Equation Approach

Consider the stochastic process $\{y_k\}$ in (1), and define at time k the future and past of the process, respectively, as

$$Y_k^+ = \begin{bmatrix} y_k \\ y_{k+1} \\ \cdot \\ \cdot \\ \cdot \end{bmatrix} \ , \qquad Y_k^- = \begin{bmatrix} y_{k-1} \\ y_{k-2} \\ \cdot \\ \cdot \\ \cdot \end{bmatrix} \tag{3}$$

Let $S(Y_k^+)$ and $S(Y_k^-)$ be the Hilbert spaces obtained by taking the closed linear span of the random variables in Y_k^+ and Y_k^-, respectively. (In general, $S(\cdot)$ will be the Hilbert space generated by the closed linear span of the elements of "$\cdot$".) The inner product between two random variables w and z is $E(wz)$, where E denotes mathematical expectation. Orthogonal projection in these Hilbert spaces is equivalent to conditional expectation, i.e. $w \mid z = E(w \mid z)$, where "$|$" on the left hand side of this equation denotes the orthogonal projection of w onto $S(z)$. This projection is given explicitly by

$$w \mid z = E(wz^T)E(zz^T)^{-1}z \tag{4}$$

where w and z are zero-mean random variables or random vectors. If $w = Y_k^+$ and $z = Y_k^-$, then the projection $Y_k^+ \mid Y_k^-$ in fact defines the state space of a Markovian model of the form (1) [5]. Thus this projection is central to any stochastic realization algorithm. From (4), it can be seen that this projection is given in terms of the following matrices formed from the covariance sequence:

$$Y_k^+ \mid Y_k^- = HR^{-1}Y_k^- \tag{5}$$

where

$$H = E[Y_k^+(Y_k^-)^T] = \begin{bmatrix} r_1 & r_2 & r_3 & \cdot \\ r_2 & r_3 & r_4 & \cdot \\ r_3 & r_4 & r_5 & \cdot \\ \cdot & \cdot & \cdot & \cdot \end{bmatrix} \tag{6}$$

$$R = E[Y_k^-(Y_k^-)^T] = \begin{bmatrix} r_0 & r_1 & r_2 & \cdot \\ r_1 & r_0 & r_1 & \cdot \\ r_2 & r_1 & r_0 & \cdot \\ \cdot & \cdot & \cdot & \cdot \end{bmatrix} = LL^T$$

and L is the Cholesky factor (square root) of R. Recall that this exposition started with the assumption that $\{y_k\}$ was a finite-dimensional process which had a pth-order Markovian representation. This is equivalent to the Hankel matrix H having rank p. In this case, both the process $\{y_k\}$ and the covariance sequence $\{r_j\}$ will be referred to as p^{th}-order.

From deterministic realization theory [6,7], there exists an (F_c, g_c, h_c) whose impulse response equals the covariance sequence r_j, $j = 1, 2, \cdots$. Such a realization will be called a *covariance realization.* It is known that a given covariance realization corresponds to a particular factorization of the Hankel matrix as $H = OC$. (See the realization formulas in Appendix A.) It is also known that a covariance realization can be uniquely associated with a stochastic realization, as shown in the following result.

Lemma 1

Given a covariance realization (F_c, g_c, h_c) associated with the covariance sequence $\{r_j\}$ of a stochastic process $\{y_k\}$, there exists an innovations model for $\{y_k\}$ with $(F, h) = (F_c, h_c)$ and $g = (g_c - FPh^T)\rho^{-1}$.

Proof

See Appendix C.

The above lemma provides a method for solving the stochastic realization problem. The first step is to find a covariance realization and set $(F, h) = (F_c, h_c)$. The second step is to solve an algebraic Riccati equation for P, the state covariance matrix of the innovations model. This equation is derived by multiplying the first equation in (1) by x_{k+1}^T and taking the expectation of both sides to obtain

$$P = FPF^T + g\rho g^T \tag{7}$$

Then, multiply the second equation in (1) by y_k and take expected values to obtain

$$r_0 = hPh^T + \rho \tag{8}$$

Finally, by substituting (8) and the expression for g given in the lemma into (7), we obtain the following algebraic Riccati equation

$$P = FPF^T + (g_c - FPh^T)(r_0 - hPh^T)^{-1}(g_c - FPh^T)^T \tag{9}$$

Any positive-definite solution P to (9) results in a stochastic realization; however, the minimal solution to (9) is the one that gives the innovations model*. The Riccati equation approach, which is detailed in [1], is summarized by

Algorithm 1

1) Given a covariance sequence r_j, $j = 0, 1, \cdots, L$, form the covariance Hankel H and the matrix R using (6).
2) Factor H as $H = OC$ and obtain the covariance realization (F_c, g_c, h_c) using the formulas in Appendix A.
3) Solve the Riccati equation (9) for P.
4) Then the innovations model of the form (1) is given by $F = F_c$, $h = h_c$, $\rho = r_0 - hPh^T$, $g = (g_c - FPh^T)\rho^{-1}$ with state covariance matrix P.

It was stated earlier that this approach was valid for any finite number of covariance lags $L \geq 2p$, where p is the dimension of the resulting Markovian model. The reason is that the covariance realization is uniquely specified by $2p$ or more covariance lags. This is a result from deterministic realization theory [8]. Since the same covariance

* A positive-definite matrix A is said to be greater than a positive-definite matrix B if $A - B$ is positive-definite.

realization is obtained for any number of covariance lags greater than $2p$, the same Markovian realization will also be obtained. Thus the Riccati equation approach to stochastic realization will give the same result for a pth-order covariance sequence of any length $L \geq 2p$.

Factorization Approach

If an *infinite,* p^{th}-order covariance sequence is available, then a stochastic realization can be obtained simply by the factorization of a matrix. Since this approach uses an infinite covariance sequence, it cannot be used directly in practice. However, it is important for the following two reasons. First, this approach is related to a well-known model reduction technique as shown in the next section. Second, it is possible to scale the data in such a way that the infinite data results are valid for a finite covariance sequence. These two facts will be combined to obtain a new stochastic realization algorithm. The basis for the factorization approach is the following

Theorem 1

Let (F_c, g_c, h_c) be the covariance realization corresponding to the factorization of the infinite rank p covariance Hankel as $H = OC$. Then the innovations model for $\{y_k\}$ for which $(F, h) = (F_c, h_c)$ is obtained by choosing the state vector as $x_k = CR^{-1}Y_k^-$. In this case, the state covariance matrix of the innovations model is $P = CR^{-T}C^T$.

Proof

See Appendix C.

The state vector of any innovations model must always be a linear function of Y_k^-. The theorem gives this linear function in terms of a factorization of the covariance Hankel in such a way that the innovations model and the covariance realization have the same F and h. This is the same as in the Riccati equation approach, but in this case, the state covariance matrix is also specified in terms of the factorization. Thus no Riccati equation has to be solved, but P can be used directly to find ρ and g. The factorization approach may be summarized by

Algorithm 2

1) Given an infinite covariance sequence $\{r_j\}$, $j = 0, 1, \cdots$, form the covariance Hankel H and the matrix R using (6).
2) Factor H as $H = OC$. Obtain the covariance realization (F_c, g_c, h_c) and the matrix $P = CR^{-T}C^T$.
3) The innovations model of the form (1) is given by $F = F_c$, $h = h_c$, $\rho = r_0 - hPh^T$, $g = (g_c - FPh^T)\rho^{-1}$ with state covariance P.

DETERMINISTIC BALANCING AND STOCHASTIC MODEL REDUCTION

The factorization approach outlined in the last section will yield different stochastic realizations depending on the actual factorization of H in Step 2. In order to apply these results to covariance sequences that have been perturbed by estimation errors, the resulting stochastic realizations must be amenable to model reduction. In this section, a particular factorization of H is presented which has this desirable property.

Deterministic Balancing

The stochastic model reduction technique presented later in this section will be shown to be equivalent to deterministic balancing. Thus the deterministic theory is briefly reviewed here. Let the matrices in (1) define the following *deterministic* realization:

$$\begin{aligned} x_{k+1} &= Fx_k + gw_k \\ y_k &= hx_k \end{aligned} \tag{10}$$

where w_k is a deterministic input. Note that the feedthrough term in the output equation of (10) is omitted since it is irrelevant as far as model reduction goes. The model reduction technique used in this chapter is that of *balancing* [9-11]. Balanced realizations can be obtained by an appropriate factorization of the infinite Hankel matrix H formed from the impulse response; that is, the response of (10) when w_k is the Kronecker delta function. For any factorization $H = OC$, the observability and controllability Gramians are defined, respectively, as

$$G_o = O^T O \quad , \quad G_c = CC^T \tag{11}$$

A balanced realization is defined as one for which G_o and G_c are simultaneously diagonal, and equal. For the purpose of model reduction, it is only necessary that the Gramians be simultaneously diagonal [11]. Realizations with this property are said to be *principal axis* realizations [12], although we will continue to use the term "balanced" in this case. Reduced-order models can be obtained as subsystems of such realizations, and this approach corresponds to retaining the strongly observable and controllable part of the system [9].

Given an infinite impulse response sequence, a balanced realization may be obtained by the following singular value decomposition (SVD) [13] of the Hankel matrix:

$$H = (U\Sigma^{\frac{1}{2}})(\Sigma^{\frac{1}{2}}V^T) = OC \tag{12}$$

Since U and V have orthogonal columns, it is clear that $G_o = G_c = \Sigma$. Thus the factorization in (12) corresponds to a balanced realization, where the matrices are given by the realization formulas in Appendix A.

Stochastic Model Reduction

Recall from (5) that the state space of a stochastic realization is spanned by $HR^{-1}Y_k^-$. If H is of rank p, then the dimension of the state space is p. In order to obtain an rth order model, $r < p$, it is necessary to find a matrix G such that GY_k^- nearly spans the state space. Arun, Bhaskar Rao, and Kung [14,15] have suggested choosing G to minimize the following criterion:

$$\text{trace of covar}\,(Y_k^+ - (Y_k^+ \mid GY_k^-)) \tag{13}$$

where the covariance of a vector random variable v_k is denoted covar $(v_k) = E\,[v_k\, v_k^T]$. They call the resulting algorithm the unweighted principal components (UPC) algorithm. The first step in the UPC algorithm is to perform the following spectral decomposition:

$$HR^{-1}H^T = U\Sigma U^T \tag{14}$$

The full (p^{th}) order model is then given by

$$\begin{aligned} F &= U^L\, U^T \\ h &= \text{first row of } U \\ P &= \Lambda \\ \rho &= r_0 - h\Lambda h^T \\ \Lambda - F\Lambda F^T &= \rho g g^T \end{aligned} \tag{15}$$

It should be mentioned that an algorithm for transforming an arbitrary stochastic realization to principal component (deterministically balanced) form was derived in [16].

To relate the UPC algorithm to the factorization approach mentioned earlier, it is necessary to work with the "square root" of (14), i.e.

$$HL^{-T} = U\Sigma V^T \tag{16}$$

which yields the following factorization of the Hankel matrix:

$$H = (U)(\Sigma V^T L^T) = OC \tag{17}$$

The resulting stochastic realization can be obtained from step 3 of the factorization approach, or from (15), since the matrix U in (17) is the same as that in (14), and $\Sigma^2 = \Lambda$. Note that Step 3 provides an explicit formula for g, while (15) does not. However, an advantage of using (14) and (15) is that they can be applied to sample covariance sequences that are not positive definite. This condition often arises when the covariance sequence is estimated from a short data record.

The equivalence between UPC and deterministic balancing was established in [17], and is summarized by the following

Theorem 2

Let $\{y_k\}$ be a scalar stochastic process with covariance sequence $\{r_j\}$, $j = 0, 1, \cdots$. Then the UPC algorithm applied to $\{r_j\}$ yields a realization (F, g, h) which is balanced in the deterministic sense.

The consequence of this theorem for stochastic model reduction is as follows: any innovations model (F, g, h) for a stochastic process $\{y_k\}$ can be transformed to balanced form by deterministic balancing. The state variables of the balanced model are the principal components of the state space $S(Y_k^+ \mid Y_k^-)$. Using either the results of deterministic balancing or the UPC algorithm, a reduced-order model is obtained by taking a subsystem of the original model. This suggests the following interpretation: retaining the strongly observable-controllable part of the innovations model in the deterministic setting corresponds to extracting the principal components of the state space in the stochastic setting. In other words, whether the input to a system is white-noise or a deterministic impulse, the "balanced basis" is useful for distinguishing the relative importance of each state variable.

The above theorem does not have immediate application to the approximate stochastic realization problem because it is an infinite data result. Indeed, the factorization approach to deterministic balancing presented in the last section is also an infinite-data result. However, the theorem indicates that any high-order realizations that are obtained as an intermediate step in the approximate realization problem should be balanced before their order is reduced.

It should be mentioned that there are at least two types of balancing in the literature. Deterministic balancing as defined in the last section was originally motivated by the deterministic model reduction problem, but it also has application to the stochastic model reduction problem as shown by Theorem 2. Desai and Pal [19,20] have introduced a technique called stochastic balancing which was motivated by the stochastic model reduction problem. Stochastic balancing is based on a canonical correlation analysis of the future and past of a process, and is *not* equivalent to a deterministic balancing of an innovations model. A comparison of the UPC algorithm (deterministic balancing) and stochastic balancing is given in [21]. In this chapter, the term balancing should be taken to mean deterministic balancing.

Recall from the previous section that the Riccati equation approach is valid for a finite number of covariance lags. Thus it can be used together with the above theorem to suggest a finite-data approximate realization algorithm as follows:

Algorithm 3

1) Perform the factorization in (16) for matrices derived from a finite covariance sequence. Identify a high order covariance realization.
2) Solve the algebraic Riccati equation (9) and use Step 4 of Algorithm 1 to obtain a high-order stochastic realization.

3) Compute the similarity transformation which balances (F, g, h) obtained in Step 3 [18, see Appendix B]. Apply this transformation to the stochastic realization.
4) An approximate realization can then be obtained as a subsystem of the balanced realization.

Although it provides a way of implementing the infinite-data results of the theorem using a finite covariance sequence, the above algorithm has two problems. The first is that it requires more computation than previous algorithms (an SVD *and* a Riccati solution must be obtained). The second, and far more serious problem, is that the balancing transformation can not always be computed. The difficulty is that the high-order realization obtained in Step 2 is an exact *partial realization* of the given covariance sequence, and as such, is prone to being unstable [22]. If the realization is unstable, then it cannot be balanced, since the Gramians do not exist in this case. It should be noted that the instability is clearly a result of estimation errors, since a partial realization obtained from exact covariance data will always be stable. It is shown in the next section that it is possible to scale the data in such a way that a slight variation of Algorithm 3 can be applied to any sample covariance sequence.

MODELING FROM PERTURBED, FINITE COVARIANCE SEQUENCES

Given N observations of a random process $\{y_k\}$, $k = 0, \cdots, N-1$, it is first necessary to estimate L lags of the covariance sequence. Although there are many ways of doing this, we will simply use the following unbiased estimator:

$$r_j = \frac{1}{N-j} \sum_{i=0}^{N-j} y_i \, y_{i+j} \tag{18}$$

In this section, two algorithms are presented which can be used with perturbed, finite covariance sequences. The basis for both algorithms is the following easily verified fact: if the covariance sequence has decayed to zero by lag L (i.e. covariance lags greater than L equal zero with respect to the precision of the computer that is performing the calculations), then the infinite covariance sequence (as represented by the computer) can be obtained by simply padding with zeros. In fact, only $L-2$ zeros have to be added to a given sequence in order to form the following finite matrices:

$$H_{L,L-1} = \begin{bmatrix} r_1 & r_2 & \cdot & \cdot & r_{L-1} \\ r_2 & r_3 & \cdot & \cdot & r_L \\ \cdot & \cdot & \cdot & \cdot & 0 \\ \cdot & \cdot & 0 & \cdot & \cdot \\ r_L & 0 & 0 & \cdot & 0 \end{bmatrix}, \quad R_{L-1} = \begin{bmatrix} r_0 & r_1 & \cdot & \cdot & r_{L-2} \\ r_1 & r_0 & \cdot & \cdot & r_{L-3} \\ r_2 & r_1 & \cdot & \cdot & \cdot \\ \cdot & \cdot & \cdot & \cdot & \cdot \\ r_{L-2} & r_{L-3} & \cdot & \cdot & r_0 \end{bmatrix} = L_{L-1} L_{L-1}^T$$

where the matrix L_{L-1} is the *upper triangular* Cholesky factor. This upper-lower factorization of R is needed in (20.a) below.

Because of the lower triangle of zeros in $H_{L,L-1}$, it can be embedded in a matrix of zeros without destroying the Hankel structure; that is, the infinite Hankel matrix formed from the covariance sequence padded with an infinite number of zeros is

$$H = \begin{bmatrix} H_{L,L-1} & 0 \\ 0 & 0 \end{bmatrix} \tag{19}$$

In this case, the infinite-data factorization in (16) reduces to

$$HL^{-T} = \begin{bmatrix} H_{L,L-1}L_{L-1}^{-T} & 0 \\ 0 & 0 \end{bmatrix} = \begin{bmatrix} U\Sigma V^T & 0 \\ 0 & 0 \end{bmatrix} \tag{20.a}$$

$$= \begin{bmatrix} U \\ 0 \end{bmatrix} \begin{bmatrix} \Sigma \end{bmatrix} \begin{bmatrix} V^T & 0 \end{bmatrix} \tag{20.b}$$

where $U\Sigma V^T$ is the finite SVD of $H_{L,L-1}L_{L-1}^{-T}$. The *infinite* SVD of HL^{-T} is given in (20.b), and is just the finite SVD appropriately padded with zeros. Because of these zeros, the infinite-data realization formulas which use the infinite SVD give exactly the same results as the finite-data realization formulas in Appendix A which use the finite factorization. This fact will be exploited in the algorithms that follow.

It has just been shown that if the estimated covariance sequence has decayed to zero, then the infinite-data results can be obtained from finite-data factorizations. However, the covariance sequence will not have decayed to zero in general, especially if it has been estimated from a short data record. Thus it is necessary to scale the data to force the covariance sequence to zero. It is possible to do this scaling in two different ways, leading to two new algorithms. In both cases, it is possible to apply the inverse scaling after the model reduction has been performed.

It should be mentioned that any scaling that is done before model reduction is performed changes the nature of the approximation inherent in the reduction procedure. However, the purpose of model reduction in this case is simply to remove the effects of the estimation error, and *not* to approximate signal components. It has been found experimentally that the balancing procedure works well even after scaling.

Scaling the Covariance Realization

The first algorithm is based on scaling the estimated covariance sequence $\{r_j\}$ by a geometric progression $\{\mu^j\}$ where μ is a positive real number. The scaled covariance sequence is $\{\overline{r}_j\}=\{\mu^j r_k\}$, and μ is chosen so that this sequence has effectively decayed to zero by lag L. The effect of this scaling is to move the poles and zeros of the covariance realization radially inward by the factor μ. The inverse scaling can easily be applied to the reduced-order covariance realization, as shown below. Then, the Riccati equation must be solved to obtain the stochastic realization. Before the details of this algorithm are given, a two-step procedure to compute μ is given. The first step is to compute a "bounding exponential" for the given covariance sequence, i.e. compute

$$\alpha = \max_j \exp\left\{ \frac{1}{j} \ln \frac{|r_j|}{r_0} \right\} \quad , \quad j = 1 , \cdots , L \tag{21}$$

This choice of α guarantees that $r_0 \alpha^j \geq |r_j|$, $j = 1 , \cdots , L$. The second step is to calculate a value of μ which reduces the magnitude of the bounding exponential to an acceptably small level at lag L . This is accomplished by solving the following equation for μ:

$$r_0 (\alpha \mu)^L = \epsilon \tag{22}$$

where ϵ is a number related to the precision of the computer. The algorithm can now be summarized below. Simulation results for this algorithm are given in [23].

Algorithm 4

1) Estimate $L+1$ covariance lags using (18). Calculate α and μ from (21) and (22), and let $\beta = \dfrac{1}{\mu}$.
2) Form the weighted sequence $\overline{r}_j = \mu^j r_j$, $j = 0 , \cdots , L$, and pad it with $L-2$ zeros. Form the matrices H , R , and L and perform the SVD in (20.a).
3) Group this factorization as $H = (U)(\Sigma V^T L^T) = OC$, and use the realization formulas in Appendix A to obtain the balanced covariance realization (F , g_c , h). The reduced-order covariance realization is then obtained by taking a scaled subsystem of this model as follows: $(\beta F_{11} , \beta g_{c_1} , h)$. P , g , and ρ are obtained by solving the Riccati equation.

Note that the scaled covariance sequence decays to zero, and so the finite SVD is equivalent to an infinite SVD. From the discussion on the factorization approach (Algorithm 2), an infinite factorization is equivalent to solving a Riccati equation. However, since scaling is applied to the covariance sequence in the above algorithm, the inverse scaling must be applied to the (reduced-order) covariance realization. This scaling induces a nontrivial change in the state covariance matrix. Since this matrix does not merely scale, it must be computed from the Riccati equation. This additional computation can be avoided by scaling the data itself, before estimating the covariance sequence. This is equivalent to scaling the stochastic realization, as is shown next.

Scaling the Stochastic Realization

The details of this approach are a topic of current research. The point of this section is to show that a scaled stochastic realization can be obtained by a nonstationary scaling of the stochastic process. The resulting nonstationary process can be modeled using the state-space framework developed in this chapter; the realization formulas are given below. The open problems are the infinite-data extension of this approach, and its relationship to balancing.

To develop this idea, we first consider the following stochastic realization for the stationary process $\{y_k\}$ which is initialized at time $k=0$ with steady-state initial conditions

$$x_{k+1} = Fx_k + gu_k \quad , \quad x_0 = X \tag{23}$$
$$y_k = hx_k + u_k \quad , \quad k \geq 0$$

where the random vector X is Gaussian, zero mean with variance P, and P is the steady-state state covariance matrix associated with (23), i.e. the solution to the Lyapunov equation (7). (23) can be solved to express y_k directly in terms of the input sequence and the initial state as follows:

$$y_k = hF^k x_0 + \sum_{j=0}^{k-1} hF^{k-j-1} gu_j \tag{24}$$

We now consider the effect of scaling $\{y_k\}$ by a geometric progression, i.e. we consider a finite observation sequence, and define

$$\overline{y}_k = \mu^k y_k \quad , \quad k = 0, \cdots, N-1 \tag{25}$$

It is easy to see that (24) can be modified to provide an expression for $\overline{y}_k$,

$$\overline{y}_k = h(\mu F)^k x_0 + \sum_{j=0}^{k-1} h(\mu F)^{k-j-1}(\mu g)(\mu^j u_j) \tag{26}$$

and that this expression is the solution to the following state-space equations:

$$\overline{x}_{k+1} = \overline{F}\,\overline{x}_k + \overline{g}\,\overline{u}_k \quad , \quad \overline{x}_0 = X \tag{27}$$
$$\overline{y}_k = \overline{h}\,\overline{x}_k + \overline{u}_k$$

where $\overline{u}_k = \mu^k u_k$, $\overline{y}_k = \mu^k y_k$, $\overline{F} = \mu F$, $\overline{g} = \mu g$, and $\overline{h} = h$.

Thus the scaled output process can be generated by passing the scaled input process through the scaled system. It should be noted that $\{\overline{u}_k\}$ and $\{\overline{y}_k\}$ are both *nonstationary* processes; however, $(\overline{F}, \overline{g}, \overline{h})$ is a time-invariant system. The fact that the stochastic process is nonstationary means that the Toeplitz and Hankel matrices R and H which define the predictor space must be defined in terms of their nonstationary counterparts R_k and H_k defined below. However, the fact that the underlying stochastic realization is time invariant means that the realization equations only have to be solved at a single, arbitrary index k. This is in contrast to the modeling of a more general class of nonstationary process which requires solving the realization equations at *every* time index $k \geq 0$ [24].

The realization equations for $\{\overline{y}_k\}$ are developed in a manner similar to the stationary case. Here, however, we explicitly consider only the n-step future and past, as defined below:

$$\overline{Y}_k^{\,+} = \begin{bmatrix} \overline{y}_k \\ \overline{y}_{k+1} \\ \cdot \\ \cdot \\ \cdot \\ \overline{y}_{k+n} \end{bmatrix}, \quad \overline{Y}_k^{\,-} = \begin{bmatrix} \overline{y}_{k-1} \\ \overline{y}_{k-2} \\ \cdot \\ \cdot \\ \cdot \\ \overline{y}_{k-n} \end{bmatrix} \tag{28}$$

The projection of the future onto the past is now expressed in terms of the following two matrices:

$$H_k = E[\overline{Y}_k^{\,+}(\overline{Y}_k^{\,-})^T] = \begin{bmatrix} \overline{r}_{k,k-1} & \overline{r}_{k,k-2} & \cdot\ \cdot & \overline{r}_{k,k-n} \\ \overline{r}_{k+1,k-1} & \overline{r}_{k+1,k-2} & \cdot\ \cdot & \overline{r}_{k+1,k-n+1} \\ \cdot & \cdot & \cdot\ \cdot & \cdot \\ \cdot & \cdot & \cdot\ \cdot & \cdot \\ \overline{r}_{k+n,k-1} & \cdot & \cdot\ \cdot & \overline{r}_{k+n,k-n} \end{bmatrix} \tag{29}$$

and

$$R_k = E[\overline{Y}_k^{\,-}(\overline{Y}_k^{\,-})^T = \begin{bmatrix} \overline{r}_{k-1,k-1} & \overline{r}_{k-1,k-2} & \cdot\ \cdot & \overline{r}_{k-1,k-n} \\ \overline{r}_{k-2,k-1} & \overline{r}_{k-2,k-2} & \cdot\ \cdot & \cdot \\ \cdot & \cdot & \cdot\ \cdot & \cdot \\ \cdot & \cdot & \cdot\ \cdot & \cdot \\ \overline{r}_{k-1,k-n} & \cdot & \cdot\ \cdot & \overline{r}_{k-n,k-n} \end{bmatrix} = L_k L_k^T$$

where the nonstationary covariance sequence is

$$\overline{r}_{i,j} = E(\overline{y}_i\,\overline{y}_j) = \mu^{i+j}\, r_{|i-j|} \tag{30}$$

Since the process $\{y_k\}$, and hence $\{\overline{y}_k\}$, is only observed over the interval $0 \le k \le N-1$, H_k and R_k can only make use of the data in this interval. Note that H_k makes use of $\overline{r}_{k,k-L+1} = E(\overline{y}_k\,\overline{y}_{k-L+1})$. Thus we must have $k-L+1 \ge 0$. This means that $k_0 = L-1$ is the smallest time index for which we will construct H_k and R_k.

At time index $k = k_0$, a stochastic realization can be obtained from the following SVD (it will be shown shortly that the singular vectors do not depend on k):

$$H_{k_0} L_{k_0}^{-T} = U \Sigma_{k_0} V^T \tag{31}$$

To obtain the decomposition for future values of k, we simply note that for $k \ge k_0$,

$$R_k = \mu^{2(k-k_0)} R_{k_0}\,, \quad H_k = \mu^{2(k-k_0)} H_{k_0} \tag{32}$$

and so

$$L_k = \mu^{k-k_0} L_{k_0} \tag{33}$$

Thus, for $k \geq k_0$, the SVD is

$$H_k L_k^{-T} = \mu^{k-k_0} H_{k_0} L_{k_0}^{-T} = U\left(\mu^{k-k_0}\Sigma_{k_0}\right) V^T$$

In other words, the time variation results in a simple scaling of the singular values, while the singular vectors are constant. This implies that the SVD only has to be performed once (say at $k = k_0$) to obtain the stochastic realization.

Note that (31) can be rewritten as

$$H_{k_0} = (U)(\Sigma_{k_0} V^T L_{k_0}^{-T}) = OC_{k_0} \tag{34}$$

Since O is time-invariant, the realization formulas in Appendix A show that $\bar{F}$ and $\bar{h}$ are constant. Since C_{k_0} is a function of time, the input vector of the covariance realization is also time varying; however, when $\bar{g}$ is obtained from g_c using the formula given in Lemma 1, the time variation cancels out, leaving a time-invariant $\bar{g}$.

The purpose of this section was to provide an introduction to the idea of scaling the stochastic realization. Such details as relating the finite factorizations to infinite factorizations and balancing the scaled realization remain to be worked out.

ACKNOWLEDGMENTS

The author is grateful to Prof. B.W. Dickinson for many helpful discussions, and for assisting in producing the final manuscript; the Department of Electrical Engineering at Princeton University provided the text processing facilities. Thanks are also extended to I. Ozguc for performing computer simulations related to this work. This work was supported by the National Science Foundation through Grant ECS-8404742.

APPENDIX A - Deterministic Realization Formulas

Infinite Data Case

Let h_k, $k = 1, 2, \cdots$ be the impulse response of the following asymptotically stable system

$$x_{k+1} = Fx_k + gu_k \tag{A.1}$$
$$y_k = hx_k$$

It is easy to show that when u_k is a Kronecker delta function, the output $y_k = hF^{k-1}g$, $k \geq 1$. Let H be the infinite Hankel matrix whose i,j element is $h_{i+j-2} = hF^{i+j-2}g$. This implies that the Hankel matrix can be factored as

$$H = \begin{bmatrix} h \\ hF \\ hF^2 \\ \cdot \\ \cdot \\ \cdot \end{bmatrix} \begin{bmatrix} g & Fg & F^2g & . & . & . \end{bmatrix} = OC \tag{A.2}$$

where O and C are both full rank. Thus, knowing only the factorization (A.2), one can recover the state-space model (A.1) as follows:

$$h = \text{first row of O} \tag{A.3}$$

$$g = \text{first column of } C$$

Now define $O^{\uparrow}$ and $\overleftarrow{C}$ as

$$O^{\uparrow} = \begin{bmatrix} hF \\ hF^2 \\ \cdot \\ \cdot \\ \cdot \end{bmatrix}, \quad \overleftarrow{C} = \begin{bmatrix} Fg & F^2 g & . & . & . \end{bmatrix} \tag{A.4}$$

Then from (A.4) and (A.2), it is clear that

$$OF = O^{\uparrow}, \quad FC = \overleftarrow{C} \tag{A.5}$$

The matrix F can be obtained by solving either equation in (A.5) as follows:

$$F = O^L O^{\uparrow} = \overleftarrow{C} C^R \tag{A.6}$$

where "L" and "R" denote left and right inverse, respectively, and can be computed in general as

$$O^L = (O^T O)^{-1} O^T, \quad C^R = C(CC^T)^{-1} \tag{A.7}$$

If

$$F = \begin{bmatrix} F_{11} & F_{12} \\ F_{21} & F_{22} \end{bmatrix}, \quad g = \begin{bmatrix} g_1 \\ g_2 \end{bmatrix}, \quad h = \begin{bmatrix} h_1 & h_2 \end{bmatrix}$$

then the reduced-order model is simply (F_{11}, g_1, h_1).

Finite Data Case

Suppose only a finite portion of the impulse response in the previous section is available. To be specific, assume the number of data points is even and equal to $2p$. The Hankel matrix with $p+1$ rows and p columns is formed, and factored into observability and controllability matrices as follows:

$$H_{p+1,p} = O_{p+1,n} C_{n,p} \tag{A.8}$$

where n is the rank of the Hankel matrix.

It is clear that g and h are still defined by (A.3). To obtain F, first define O_1 to be the first p rows of O, and O_2 to be the last p rows of O. Then

$$O_1 F = O_2 \tag{A.9}$$

or

$$F = O_1^L O_2 \tag{A.10}$$

Noisy Finite Data Case

Suppose the impulse response has been corrupted by additive noise. In that case, the rank of $H_{p+1,p}$ will be p with probability one. However, for reasonable signal-to-noise ratios, it is usually the case that the Hankel matrix is "close" to having rank n. This "numerical rank" is determined by the number of "dominant" singular values of H [25,26]. In this case, the SVD can be partitioned as follows:

$$H_{p+1,p} = \begin{bmatrix} U & \overline{U} \end{bmatrix} \begin{bmatrix} \Sigma & 0 \\ 0 & \overline{\Sigma} \end{bmatrix} \begin{bmatrix} V^T \\ \overline{V}^T \end{bmatrix} \tag{A.11}$$

where Σ contains the n dominant singular values and the singular vectors are partitioned accordingly. Then the optimal (in Frobenius norm) rank n approximation to H is

$$\hat{H} = (U)(\Sigma V^T) = \hat{O}\hat{C} \tag{A.12}$$

In this case, there does not exist a matrix F which shifts the matrix $\hat{O}$, since $\hat{H}$ is not a Hankel matrix in general. However, a least-squares problem can be set up as follows:

$$\min_F \| \hat{O}_1 F - O_2 \|_2^2 \tag{A.13}$$

and the solution is

$$F = \hat{O}_1^L \hat{O}_2 \tag{A.14}$$

Note that this is the same equation as in the noise-free case (A.10), but that the interpretation is somewhat different. Also, the vectors g and h can be obtained from (A.3) as before [27].

To summarize, in the noisy data case, the noise-free realization formulas are used on a reduced-rank approximation to the Hankel matrix. It is indeed fortunate that the tool for determining the numerical rank of H is the same tool that is used in obtaining balanced realizations, namely, the SVD.

APPENDIX B - Computation of Balancing Transformations

Any stable realization (F, g, h) can be transformed to balanced form by an appropriate similarity transformation. The procedure is to first compute the controllability and observability Gramians as the solutions to the following Lyapunov equations

$$G_c = F G_c F^T + g g^T$$

$$G_o = F^T G_0 F + h^T h$$

Then the similarity transformation is computed as follows [18]:

1) Perform a Cholesky factorization of G_c to obtain $G_c = LL^T$.
2) Perform an eigenvalue/eigenvector decomposition of the symmetric matrix $L^T G_o L$ to obtain $L G_0 L = U \Sigma^2 U^T$.
3) The balancing transformation is $T = \Sigma U^T L^{-1}$.

The balanced realization (F_b, g_b, h_b) is then given by

$$F_b = TFT^{-1} \quad , \quad g_b = Tg \quad = \quad h_b = hT^{-1}$$

and the Gramians of the system are both equal to Σ.

APPENDIX C - Proofs

Proof of Lemma 1

From the definition of the covariance realization,

$$r_j = h_c F_c^{j-1} g_c \tag{C.1}$$

and from (1), the covariance sequence of the output of an innovations model is seen to be

$$r_j = hF^{j-1}(FPh^T + \rho g) \tag{C.2}$$

The lemma is established by equating (C.1) and (C.2).

Proof of Theorem 1

Note that $x_k | \{Y_k\} = x_k$ so that this choice of state does indeed give an innovations model. We need to show that it gives the innovations model with $(F, h) = (F_c, h_c)$.

$$E[x_{k+1} x_k^T] = E[(Fx_k + gu_k) x_k^T]$$

$$CR^{-1}\overleftarrow{R} R^{-T} C^T = FE[x_k x_k^T] = FP$$

$$\overleftarrow{C} R^{-T} C^T = FCR^{-1}RR^{-T} C^T$$

$$F_c CR^{-T} C^T = FCR^{-T} C^T = FP \quad (\text{i.e. } P = CR^{-T} C^T) \tag{C.3}$$

Note that since $R = LL^T$, $CR^{-T} C^T = (CL^{-T})(CL^{-T})^T$, and since CL^{-T} has rank p, $CR^{-T} C^T$ is nonsingular. So by (C.1), $F = F_c$. Now let $e_1^T = (1, 0, 0, \cdots)$. Then

$$E[y_k x_k^T] = E[(hx_k + u_k) x_k^T]$$

$$E[y_k (Y_k^-)^T R^{-T} C^T] = hCR^{-T} C^T e_1^T HR^{-T} C^T = hCR^{-T} C^T$$

$$e_1^T OCR^{-T} C^T = hCR^{-T} C^T$$

and it has been shown above that $CR^{-T} C^T$ is nonsingular, thus

$$e_1^T O = h = \text{first row of O} = h_c$$

REFERENCES

1. P. Faurre, In: *System Identification: Advances and Case Studies* (Eds. R.K. Mehra and D.G. Lainiotis), Academic Press, New York, 1976, pp. 1-25.

2. R.J. Vaccaro and B.W. Dickinson, In: *Proc. 19th Annual Asilomar Conf.*, Pacific Grove, CA, November, 1985.
3. M. Gevers and W.R.E Wouters, *Journal A*, vol. 19, pp. 90-110, 1978.
4. B.D.O. Anderson and J.B. Moore, *Optimal Filtering*, Prentice-Hall, Englewood-Cliffs, 1979.
5. H. Akaike, *IEEE Trans. Automat. Contr.*, vol. AC-19, pp. 667-674, 1974.
6. B.L. Ho and R.E. Kalman, In: *Proc. 3rd Allerton Conf.*, Monticello, Illinois, pp.449-459, 1965.
7. J. Rissanen, *SIAM J. Contr.*, vol. 9, pp. 420-430, 1971.
8. R.E. Kalman, In: *Acta Polytechnica Scandinavica*, Mathematics and Computer Science Series No. 31, pp. 9-32, 1979.
9. B.C. Moore, *IEEE Trans. Automat. Contr.*, vol. AC-26, pp. 17-32, 1981.
10. L. Pernebo and L.M. Silverman, *IEEE Trans. Automat. Contr.*, vol. AC-27, pp. 382-387, 1982.
11. R.J. Vaccaro, Ph.D. Dissertation, Princeton University, Princeton, NJ, 1983.
12. K.V. Fernando and H. Nicholson, *IEEE Trans. Automat. Contr.*, vol. AC-28, pp. 228-231, 1983.
13. V.C. Klema and A.J. Laub, *IEEE Trans. Automat. Contr.*, vol. AC-25, pp. 164-176, 1980.
14. K.S. Arun, D.V. Bhaskar Rao, and S.Y. Kung, In: *Proc. 22nd IEEE Conf. Decision Contr.*, San Antonio, TX, pp. 1353-1355, 1983.
15. K.S. Arun and S.Y. Kung, In: *Proc. IEEE Spectral Estimation Workshop II*, pp. 161-167, 1983.
16. S. Fujishige, S. Nagai, and Y. Sawaragi, *Int. J. Contr.*, vol. 22, pp. 807-819, 1975.
17. R.J. Vaccaro, *IEEE Trans. Automat. Contr.*, vol. AC-30, pp. 921-923, 1985.
18. A.J. Laub, In: *Proc. 1980 JACC*, FA8-E, 1980.
19. U.B. Desai and D. Pal, In: *Proc. 16th Annual Conf. Inform. Sci., Syst.*, Princeton Univ., Princeton, NJ, 1982; also in *Proc. 22nd IEEE Conf. Decision, Contr.*, Orlando, FL, 1982.
20. U.B. Desai and D. Pal, *IEEE Trans. Automat. Contr.*, vol. AC-29, pp. 1097-1100, 1984.
21. J.A. Ramos and E.I. Verriest, In: *Proc. American Contr. Conf.*, San Diego, CA, pp. 150-155, 1984.
22. C.I. Byrnes and A. Lindquist, *Syst. Contr. Letters*, vol. 2, pp. 99-105, 1982.
23 R.J. Vaccaro, *IEEE Proceedings (Letters)*, to appear, 1986.
24. Y. Baram, *IEEE Trans. Automat. Contr.*, vol. AC-26, pp. 1225-1231, 1981.
25. G.H. Golub, V.C. Klema, and G.W. Stewart, Stanford Univ. Tech. Rep. STAN-CS-76-559, Computer Sci. Dept., Stanford Univ., Palo Alto, CA, 1976.
26. H.P. Zeiger and A.J. McEwen, *IEEE Trans. Automat. Contr.*, p. 153, 1974.
27. S.Y. Kung, In: *Proc. 12th Asilomar Conf.*, Pacific Grove, CA, pp. 705-714, 1978.

6

MODEL REDUCTION VIA BALANCING, AND CONNECTIONS WITH OTHER METHODS

Erik I. Verriest
School of Electrical Engineering
Georgia Institute of Technology
Atlanta, Georgia 30332

ABSTRACT

This paper starts with a rather philosophical viewpoint on the concepts of modeling, model reduction, and randomness. The theory of open-loop deterministic balancing is introduced as a particular implementation of a model reduction scheme. The discussion focusses on the choice of the criterion. Thus motivated, it is shown that similar ideas can be employed in the reduction of optimally controlled systems under the presence of noise, leading to the LQG-balanced realizations. This connects to the stochastic balanced realizations. Finally, different stochastic realization algorithms are cast in the common framework of the RV-coefficient, and the deeper geometric significance of this measure is explored.

1 INTRODUCTION:

1.1 Modeling and Model Reduction

Until recently, modeling has been to a large extent a heuristic and unrigorous process where ad-hoc procedures abounded. For this reason, further attention and research to this problem has been more than welcome. In effect, the first half of the eighties has seen a proliferation in modeling and model reduction methods which are firmly based on mathematical rigor. (e.g. [1], [2], [3])

The dichotomy between modeling and model reduction is rather weak and different authors may provide different definitions. Perhaps the most intuitive notion is to let modeling be the process whereby an abstract mathematical model is matched to the physical reality; and model reduction the process whereby a simpler mathematical model is derived from an existing mathematical model. In this regard modeling is what is usually called "Identification", while model reduction belongs to the realm of Approximation Theory.

Keeping in mind that the physical world and thus all real-life systems are the basic entities, perhaps eluding a description as a whole, one can only abstract some aspects of its behavior and model these properties in a formal theory. In what follows then, it will be assumed that the "physical reality" is that what allows observation. Modeling thus infers a procedure which formalizes in a mathematical abstraction some aspects of the behavior of the physical entity. Clearly such a formalization cannot be unique.

Together with the mathematical abstraction (model) one must give its scope, i.e. which aspects of the physical system it models. Models inherently have their limitations. A linear small signal model of a transistor for instance, no matter how accurate its parametrization, will be unable to predict the switching properties of digital transistor circuits. Clearly then, the scope of the model should be matched to whatever one expects from the model. In the study of the kinematics of machinery, there is no need to apply the theory of relativity, but in the study of particle accelerators, the classical theory no longer suffices.

Once the scope of the model has been laid down, one must determine the accuracy of that model. How well does it describe the domain-aspect of the physical reality? A better model is obviously the one that, given the same domain, predicts the behavior of the physical system more accurately. Biped motion can be crudely modeled with a ball-and-stick model, where for instance each stick is rigid, and perhaps massless. A better model would be the one incorporating the distributed nature of the masses, actuators, etc.

Also, one must be able to explain when a given model is more accurate, or better than another one. More specifically, this consists in finding a measure for the accuracy, or more abstractly a suitable topology, with a meaningful physical interpretation, so that the approximation problem for models, within the same scope, is well defined.

Thirdly, another definitely more practical aspect of a formal theory is its complexity. Roughly speaking, complexity refers to the number of ad-hoc rules (postulates) that the theory requires, as well as the smallest number of parameters that need to be specified a priori in order to obtain uniquely predictable (computable) answers within the model. In a Newtonian mechanistic model the whole universe would be predictible given the initial position, velocity and mass of each particle constituting the universe. In this theory there is one basic postulate: the (Newtonian) universal law of gravity, but the parameter set is ... well, very big indeed. Such a model would clearly be impractical, if not unfeasible, if one were interested in studying the dynamics of the solar system, or the kinetics of a gas in a containter.

To summarize, every formal modeling theory should be accompanied by these three quantifiers:

-its domain of validity

-its predictability or accuracy

-its complexity

Hence, there exists only a partial ordering between models, and blank statements as "Model A is better than model B." definitely do not make any sense without any indication of these three quantifiers. Even given the quantifiers, different models may simply not be comparable. Whether one favors a general model of large scope, or several specialized ones of lower scope and complexity is now more a matter of personal taste. Of course the particular purpose or objective of the model should influence such a choice.

Model reduction problems aim at reducing the complexity although there generally is a trade off with the accuracy and the scope. Within the established mathematical framework, this resorts to finding a more attractive subset of the space which is dense (with respect to the topology) in the given space, as for instance in polynomial approximation. Alternatively, it may mean the search for a lower dimensional subspace of some given space, e.g. finite element approximations of distributed systems. At any rate, the modelers dream is to come up with a mathematical model, which is suitably small in order to allow computable predictions of the reality.

Model reduction can be accomplished in many ways: For example, suppose that one has a large dimensional system, perhaps of weakly interacting subsystems. Existing techniques find a lower dimensional model, e.g. by aggregation. There is no doubt that the result will be a simpler model. On the other hand, one could take the opposite approach, and let the number of weakly interacting subsystems approach infinity, only to realize a statistical or probabilistic description of the system. Such a probabilistic description may result in a fewer number of parameters (e.g. first and second order moments). In fact, this is exactly the approach of statistical dynamics. Again, which approach is favorable will depend on the purpose of the model. If one is only interested in the average behavior of the system, then the statistical description may be preferrable. One does not need to know the detailed trajectories of the gasmolecules in order to understand the workings of an internal combustion engine.

1.2 Stochastic Models and the Origins of Randomness

In the previous paragraph, we already hinted at building statistical models. The observed data set on which one tries to model some behavior, typically shows fluctuations. These fluctuations arise from two origins:

i) Some variables (parameters) of the system may be random. In this sense the resulting probabilities are unambiguously defined, i.e. the randomness is imposed from the "outside". (e.g. random boundary conditions).

ii) Randomness can be introduced in an arbitrary way, to reflect our incomplete knowledge of an exact description of a system. For instance this can be due to uncertainties of a real probabilistic nature (e.g. quantum uncertainty). This uncertainty further arises when the number of variables is so large that a correct description would be practically impossible. Randomness is then used to replace a knowledge which is too detailed to be useful in practice.

A practical methodology for discarding information can then be organized as follows:

- Retain only a few simple features which seem relevant to the problem. (e.g. based on the different physical consequences that result from the different ways of complexity reduction).
- Give a probabilistic description. (This allows statistical predictions, despite the incomplete information).
- Compute observed quantities from within this model and compare these with experimental results. Here the "scope" and the "accuracy" are tested, thus allowing "feedback" or interaction in the modeling procedure.

A fundamental assumption is the MARKOV assumption, which is justifiable as follows:

The large set of variables, giving an exact complete microscopic description of the system can be divided in two classes, according to their relaxation times. If a first set {x} has relaxation times, much greater than all the other variables in the second class, then the timescale of the description (amounting to the scope of the model), is chosen intermediate to the long and short relaxation times. Hence, all memory effects are accounted for by the variables {x}, and it is adequate to assume that they form a Markov process.

Another frequently made assumption is that of STATIONARITY, implying that

i) all external influences on the system are time-independent on the chosen timescale.

ii) the classification of all variables in "fast" and "slow" is preserved during the evolution of the system.

2 OPEN LOOP BALANCING

2.1 Reachability and Observability

A state space model of a continuous time (the theory for discrete time systems is very similar and omitted) linear system with n inputs and p outputs is characterized by a triple of matrices (F,G,H)

$$F \in \mathbb{R}^{nxn}; \quad G \in \mathbb{R}^{nxm}; \quad H \in \mathbb{R}^{pxn}$$

where n is the order of the system. In general, the matrices are indexed by the reals $\mathbb{R}$.

For continuous time systems the relations are

$$\dot{x}\,(t) = F(t)\,x(t) + G(t)\,u(t) \tag{2.1}$$

$$y(t) = H(t)\,x(t) \tag{2.2}$$

If F, G and H are invariant with time, it is well known [4] that the reachability and observability of the system are determined by the fullrankness of the reachability and observability matrices, respectively $[G, FG,\ldots, F^{n-1}G]$ and $[H', F'H',\ldots F'^{n-1}H']$. However, the rankdefect of a matrix is very difficult to determine numerically because of the finite precision arithmetic of all computers. Moreover, these criteria do not provide any means to attach a measure of the degree of observability or reachability of the given system. A quantitative measure of the reachability ($\mathbb{R}$) or observability ($\mathbb{O}$) in some interval (t_o, t_1) is obtained via the (weighted) Gramian matrices, defined as: (Φ (.,.) is the transition matrix of F)

$$\mathbb{O}_W[t_o, t] = \int_{t_o}^{t} \Phi'(\tau, t_o)H'(\tau)W(\tau)H(\tau)\Phi(\tau, t_o)\, d\tau \tag{2.3}$$

$$\mathfrak{R}_M[t_o, t] = \int_{t_o}^{t} \Phi(t,\tau)G(\tau)M^{-1}(\tau)G'(\tau)\Phi'(t, \tau)\, d\tau \qquad (2.4)$$

Note that these matrices are well defined also in the timevarying case, as long as the integrals converge. The matrices W(t) and M(t) are assumed to be (positive or negative) definite, (usually identity). An interesting interpretation of these Gramians as weighting matrices for energies and uncertainties is given in the following subsection. In fact, this interpretation forms the basis for the model reduction algorithms to be introduced in the next subsection. The last subsection then describes the properties of the so-called balanced realizations, which were first introduced by Moore [5] for the time-invariant case.

2.2 Interpretation of the Gramians.

2.2.1 Deterministic

We start with simple thought experiments. Assume that the relevant input and output signals are in L_2. Let the system be in the state x_o initially. The output of the undriven systems is

$$y(t) = H(t)\Phi(t, t_o)\, x_o \qquad (2.5)$$

In general, the weighted L_2-norm is a particular measure of the "strength" in the signal, even though there may not be an underlying energy in a physical sense. The cross terms measure the degree of "interference" between the different components. Also, it is always possible to renormalize or take linear combinations of the existing output signals that have a more direct physical interpretation in terms of energy. Equivalently, one can define a weighting matrix W for the outputs, thus effectively measuring the "energy" as a weighted L_2-norm. With this generalization, the available W-measured output-energy U_w in the interval t_o to t_f for a system in state x_o at time t_o is given by,

$$U_W = \int_{t_o}^{t_f} y(t)' \, W(t) \, y(t) \, dt$$

$$= \int_{t_o}^{t_f} x_o' \, \Phi' \, (t, t)H'(t)W(t)H(t)\Phi(t, t)x_o \, dt \qquad (2.6)$$

$$= x_o' \, \mathfrak{O}_W \, [t_o, t_f]x_o$$

The generalized Observability Gramian $\mathfrak{O}_W \, [t_o, t_f]$ is a weighting matrix for the output L_2-measure given the initial state. If the system is observable (i.e. $\mathfrak{O}_W$ nonsingular), the state x_o can be recovered as (assuming that the system is undriven in $[t_o, t_f]$)

$$x_o = (\mathfrak{O}_W \, [t_o, t_f])^{-1} \int_{t_o}^{t_f} \Phi(t, t_o)H(t)'W(t)y(t) \, dt \qquad (2.7)$$

Consider now the dual problem of determining the inputs which drive the system from the zero state at t_o to any arbitrary state x_f at t_f. If the matrix $\mathfrak{R}_M[t_o, t_f]$ is nonsingular, then a particular input achieving this is

$$u(t) = M(t)^{-1} \, G(t) \, \Phi(t_f, t) \, (\mathfrak{R}_M \, [t_o, t_f])^{-1} \, x_f \qquad (2.8)$$

The optimality properties of this input are well-known [6]. It is the input with the least amount of "energy", as measured in a M-weighted L_2-norm.

$$||u||_M^2 = \int_{t_o}^{t_f} u(t)' \, M(t)u(t) \, dt \qquad (2.9)$$

The corresponding minimal energy is

$$U_R = x_f' (\mathfrak{R}_M [t_o, t_f])^{-1} x_f \qquad (2.10)$$

Again we see the role played by the M-weighted Gramian matrix. Its inverse appears as a weighting matrix for the minimal steering effort to the state x_f from the zero state.

2.2.2 Stochastic.

Here also we start with two thought experiments. One characterizing "uncertainties" relating to the inputs to the sytstem, the other one relating the state uncertainty to the outputs.

Let the system be driven by a white gaussian (vector) input signal, of zero mean, and covariance matrix Q(t). Assuming that this input is uncorrelated with the initial state of the system, the state covariance matrix P(t) at time t is given by

$$P(t) = \Phi(t, t_o)\, P(t_o)\, \Phi'(t, t_o) + \mathfrak{R}_{Q^{-1}}[t_o, t] \qquad (2.11)$$

The first term equals the covariance $\Pi(t, t_o)$ for the free-running undisturbed system. The second term is the generalized Q^{-1}-weighted Reachability Gramian (2.4) for $M = Q^{-1}$. It is a measure of the uncertainty induced in the state by a maximally random input. The disturbability of the state (as measured by the covariance) in the direction d by a white Gaussian input is given by

$$d'P(t)d = \text{Tr}\, DP(t) \qquad (2.12)$$

where $D = dd'$ and Tr is the trace function. The expected value of the A-weighted state "energy" in the realization is

$$\begin{aligned} E\, x(t)'\, A(t)\, x(t) &= \text{Tr}\, A(t)P(t) \qquad (2.13) \\ &= \text{Tr}\, \Pi(t, t_o) + \text{Tr}\, A(t)\, \mathfrak{R}_{Q^{-1}}[t_o, t] \end{aligned}$$

Here $\mathfrak{R}_{Q^{-1}}[t_o, t]$ appears as a weighting for D(t) and A(t) under the trace-norm. The second term in (2.13) is interpreted as the average energy increase in the states of the given realization due to the process noise with covariance Q(t).

Finally, consider the state estimation problem for a system with observation noise, but no driving terms. If the measurement noise is white with covariancematrix R(t), and, for simplicity, assumed to be uncorrelated with the initial state x_o, then the (Kalman filter) solution to the problem leads to the classical result $(S_o = P(t_o|t_o)^{-1})$

$$P(t_o|t)^{-1} = S_o + \mathfrak{O}_{R^{-1}}[t_o, t] \tag{2.14}$$

The matrix $\mathfrak{O}_{R^{-1}}[t_o, t]$ is a (matrix valued) measure for the information (or $\mathfrak{O}_{R^{-1}}^{-1}[t_o, t]$ for the uncertainty) conveyed by the observation process in (t_o, t) about the initial state x_o, and is usually referred to as the "Information matrix" in the estimation literature. In particular, if there is no prior information, $P(t_o|t)^{-1}$ is the zero matrix, and

$$P(t_o|t) = (\mathfrak{O}_{R^{-1}}[t_o, t])^{-1}$$

The above illustrates in a simple way how $\mathfrak{R}^{-1}$ and $\mathfrak{O}$ relate to (generalized) energies, while their inverses $\mathfrak{R}$ and $\mathfrak{O}^{-1}$ have to do with "uncertainties". The lower the required (minimal) energy to reach a certain state is, the more "reachable" that state is. Similarly, the higher the output energy available from the system, the more information we have about that system, and the smaller the errorcovariance of the filtered initial state. The Gramians provide, therefore, a suitable measure for the degrees of reachability and observability in a system.

With these remarks serving as a motivation, we proceed to the formal definitions.

2.3 Balanced Realizations

2.3.1 The Canonical Gramian

Under a similarity transformation of the state space form of a system, the quantitative reachability and observability properties of a realization are changed. Indeed if T(t) is the (nonsingular) transformation

$$T: (F, G, H) \to (TFT^{-1}, TG, HT^{-1})$$

then the Gramians for the new realization are

$$T: \quad \mathfrak{R}[t_o, t_f] \to T(t_f)\, \mathfrak{R}[t_o, t_f]\, T'(t_f)$$

$$T: \quad \mathfrak{O}[t_o, t_f] \to T(t_o)^{-T}\, \mathfrak{O}[t_o, t_f]\, T^{-1}(t_o)$$

It has been shown [7] that if the matrices F, G, H and the weights W and M are real analytic functions of time, and if the system is completely reachable and observable, then a similarity transformation exists such that both $\mathfrak{R}$ and $\mathfrak{O}$ are diagonal and equal. If the diagonal elements are separated, then one can define a (unique) canonical form by inducing some ordering in these elements. In the time invariant case, it is customary to order them according to decreasing magnitude. In the sequel we shall refer to these as the CANONICAL ELEMENTS, and to the Gramians in balanced form as the CANONICAL GRAMIAN. The open loop canonical Gramian will be denoted as Λ. (In the signal processing and digital filtering context, the canonical elements are also known as the second order modes [8]). The resulting realization (TFT^{-1}, TG, HT^{-1}) is then called "balanced with respect to the weights W and M". (Usually, only balancedness is considered with respect to the weights W=M=I). Algorithms for obtaining balanced realizations are based on the singular value decomposition of the Gramians in an arbitrary realization ([5], [7]). Recently more direct methods have been obtained for computing the balanced realizations for time invariant systems ([9], [10], [11], [12], [13]). The timevarying balanced realizations are an extention of the original balanced realizations for time invariant systems introduced by Moore, and were introduced in [7]. In view of the interpretations of the Gramians developed in the previous section, it is clear that in each coordinate direction of the balanced state space the degree of reachability and observability is the same.

2.3.2 Model Reduction via Balancing

In order to fix the ideas on how this might be used as a criterion for model reduction, consider a nonminimal time-invariant state space realization of a system. It is well known that such a realization is nonreachable and/or unobservable [4]. A minimal realization, having identical input-output properties if the system is initially at rest, can be obtained by removing these unobservable or nonreachable modes from the original description, e.g. via a truncation (projection of dynamics) of the standard decomposition of the nonminimal system, thus effectively deleting the unreachable and noncontrollable parts of the state space. The Gramians give now a quantitative measure for observability and reachability, rather than the binary value assigned by the (Kalman) criterium. If one were now to "delete" a component which has a high cost associated with its reachability, then this component may have very good observability properties, and therefore be very significant in the input-output description of the system. Since in fact the state intervenes only as an interface between input and output, a transformation (for instance just a scaling) can be used to yield a new representation in which this difficult-to-reach state component has become very easy to reach. The opposite would then also be true for its observability properties. Clearly, component reachability and component observability is not an absolute criterion for the importance towards the input-output or external description. What one needs to look for is invariants with respect to arbitrary state space transformations. The product of the reachability and observability Gramians transforms under T as a similarity. Hence the eigenvalues of this product are invariants.

But these eigenvalues are exactly the squares of the canonical elements. Hence it turns out that the relative importance of a (balanced) state space component with respect to the external system behavior is quantitatively determined by the "joint degree of reachability and observability" associated with this system dimension. By virtue of the interpretations, we developed in the previous section, this is exactly described by the elements of the Gramians of the balanced realization. Based on this description, the canonical elements can be used by the system analyst or control designer to decide which components to use in a reduced order model for the original system. Such a reduced order model is then obtained by "projection of dynamics". One partitions the original system (in balanced form) as

$$\begin{bmatrix} F_{11} & F_{12} \\ F_{21} & F_{22} \end{bmatrix} \qquad \begin{bmatrix} G_1 \\ G_2 \end{bmatrix} \qquad [H_1 \quad H_2]$$

where it is assumed that the canonical elements are ordered with respect to their magnitude. The reduced order model is then obtained as $(F_{11},\ G_1,\ H_1)$. It simply means that the components which were difficult to control and observe are considered as completely uncontrollable and unobservable, and subsequently, the minimal realization is obtained. In reference to section 1, our topology is derived from the trace of the canonical Gramian, under the restriction of "Projection of Dynamics". The above "projection-of-dynamics" method is also applied in the discrete case. However, it leads to some self-inconsistency, since the reduced models of the discrete balanced system are themselves not balanced. Similar interference-effects are also known in realization theory. For instancc, the reduction of a Hankel matrix via a singular value decomposition does not yield a matrix with Hankel structure in general. This has been a steady source of critique to the method.

2.3.3 Properties of Balanced Realizations

A basic property of the Gramians introduced in section 2.1 is that they satisfy the following Lyapunov equations (without loss of generality, we take M and W as the identity matrix). (t_o and t_f fixed):

$$\dot{\mathfrak{R}}\,[t_o, t] = F(t)\mathfrak{R}[t_o, t] + \mathfrak{R}[t_o, t]F'(t) + G(t)G(t)' \tag{2.15a}$$

$$\dot{\mathfrak{R}}\,[t_o, t_o] = 0 \tag{2.15b}$$

$$\dot{\mathfrak{O}}\,[t, t_f] = -F(t)'\mathfrak{O}[t, t_f] - \mathfrak{O}[t, t_f]F(t) - H'(t)H(t) \tag{2.16a}$$

$$\dot{\mathfrak{O}}\,[t_f, t_f] = 0 \tag{2.16b}$$

More general formulas for t_o and t_f depending on t have been obtained in [7]. They are of interest in "Sliding Interval" Balancing and Model Reduction. It follows that for the balanced realization of a time invariant realization with $t_o = -\infty$ and $t_f = \infty$, the canonical Gramian satisfies the symmetrical equations.

$$F\Lambda + \Lambda F' + GG' = 0 \tag{2.17}$$

$$F'\Lambda + \Lambda F + H'H = 0 \tag{2.18}$$

These equations form the basis for the derivations of a whole set of nice properties for balanced realizations (see [9], [11], and [14]). e.g., for some signature matrix E (a diagonal matrix having either +1 or -1 as diagonal elements) one can show that for SISO systems

$$EFE = F' \ ; \ EG = H' \tag{2.19}$$

3 Balancing in the LQG-sense

In the open loop case, the balanced realization led to a natural selection of reduced order models through the "projection of dynamics". Adopting this procedure for the design of a reduced order controller is very dangerous however, due to the feedback around the system. A more direct approach is needed, treating the closed loop as a whole. In fact, the degree of uncertainty (i.e., the noise covariances) and the performace index or cost-functional of the system should be taken into account for the selection of a reduced model. Skelton et al. [15-17] suggested a weighting with respect to the "component-costs". In a stochastic context, this may be undesirable, since it may lead to "bad surprises". Indeed, if the uncertainty associated with a dynamical element, with small expected cost contribution, is high, then the actual cost contribution for a sample trajectory of the stochastic system may be quite different from its (lower) expectation. This motivates the balancing with respect to the optimal deterministic controller, and the stochastic observer via the separation principle. ([18-19], [20])

The basics of the LQG-theory are well established, and can be found in many textbooks. The solution to the optimal control problem for a linear system with a quadratic performance index in the presence of white gaussian noise falls apart into the design of the deterministic controller (i.e. assuming perfect knowledge of the state of a system), and a stochastic observer for the noisy system driven by an external (but assumed known) input. This constitutes the celebrated Certainty-Equivalence Principle [6]. We shall briefly summarize the solution for this stochastic control problem. As was done in the open loop case, here also we shall try to give an interpretation to the solution, and clarify the different components in it. Several different problems are now of interest. In digital control (using fixed point arithmetic) the interest is in minimal sensitivity (with respect to the finite wordlength effects) design of the digital controller. In general control, one might be interested in a suboptimal but reduced order controller (in order to reduce the computational burden). Finally one might just be concerned with the modeling and analysis of the overall feedback system (thus including the plant) for the purpose of assessing the dominant contributions to the performance index, or uncertainty. In this case reduced order models for the combined plant and regulator are of interest.

It will be shown that the ideas of balanced realizations, when properly (re)defined are again very usefull. The LQG-balancing for continuous time systems will be motivated from the following analysis.

3.1 The LQG-terminal controller

In order to fix the ideas, consider the stochastic sytem

$$\dot{x} = F x + G u + w \quad ; \dim x = n, \dim u = m \tag{3.1}$$

$$y = H x + v \quad ; \dim y = p \tag{3.2}$$

with the initial state normally distributed:

$$x(t_o) \sim N(x_o, P_o) \tag{3.3}$$

For simplicity (but without loss of generality) we shall assume that w and v are uncorrelated zero mean white gaussian noise processes, with covariances Q and R respectively. They are further assumed to be independent from the initial conditions. Let the design objective be the minimization of a positive semi-definite quadratic performance index

$$J = E\left(x'(t_f)\, S_f\, x(t_f) + \int_{t_o}^{t_f} (x'Ax + u'Bu)dt \right) \tag{3.4}$$

It is assumed that the matrices R and B are positive definite, but otherwise arbitrary. In fact, this amounts to a slight overparametrization, but avoids some preliminary transformation. First, assuming that the states can be perfectly measured, the (deterministic) optimal closed loop system will have dynamics:

$$\dot{x} = (F - G C) x \qquad x(t_o) = x \tag{3.5}$$

$$C = B^{-1} G' S \tag{3.6}$$

$$\dot{S} = -S(F-GC) - (F-GC)' S - A - C'BC \qquad S(t_f) = S_f \tag{3.7}$$

The solution $S(t; t_f, S_f)$ of (3.7) has an interpretation as a weighting matrix for the minimum "cost-to-go" from the state $x(t)$ at time t. For $S_f = 0$, $S(t)$ is exactly the observability Gramian of the closed loop system sporting the fictious $(n+m)$-dimensional output

$$z = L x \tag{3.8}$$

$$L' = [\, -C'B^{1/2},\ A^{1/2}] \tag{3.9}$$

The performance index then is the output energy (as discussed in section 2.2) of this system. The presence of the nonzero S_f can be interpreted as the instantaneous release ("flushing") of the remaining "energy" in the system (due to a nonzero state) at time t_f over a weight matrix S_f. Equivalently, it is also a measure for the amount of information, about an a priori unknown initial condition, that this fictious output would carry, if corrupted by unit variance white gaussian noise.

Similarly, the filter error dynamics are given by

$$\dot{\tilde{x}} = (F - KH)\,\tilde{x} + M\omega \quad ; \dim \omega = n+p \tag{3.10}$$

$$M = [\, Q^{1/2}, KR^{1/2}\,] \tag{3.11}$$

where

$$\tilde{x} = x - \hat{x} \tag{3.12}$$

$$K = P H' R^{-1} \tag{3.13}$$

$$\dot{P} = (F-KH)P + P(F-KH)' + Q + KRK' \quad P(t_o) = P_o \tag{3.14}$$

and $\omega(t)$ is a white noise of unit variance. Again, $P(t; t_o, P_o)$ is a measure for the uncertainty in the closed loop system (3.10), and characterizes the "disturbability" by the noise. It is in fact the covariance $E(\widetilde{xx}')$ of the estimation error at time t, if at time t_o the error was P_o.

The Certainty Equivalence principle states that the optimal control for the system (3.1 - 3.2) where the states are not perfectly known, is given by feedback of the estimates of the states over the optimal control-gains. The overall equations are thus

$$u = - C x \tag{3.15}$$

$$\dot{\hat{x}} = F \hat{x} + G u + K (y - H \hat{x}) \quad ; \quad \hat{x}_0 = 0 \tag{3.16}$$

(The variance of the estimate $E(\hat{x}\hat{x}')$ will be denoted by Σ, while Π is used for the state variance $E\,(xx')$. By the optimality of the estimates, we have then $\Pi = \Sigma + P$. Note that the innovation $\epsilon = (y - H \hat{x})$ acts as a white Gaussian noise with covariance R. The following equivalent equations are easily derived.

$$\dot{x} = (F - G C) \tilde{x} + G C \tilde{x} + w \tag{3.17}$$

$$\dot{\hat{x}} = (F - G C) \hat{x} + K H \tilde{x} + K v \tag{3.18}$$

$$\dot{\hat{x}} = (F - G C - K H) \hat{x} + K y \tag{3.19}$$

The optional performance index (3.4) can now be evaluated in several different forms (using partial integration and the Riccati equations for S and P combined with the Lyapunov equations for Π and Σ in the closed loop)

$$J = \mathrm{Tr}\,\{\Pi_f S_f + \int_{t_o}^{t_f} (A\Pi + C'BC\Sigma)\, dt\} \tag{3.20}$$

$$= \mathrm{Tr}\,\{\Pi_o S_o + \int_{t_o}^{t_f} (SQ + C'BCP)\, dt\} \tag{3.21}$$

$$= \mathrm{Tr}\,\{P_f S_f + \int_{t_o}^{t_f} (AP + KRK'S)\, dt\} \tag{3.22}$$

Several (equivalent) interpretations follow from these equations. (3.19) and (3.15) give an open loop representation for the optimal stochastic controller, with input y and output the control u (figure 1). The equations (3.10) and (3.17) lead to a decomposition as a cascade of a system (F-KH, M, I), driven by standard white gaussian noise, connected via a "transmission" matrix GC to the system (F-GC, I, L), (figure 2). Whereas the former subsystem represents the dynamics of the estimation error $\tilde{x}$ (driven by the fictitious noise ω for which $M\omega = w-Kv$), the latter will represent the plant states x, if and only if an additional gaussian input (w) is summed at its input. This additional noise has covariance Q, and is correlated with the noise ω according to $M\,E(\omega w') = Q$.

In terms of this decomposition we define a fictitious output $z = z_1 + z_2$ where z_1 is the (n+m)-dimensional output from the second subsystem and $z_2 = \Gamma x$ where Γ is the (n+m) by n matrix $\Gamma' = [\ C'B^{1/2},\ 0\ ]$ and L is as defined in (3.9). Because the outputs z_1 and z_2 are "maximally interfering" (due to their correlation), the variance of their sum z, is actually the difference of their individual variances, which is the integrand in (3.20).

Another interesting representation (figure 3) can be derived starting from the equations (3.10) and (3.18). Again a cascade is formed, beginning with the system (F-KH, M, I) driven by standard white gaussian noise. This time the output (which is the representation of the state estimation error) drives, via the "transmission-matrix" KH, the system (F-GC, I, L). This system will have the variable $\hat{x}$ as state, if again an additional "correction" input Kv (with variance KRK') is added, having a correlation T with ω satisfying:

$$MTK' = E(M\omega)\,(Kv)' = -KRK.$$

A fictitious output $z=z_3 + z_4$ is defined which generates the integrand in the performance index. In this case, it is readily verified that this is accomplished by z_3, the output of the cascade and $z_4 = N\,\tilde{x}$, where $N' = [0,\ A^{1/2}]$. Note that in this decomposition the input to the x-subsystem is actually K times the innovations process. This is known to act as a white noise. Here z_3 and z_4 are noninterfering (uncorrelated).

3.2 Interpretation of the Cost Functional.

The two decompositions described in the last paragraph, lead to a "cost-decoupled" interpretation of the various terms. For simplicity, we shall fix the ideas on the LQG-regulator problem. The matrices F, G, H, Q, and R are all supposed to be time-

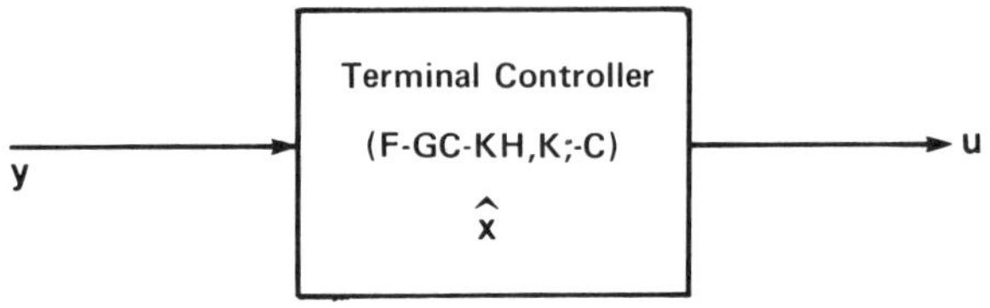

Figure 1: Open Loop representation of the optimal LQG system.

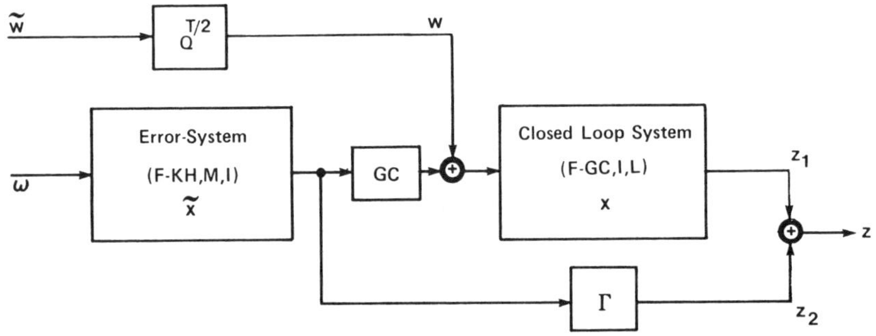

Figure 2: The $(\tilde{x}, x)$ representation of the optimal LQG system.

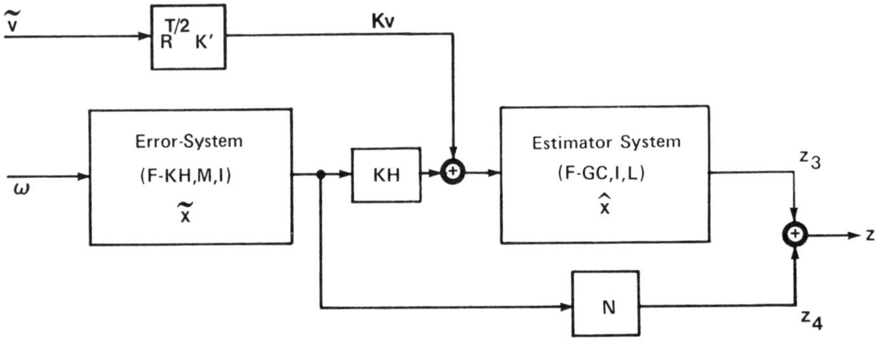

Figure 3: The $(\tilde{x}, \hat{x})$ representation of the optimal LQG system.

invariant, and it will be assumed that the statistical stationary state exists. For the steady-state regulator problem, the cost-rate, rather than the cost (which is infinite) is computed. Let in the LQG terminal controller, t_0 and t_f approach $-\infty$ and $+\infty$ respectively. Then the cost rate is the limit of the expected cost per time unit. It follows then from the equations (3.20 - 3.22) that the cost rate is

$$j = \mathrm{Tr}\{A\Pi + C'BC\Sigma\} = \mathrm{Tr}\{SQ + C'BCP\} = \mathrm{Tr}\{AP + KRK'S\} \tag{3.23}$$

where now P and S satisfy the algebraic Riccati equations, respectively

$$(F-KH)\,P + P\,(F-KH)' + MM' = 0 \tag{3.24}$$

$$(F-GC)'S + S\,(F-GC) + L'L = 0 \tag{3.25}$$

As already discussed, the following identifications follow, where the "*" entry is irrelevant:

P is the REACHABILITY gramian for the system (M, F-KH, *)

S is the OBSERVABILITY gramian of the system (*, F-GC, L)

Consider now a fictitious system, consisting of the two decoupled subsystems (F-GC, $Q^{1/2}$, L) and (F-KH, M, Γ). If both are driven by independent standard white gaussian noise, then their outputs (respectively z_1 and z_2) will be uncorrelated as well, and their expected "powers" additive. The contribution of the first is exactly Tr $\{Q\,\mathfrak{O}(F-GC, L)\}$ = Tr{QS}, while the contribution of the second is $\mathrm{Tr}\{MM'\mathfrak{O}(F-KH, \Gamma)\} = \mathrm{Tr}\{\Gamma\Gamma'\mathfrak{R}(F-KH, M)\}$ = Tr{C'BCP}. We used the fact that in the steady state, the A-weighted output power of a system (F, G, H) driven by a zero mean white gaussian input of variance Q, can be expressed as

$$\mathrm{Tr}\; H'AH\mathfrak{R}_{Q\text{-}1} = \mathrm{Tr}\; GQG'\mathfrak{O}_A \tag{3.26}$$

But note that this is exactly the system of figure 2 if one replaced the input noises by uncorrelated ones, and set the transmission matrix GC to zero. The two partial outputs are then indeed noninterfering. This leads to the INTERPRETATION that Tr {QS} is the partial cost rate in the deterministic closed loop system due to the process noise ω only, i.e., assuming full knowledge of the state x. The part Tr{C'BCP} is the cost rate due to the state estimation error. It is as if the role of the transmission matrix GC and the input correlation is to guarantee maximal destructive interference between the outputs, (so that the variance of the output is the difference of the individual variances).

Similarly, with regards to the figure 3, we have the estimator subsystem (F-GC, K, L) driven by the innovations ϵ, and with output z_3, and the estimation-error system (F-KH, M, N) driven by ω and with output z_4. The two outputs are uncorrelated, and their sum $z = z_3 + z_4$ has covariance equal to the cost rate of the optimally controlled system. A cost-equivalent decoupled form can be constructed, consisting of the system (F-GC, K, L) (the estimator system) driven by v, considered independent of the noise ω, which drives the error system (M, F-KH, N). Their output "power" contributions are respectively Tr{KRK'S} and Tr {AP}. Thus the role of the transmission matrix KH in figure 3 seems to be to effectively uncorrelate the two input noises.

The contribution TR{KRK'S} can thus be identified as the cost rate for the closed loop system under the assumption that the estimated state is the correct state. However since not $\hat{x}$, but x is the state of the closed loop system, a correction occurs due to the imperfect knowledge of the state (i.e. $\tilde{x}$). This is represented by the (independent) contribution of the error subsystem (M, F-KH, N), driven by the equivalent noise ω, with cost-rate contribution Tr{AP}.

3.3 LQG-Balanced Realizations

Since S and P transform under a similarity transformation T as T^{-T} and ST^{-1} and TPT' respectively, it is possible to transform any given realization such that in the new coordinate system,

$$P = S = \Omega$$

where Ω is diagonal, with its elements ordered in magnitude. Ω is called the CANONICAL RICCATIAN. The new realization will then be referred to as the LQG-BALANCED realization. Note that Q as well as A also transform under the similarity. (B and R are of course invariant as they relate to "external" variables).

The cost rate for the optimally regulated system is then in the balanced coordinates,

$$j = \text{Tr}\ \Omega(Q+C'BC) = \text{Tr}\ \Omega(A+KRK') \tag{3.27}$$

or, using the fact that Ω is diagonal:

$$j = \sum_{i=1}^{n} \Omega_i(Q_{ii}+\Omega_{ii}^2\,(GB^{-1}G')_{ii}) \tag{3.28}$$

$$j = \sum_{i=1}^{n} \Omega_i(A_{ii}+\Omega_{ii}^2(HR^{-1}H')_{ii}) \tag{3.29}$$

It is clear that the cost rates corresponding to the individual state components are not simply determined by the magnitudes of the elements of the canonical Riccatian, but also depend on the relative magnitudes of the diagonal elements of Q, $GB^{-1}G'$, or A and $H'R^{-1}H$.

If the system (in balanced coordinates) is partitioned into two coupled subsystems with $\Omega_1 \geq \Omega_2$, then a sufficient condition for the part corresponding with Ω_1 to have the dominant cost rate contribution, is that either of the followyng sets of inequalities are satisfied: (the indices refer to the block-entries)

$$\begin{cases} \Omega_1 \geq \Omega_2 \\ Q_{11} \geq Q_{22} \\ (GB^{-1}G')_{11} \geq (GB^{-1}G')_{22} \end{cases} \tag{3.30}$$

$$\begin{cases} \Omega_1 \geq \Omega_2 \\ A_{11} \geq A_{22} \\ (H'R^{-1}H)_{11} \geq (H'R^{-1}H)_{22} \end{cases} \tag{3.31}$$

The various terms can be interpreted as quantifying the following:

Q	:	Disturbance (noise) in the plant.
$GB^{-1}G'$	:	"Potential" of the system input to decrease the regulation cost.
A	:	Cost on the state deviations.
$H'R^{-1}H$	:	Information (about the state) gained from the measurements (observations).

The first set of inequalities expresses that the set of variates that are most disturbed by noise and for which at the same time the input-potential is high, are dominant. (The larger the input-potential, the less the cost of control). Alternatively, state variables for which the information contained in the measurements and the state-cost is highest, also contribute to the major parts in the regulation cost-rate.

If one were only interested in obtaining a simple model for the optimally regulated system, for instance with the goal of identifying the dominant contributions to the uncertainties and the costs, then the combined plant and regulator may be reduced by "projection of dynamics". The decision on the order of the reduced model can for instance be based on tresholding the ratios

$$\alpha(r) = \frac{\mathrm{Tr}\,\Omega_1}{\mathrm{Tr}\,\Omega} \quad \text{and} \quad \beta(r) = \frac{j(r)}{j(n)}$$

where

$$j(r) = \sum_{i=1}^{r} \Omega_i(Q_{ii} + \Omega_{ii}^2(GB^{-1}G')_{ii})$$

If on the other hand one wants to design a reduced order regulator for a fixed plant, the above cannot be taken over directly. It has been shown that the design of a reduced order regulator based on a reduced order model of the plant may be unsatisfactory. Also, a "projection of dynamics" approach on the full order combined plant and regulator, based on the magnitude of the elements of the canonical Riccatian alone may not guarantee the stability of the regulation of the (full order) plant with the obtained reduced regulator ([16-17], [20]). Also in the open loop case, the Λ provides insufficient information ([14], [21]).

In reference to the decompositions in figures 2 and 3, the following property of the transmission matrix is derived.

Theorem The lower (upper) triangular part of the transmission matrix GC (KH) is dominant in the balanced coordinates.

proof: Since $T = GC = GB^{-1}G'\Omega$, it follows that ΩT is symmetric. But then $\Omega_i T_{ij} = T_{ji}\Omega_j$ for all i and j. Hence, $T_{ji} = T_{ij}\ \Omega_i/\Omega_j$. Since by assumption the elements Ω_i are ordered, we get $T_{ji} \geq T_{ij}$ whenever $j \geq i$. Similarly, $X = KH = \Omega H'R^{-1}H$ and thus $X\Omega$ is symmetric, from which $X_{ji} \leq X_{ij}$ whenever $j \geq i$.

Consider figure 2. Keeping the ordering in mind for the balanced case, it follows that low uncertainty states are more perturbed by the high uncertainty states than vice versa. It further follows from the positive semi-definiteness of the "input-potential" that for $j \geq i$

$$(GB^{-1}G')_{ij}^2 \leq (GB^{-1}G')_{ii}\ (GB^{-1}G')_{jj}$$

and thus that the elements in the upper left block of $(GB^{-1}G')\Omega$ are larger in magnitude than the elements in the upper right block of T. Hence, x_2 is almost decoupled from the closed loop system (I, F-GC, L). In fact, for the same reason the upper right block of the closed loop system matrix F-GC will be close to that of F itself, so that there is almost no feedback from the x_2 subsystem. One expects, therefore, that the closed loop dynamics of the plant controlled with the reduced regulator, would have a near optimal behavior. Similar arguments work with figure 3.

4 STOCHASTIC MODELING

4.1 The Stochastic Realization Problem

Desai and Pal [22-23], extended the ideas of balancing in the LQG-sense to the stochastic realization problem. Balancing is here with respect to the state covariance matrices in the forward and the backward innovation representations. These matrices solve dual Riccati equations. The elements of the "canonical Riccatian" are connected to the canonical correlations between the past and the future observations. Arun and Kung [24] contrasted the method based on canonical correlations with a method based on Principal Components. Vaccaro showed its connection with deterministic open loop balancing [25]. Ramos and Verriest [26-27] unified the theory be showing that both the canonical correlation analysis (CCA) and the principal component analysis (PCA) are special cases of a more general optimization problem, using a new tool from multivariate statistics: the RV-coefficient introduced by Escouffier [28]. If two zero mean random vectors X_1 and X_2 (not necessarily of the same dimension) have covariance matrix

$$\mathrm{cov}(X_1,X_2) = \begin{bmatrix} \Sigma_{11} & \Sigma_{12} \\ \Sigma_{21} & \Sigma_{22} \end{bmatrix} \tag{4.1}$$

then the RV-coefficient is defined as

$$RV(X,Y) = \frac{T_r(\Sigma_{12}\Sigma_{21})}{\sqrt{T_r(\Sigma_{11}{}^2)\ \ T_r(\Sigma_{22}{}^2)}} \geq 0 \tag{4.2}$$

This measure shares many of the properties of a correlation coefficient, but is not one itself. (It is the square of the correlation if X and Y are scalar). It also allows the computation of a "figure of merit" for each algorithm in a consistent way.

This formalism is applied to stochastic realization theory as follows. Given the correlation sequence $\{\Lambda_k\}$ of a discrete time stationary stochastic sequence $\{y_k\}$, the forward and backward predictor subspaces are

$$X_k = \text{Span}\,(Y_k^+ \mid Y_k^-) \qquad (4.3)$$
$$Z_{k-1} = \text{Span}(Y_k^- \mid Y_k^+)$$

where $\underline{Y}^+$ and $\underline{Y}^-$ respectively correspond to the "future" and the "past" of the process. Here (A|B) denotes the projection of span(A) onto span(B). These two spaces form the information interface between the past and the future. Defining as usually

$$\hat{H}_k = E\{Y_k^+(Y_k^-)'\}\ ,\ R_k^+ = E\{Y_k^+(Y_k^+)'\}\ ,\ R_k^- = E\{Y_k^-(Y_k^-)'\} \qquad (4.4)$$

then CCA is equivalent to the problem of finding transformations L and M such that RV $(L'\underline{Y}^+,\ M'\underline{Y}^+)$ is maximized, subject to the constraints that $L'(R^+)L$ and $M'(R^-)M$ are diagonal. The PCA is equivalent to the problem of maximizing RV $(\underline{Y}^+,\ M'\underline{Y}^-)$ over M under the constraint $M'(R^-)M$ diagonal. The two methods are also referred to as the one-sided and the two-sided stochastic realization problem.

4.2 Geometrical Interpretation: Correlation between Subspaces

Let H be a Hilbert space. The set of all closed subspaces of H has the structure of an orthocomplemented complete lattice, also called a logic. The lattice of all closed subspaces of H and the lattice Proj H of all orthoprojectors on H are isomorphic.

In his study of the mathematical foundations of quantum mechanics, Mackey posed the problem of finding all positive measures on the closed subspaces of a Hilbert space. Such a measure must have the property that for any countable collection $\{S_i\}$ of mutually orthogonal closed subspaces the mapping is σ-additive, i.e.,

$$\sum_i \mu(S_i) = \mu(\sum_i S_i) \tag{4.5}$$

A measure satisfying the above property is for instance obtained by selecting a vector v in the Hilbert space H, and letting for each subspace A of H

$$\mu_v(A) = \|P^A(v)\|^2 \tag{4.6}$$

where P^A is the projection operation on A. Clearly, finite convex combinations of such measures also satisfy the conditions for such measures, and passing to the limit, any positive semidefinite trace class operator T also defines such a measure via

$$\mu(A) = Tr(TP^A) \tag{4.7}$$

Gleason [29] has shown that is a separable Hilbert space of dimension at least three, every measure on the closed subspaces can be represented as above, with T a positive definite operator of trace class.

Consider now a tensor product Hilber space $R^P \otimes H$, and let $\{\Psi_i\}$ be a complete Orthonormal Set (CONS) in H. Any vector x in this tensor product space has then a decomposition

$$x = \sum_i |x_i> <\Psi_i| \quad ; \quad x_i \in R^P \text{ for all i} \tag{4.8}$$

The vector x will be referred to as a "prior". Let for all A in Proj R^P : $\mu_i(A) = \|P^A(x_i)\|^2$ and define a "superposition of measures" on Proj R^P as $\mu_x = \sum_i \mu_i \alpha_i$ for some square summable positive weights $\{\alpha_i\}$. By Gleason's theorem [29] it follows that there exists an operator $T_x : R^P \to R^P$ such that

$$\mu_x(A) = \mathrm{Tr}\ T_x\ P^A \tag{4.9}$$

This operator is characteristic for the given vector x in $R^P \otimes H$ (in fact, a "sufficient statistic"), and one can think of T (or μ) as "conditioned" by the vector x. Since

$$T_x = \sum_i \alpha_i^2 |x_i\rangle \langle x_i| = xx' \tag{4.10}$$

it can be interpreted as a Gramian or covariance operator.

The measure $\mu_x(A)$ gives a numeric value to the closeness of A to R^P, given the prior x [30]. Define also the extended projectors $\tilde{P}^B \in \mathrm{Proj}\ (R^P \otimes H)$ by

$$\tilde{P}^B(x) = \sum_i \alpha_i P^B |x_i\rangle \langle \Psi_i| \tag{4.11}$$

They allow now the definition of the "variance" and "covariance" of subspaces in Proj R^P, [31]. The "posterior" variance of $A \in \mathrm{Proj}\ R^P$ given x is then the operator from R^P to R^P

$$(\tilde{P}^A x)\ (\tilde{P}^A x)' = \sum_i P^A |x_i\rangle \langle x_i| P^A = P^A T_x P^A \tag{4.12}$$

and the covariance

$$(\tilde{P}^B x)\ (\tilde{P}^A x)' = \sum_i P^B |x_i\rangle \langle x_i| P^A = P^B T_x P^A \tag{4.13}$$

This is simply interpreted as the restriction to B of the mapping T_x restricted to the subspace A, and displays the coupling or interface between A and B given x. In order to attach a numerical value to this interface, any norm on the various restrictions $P^B T_x P^A$ can be chosen. With "correlation" defined in the usual way, the choice of the Frobenius norm will lead to the RV-coefficient for the exact realization problem if $H = L_2(\Omega, B, P)$ and for the "real data" problem if $H = R^N$ for some integer N.

5 CONCLUDING REMARKS

Our discussion on the open loop balanced realizations has focussed on the continuous time case. The ideas and properties of the discrete balanced realizations are very similar. The ideas of balancing have also been used by this author to obtain improved discretization methods [32], and a new digital filter design methods [33]. Applications to bilinear systems have been developed in [34] and [35]. The treatment of the stochastic realization related problems was restricted to the discrete time case. This avoided many measure theoretic problems, that are mainly burdensome details rather than clarifying principles. Extensions to nonstationary problems [36] are possible. For some other interesting results, we refer to [37-38].

By determining well motivated measures for the correlation between subspaces of a Hilbert space, based on available priors, it was possible to unify several tools from multivariate analysis and their application in realization theory.

It is hoped that the work initiated here may be an inspiration to study further the intricate connections between modeling, approximation theory and realization theory, for many problems remain.

REFERENCES

[1] Caines, P. E., "On the Scientific Method and the Foundation of System Identification," *Int'l Symp. Math. Thy. of Networks and Systems*, Stockholm, Sweden, June 1985, (Proceedings in print).

[2] Rissanen, J., "A Universal Prior for Integers and Estimation by Minimum Description Length," *Ann. of Stat. (1983)*, Vol. 11, 2, 416-431.

[3] Larimore, W. E., "Predictive Inference, Sufficiency, Entropy and an Asymptotic Likelihood Principle," *Biometrika (1983)*, Vol. 70, 1, 175-181.

[4] Kailath, T., Linear Systems, *Prentice Hall*, 1980.

[5] Moore, B. C., "Principal Component Analysis in Linear Systems: Controllability, Observability, and Model Reduction," *IEEE Trans. Automatic Control*, AC-26, No. 5 (1) 17-32, January 1981.

[6] Bryson, A. E., and Ho, Y. C., Applied Optimal Control, *Ginn and Comp.*, 1969.

[7] Verriest, E. I. and Kailath, T., "On Generalized Balanced Realizations," *IEEE Trans. Automatic Control*, AC-28, no. 8, pp. 833-844, 1983.

[8] Mullis, C. T., and Roberts, R. A., "Synthesis of Minimum Roundoff Noise Fixed Point Digital Filters," *IEEE Trans. Circuits Syst.*, Vol. CAS-23, No. 9, pp. 551-562, 1976.

[9] Verriest, E. I., "The Structure of Multivariable Balanced Realizations," *Proc. IEEE International Symposium Circuits Syst. 1983*, Newport Beach, CA, pp. 110-113.

[10] Sveinsson, J. R., and Fairman, F. W., "Minimal Balanced Realizations of Transfer Function Matrices using Markov Parameters," *IEEE Trans. A.C.*, Vol. AC-30, No. 10, pp. 1014-1016, October 1985.

[11] Fairman, F.W., Mahil, S. S., and De Abreu, J. A., "Balanced Realization Algorithm for Scalar Continuous-Time Systems having Simple Poles," *Int. J. Systems Sci.*, Vol. 15, No. 1, pp. 685-694, 1984.

[12] Kabamba, P. T., "Balanced Forms: Canonicity and Parametrization", *IEEE Trans. A.C.*, Vol. AC-30, No. 11, pp. 1106-1109, November 1985.

[13] Young, N. J., "Balanced Realizations via Model Operators," *Int. J. Control*, Vol. 42, No. 2, pp. 369-389, 1985.

[14] Kabamba, P. T., "Balanced Gains and Their Significance for L^2 Model Reduction," *IEEE Trans. A.C.*, Vol. AC-30, No. 7, pp. 690-693, July 1985.

[15] Skelton, R. E. and Yousuff, A., "Component Cost Analysis of Large Scale Systems," *Int. J. Control,* Vol. 37, No. 2, pp. 285-304, 1983.

[16] Yousuff, A., and Skelton, R. E., "A Note on Balanced Controller Reduction," *IEEE Trans. A.C.*, VOl. AC-29, No. 3, pp. 254-257, March 1984.

[17] Yousuff, A., and Skelton, R.E., "Controller Reduction by Component Cost Analysis," *IEEE Trans. A.C.*, Vol. AC-29, No. 6, pp. 520-530, June 1984.

[18] Verriest, E. I., "Low Sensitivity Design and Optimal Order Reduction for LQG - Problem," *Proc. 24th Midwest Symp. Circ. Syst.*, Albuquerque, New Mexico, pp. 365-369, (June 1981).

[19] Verriest, E. I., "Suboptimal LQG-Design via Balanced Realizations," *Proc. 20th IEEE Conf. Dec. Control*, San Diego, CA, pp. 686-687, December 1981.

[20] Jonckheere, E. A., and Silverman, L. M., "A New Set of Invariants for Linear Systems Applications to Reduced Order Compensation Design," *IEEE Trans. A.C.*, pp. 953-964, October 1983.

[21] Therapos, C. P., "On the Selection of the Reduced Order via Balanced State Representations," *IEEE Trans. A.C.*, Vol. AC-29, No. 11, pp. 1019-1021, November 1984.

[22] Desai, U. B. and Pal, D., "A Realization Approach to Stochastic Model Reduction and Balanced Stochastic Realizations," *Proc. 21st IEEE Conf. on Decision and Control*, pp. 1105-1111, 1982.

[23] Desai, U. D., and Pal, D. A., "A Transformation Approach to Stochastic MOdel Reduction," *IEEE Trans. A.C.*, Vol. AC-29, No. 12, pp. 1097-1100, December 1984.

[24] Vaccaro, R. J., "Deterministic Balancing and Stochastic Model Reduction," *IEEE Trans. A.C.*, Vol. AC-30, No. 9, pp. 921-923, September 1985.

[25] Arun, K. S. and Kung, S. Y., "A New SVD Based Algorithm for ARMA Spectral Estimation," *Proc. 1983 IEEE-ASSP Spectral Estimation Workshop*, Tampa, Florida, November 1983.
[26] Ramos, J. A., and Verriest, E. I., "A Unifying Tool for Comparing Stochastic Realization Algorithms and Model Reduction Techniques," *Proc. 1984 Automatic Control Conf.*, San Diego, CA.
[27] Ramos, J. A., "A Stochastic Approach to Streamflow Modeling," *Ph.D. Dissertation, School of Civil Engineering*, Georgia Institute of Technology, 1985.
[28] Escoufier, Y., "Le Traitement des Variables Vectorielles," *Biometrics 29*, 751-760, (1973).
[29] Gleason, A. M., "Measures on the Closed Subspaces of a Hilbert Space," *J. Math. Mech. 6*, pp. 885-893, 1957.
[30] Verriest, E. I., "Model Reduction via Projection Methods," *Proc. 1985 Conf. Math. Thy. in Networks Syst.*, (A. Lindquist and C. Byrnes, editors) Amsterdam, North Holland, 1986.
[31] Verriest, E. I., "A Unified Theory of Model Reduction via Gleason Measures," *Mathematics in Signal Processing*, (T. S. Durani, editor), Oxford University Press, (1985).
[32] Verriest, E. I., "Reachability, Observability and Discretization," *Proc. 22nd IEEE Conf. Dec. Control*, San Antonio, Texas, pp. 854-855, December 1983.
[33] Verriest, E. I., "Digital Filter Design Based on a High Fidelity Discretization Procedure," *Proc. Conf. Inf. Sci. and Syst.*, Johns Hopkins, M.D., March 1983.
[34] Hsu, C. S., Desai, U. B. and Crawley, C. A., "Realization Algorithms and Approximation Methods of Bilinear Systems," *Proc. 22nd IEEE Conf. Dec. Control*, (Dec. 1983), pp. 783-788.
[35] Verriest, E. I., "Approximations and Order Reduction in Nonlinear Models Using an RKHS Approach," *Proc. Conf. Inf. Sci. and Systems*, Princeton, NJ, pp. 197-201, March 1984.
[36] Baram, Y., "A Geometric Approach to Stochastic Model Reduction by Canonical Variables," *IEEE Trans. A.C.*, Vol. AC-29, No. 4, pp. 358-359, April 1984.
[37] Harshavardhana, P., Johckheere, E. A., "Spectral Factor Reduction by Phase Matching: The Continuous-Time Single-Input Single-Output Case," *Int. J. Control*, Vol. 42, No. 1, pp. 43-63, 1985.
[38] Bacon, B. J., and Frazho, A. E., "A Hankel Matrix Approach to Stochastic Model Reduction," *IEEE Trans. A. C.*, Vol. AC-30, No. 11, pp. 1138-1140, November 1985.

7

THE SCATTERING MATRIX ASSOCIATED WITH A STATIONARY STOCHASTIC PROCESS: SYSTEM THEORETIC PROPERTIES AND ROLE IN REALIZATION*

Y. AVNIEL

Department of Electrical and Computer Engineering
Drexel University, Philadelphia, PA 19104

ABSTRACT

To a multivariate stationary stochastic process, we associate a scattering matrix S, which measures the interaction between the past and future of the process. This matrix valued function can be viewed as the generalized phase function associated with the spectral density. We develop a realization theory in the spectral domain and show that S can be viewed as the analog of the frequency response function in systems.

1. INTRODUCTION

The work of Adamjan-Arov on scattering [1], and their subsequent investigations of Hankel operators [2-4], have by and large not been used for the study of stationary discrete time stochastic processes (stationary sequences). Our work is a study of the ramifications of their work and the work of Nagy-Foiaş on dilation theory [16], in the context of stationary sequences.

Connections between deterministic discrete time linear systems and Nagy-Foiaş dilation theory were explored by Helton [12]. In his treatment the frequency response function is taken as the scattering matrix. However, if the frequency response function is rational but not inner, one arrives at an infinite dimensional (Nagy-Foiaş) realization--a highly undesirable fact from a systems theoretic view.

In this study, we associate with every centered regular full rank p-dimensional process $\{\underline{y}(n)\}_{-\infty}^{\infty}$ a scattering matrix S which measures the interaction between the past and future of the process $\underline{y}$. This scattering matrix is derived by using Lax-Phillips construction of the abstract scattering operator in [14] (see also [1]). In the one-dimensional case, it is essentially the phase function of the outer factor of the spectral density

*Research funded by Air Force Office of Scientific Research (AFOSR-82-0135B) and Army Research Office (DAAG29-84-K-0005), administered through the Laboratory for Information and Decision Systems, Massachusetts Institute of Technology, Cambridge, MA 02139.

(and hence can be considered as the phase function associated with the density). For the multivariate case our scattering matrix S can be viewed as the generalized phase (matrix valued) function associated with a density.

For stationary sequences the probabilistic information is available in the form of the spectral density $f_{\underline{yy}}$ (or equivalently the covariance function) and there is no obvious candidate which plays the role of an input-output map; indeed, there is no notion of inputs. The main thrust of this work is to show that S can be viewed as the analog of the frequency response function. Loosely speaking, we use scattering to bring stationary sequences into the domain of classical systems theory.

In scattering theory all pertinent information about the scatterer is contained in the scattering matrix. For completely non-deterministic processes the scattering matrix determines the density up to congruency (up to a real scalar multiple in the one-dimensional case).

In deterministic systems realization theory, for a causal frequency response function ϕ we have the general principle [12] that the Hankel operator H_ϕ (the composition of the reachability and observability maps) determines the (causal) ϕ up to an additive constant. Assuming $\phi(0) = [0]$, the general principle can be rephrased by saying that H_ϕ admits a unique causal lifting to the Laurent operator L_ϕ (input-output map). In the case of stationary sequences, the Hankel operator H_S admits a unique norm preserving lifting L_S. In the terminology of [3], S is the unique minifunction for H_S. This property can be given a variational characterization, namely, S is the unique solution to

$$\min\{\|F\|_\infty : F \in L_\infty(\mathcal{B}(\mathbb{C}^p)) \ , \quad c_k(F) = c_k(S) \ , \quad k = -1,-2,\dots\}$$

where $c_k(F)$ is the k-th Fourier coefficient of F.

That the scattering matrix can be viewed as an analog to the frequency response function in systems is further strengthened by the role we show it plays in realization theory for stationary sequences. Ruckebusch [19] and Lindquist-Picci [15] develop a realization theory for stationary sequences using the notion of a splitting subspace. Our approach seems simpler and more natural from a systems viewpoint than the one taken in [15]. Using dilation theory and translating the realization problem to the spectral domain highlights the analog role to the frequency response function played by S. Moreover, Fuhrmann's degree theory [10] arises naturally and the minimality question is answered much in the same way as in systems (see also [8]).

Model reduction for $\underline{y}$ in which H_S plays a key role was obtained in [5].

2. NOTATION

$\mathcal{Z}$ stands for the set of integers; $\delta(n)$, the indicator function of $\{0\}\subset\mathcal{Z}$; $\mathcal{C}$, the complex numbers; and for $a\in\mathcal{C}$, $\bar{a}$ denotes the complex conjugate of a. For a matrix $A = (a_{ij})_{i,j=1}^{p}$ we denote by A^* the Hermitian conjugate of $A : A^* = (b_{ij})_{i,j=1}^{p}$ $b_{ij} = \bar{a}_{ji}$, and by A' its transposition. For a family of subsets $\{M_j\}_j$ of a Hilbert space H, we denote by $\bigvee_j M_j$ the smallest closed linear manifold (subspace) that includes each M_j, and by $\bigwedge_j M_j$ the greatest subspace contained in each of them (their intersection). $\overline{M}_j$ denotes the closure of M_j in H. For subspaces M,N of H, $M\ominus N$ denotes the orthogonal complement of N in M. For a countable family $\{M_j\}_j$ of mutually orthogonal subspaces: $M_i\perp M_j$ $i\neq j$, we let $\sum_j\oplus M_j$ be their orthogonal sum. P_M stands for the orthogonal projection of H onto the subspace M. For a bounded linear operator $A : H_1 \to H_2$ of Hilbert space H_1 into H_2, we denote by $[A]$ the matrix of A with respect to specified orthonormal bases in H_1,H_2. $A|M$ stands for the restriction of A to the subspace $M\subset H_1$. $\mathcal{B}(H)$ denotes the space of all bounded linear transformations on H.

By $l_2(-\infty,\infty;N)$ we denote the usual Hilbert space of sequences $\{h_j\}_{j=-\infty}^{\infty}$ with values in (the Hilbert space) N for which $\sum_j\|h_j\|_N^2 < \infty$. $l_2(0,\infty;N)$, $l_2(-\infty,-1;N)$ are seen naturally as subspaces of $l_2(-\infty,\infty;N)$. L_2,L_∞ will denote respectively the Lebesgue spaces on the circle $T = \{e^{i\lambda} : \lambda\in[-\pi,\pi]\}$ (with respect to the normalized Lebesgue measure $d\lambda/2\pi$) of square integrable, essentially bounded complex valued functions. Each function can be viewed as defined on $[-\pi,\pi]$. Similarly one defines the spaces $L_2(\mathcal{C}^p)$, $L_\infty(\mathcal{C}^p)$ of functions f taking values in $\mathcal{C}^p$ for which $\|f(\cdot)\|_{\mathcal{C}^p} \in L_2$, $\|f(\cdot)\|_{\mathcal{C}^p} \in L_\infty$ respectively. $L_\infty(\mathcal{B}(\mathcal{C}^p))$ is defined analogously for weakly measurable, $\mathcal{B}(\mathcal{C}^p)$ valued functions f for which ess. $\sup\{\|f(e^{i\lambda})\|_{\mathcal{B}(\mathcal{C}^p)} : \lambda\in[-\pi,\pi]\} < \infty$. $\mathcal{F} : \ell_2(-\infty,\infty;\mathcal{C}^p) \to L_2(\mathcal{C}^p)$ will denote the Fourier transform operator.

$H_2^{\pm}$ are the subspaces of L_2 defined by

$$H_2^+ = \left\{f\in L_2 : \frac{1}{2\pi}\int_{-\pi}^{\pi} f(e^{i\lambda})e^{-in\lambda}\,d\lambda = 0,\quad n=-1,-2,\ldots\right\}$$

$$H_2^- = \left\{f\in L_2 : \frac{1}{2\pi}\int_{-\pi}^{\pi} f(e^{i\lambda})e^{-in\lambda}\,d\lambda = 0,\quad n=0,1,2,\ldots\right\},$$

and we have the orthogonal decomposition $L_2 = H_2^+ \oplus H_2^-$. Each $f \in H_2^+$ having a Fourier series

$$f(e^{i\lambda}) \sim \sum_0^\infty a_n e^{in\lambda}$$

generates the function

$$g(z) = \sum_0^\infty a_n z^n$$

belonging to the Hardy class H_2 of functions $g(z)$ holomorphic in $|z| < 1$ and such that

$$\|g\|_{H_2} = \sup_{0<r<1} \left[\frac{1}{2\pi} \int_{-\pi}^{\pi} |g(re^{i\lambda})|^2 d\lambda \right]^{\frac{1}{2}} < \infty .$$

Moreover, the (a.e. existing) radial limit $g(e^{i\lambda})$ of $g(z)$ equals $f(e^{i\lambda})$ a.e. and $\|f\|_{L_2} = \|g\|_{H_2}$. The function $g(z)$ is seen as the analytic extension of $f \in H_2^+$ to the unit disc $|z| < 1$ and is denoted by $f(z)$. We identify H_2^+ with H_2 and denote them commonly by H_2. Using the conjugation with respect to the unit circle ($z \to \frac{1}{\bar{z}}$), by the reflection principle, for $f \in H_2 \subset L_2$ the function $\bar{f}$ defined by $\bar{f}(e^{i\lambda}) = \overline{f(e^{i\lambda})}$ has an analytic extension to $|z| > 1 : \overline{f(\frac{1}{\bar{z}})}$, which we again denote by $\bar{f}$. The space $\bar{H}_2 = \{f \in L_2 : \bar{f} \in H_2\}$ is the space of functions $f \in L_2$ having an analytic extension to the exterior of the disc $f(z)$ and we have

$$\|f\|_{L_2} = \sup_{\rho>1} \left[\frac{1}{2\pi} \int_{-\pi}^{\pi} |f(\rho e^{i\lambda})|^2 d\lambda \right]^{\frac{1}{2}} .$$

$f \in \bar{H}_2$ are called conjugate analytic.

Analogously for the Banach space L_∞ we have the subspace $H_\infty = H_\infty^+ \subset L_\infty$ of functions $f \in L_\infty$ having an analytic extension $f(z)$ to $|z| < 1$ with

$$\|f\|_{L_\infty} = \sup_{|z|<1} |f(z)| = \|f\|_{H_\infty} .$$

Similarly, for the Hilbert space $L_2(C^p)$ we have the subspaces $H_2^+(C^p) = H_2(C^p)$, $H_2^-(C^p)$ with the orthogonal decomposition $L_2(C^p) = H_2(C^p) \oplus H_2^-(C^p)$. In $L_\infty(\mathcal{B}(C^p))$, again $H_\infty(\mathcal{B}(C^p))$ is defined as the subspace of functions in $L_\infty(\mathcal{B}(\mathbb{C}^p))$ whose negatively indexed (matrix valued) Fourier coefficients vanish. For $\Theta \in H_\infty(\mathcal{B}(\mathbb{C}^p))$ the function Θ^* defined by $\Theta^*(e^{i\lambda}) = [\Theta(e^{i\lambda})]^*$ is identified with its analytic extension $\Theta^*(\frac{1}{\bar{z}}) = \left[\Theta(\frac{1}{\bar{z}})\right]^*$ to $|z| > 1$.

A function $f \in H_\infty$ is called inner if $|f(e^{i\lambda})| = 1$ a.e. . Similarly for $\Theta \in H_\infty(\mathcal{B}(\mathbb{C}^p))$ if $\Theta(e^{i\lambda})$ is unitary a.e. . $f \in H_2$ is called outer if $\bigvee_{n \geq 0} \{\chi^n f\} = H_2$ where χ denotes the function on T defined by $\chi(e^{i\lambda}) = e^{i\lambda}$. For $\phi \in L_\infty(\mathcal{B}(\mathbb{C}^p))$ the Toeplitz operator $T_\phi : H_2(\mathbb{C}^p) \to H_2(\mathbb{C}^p)$ is defined by $T_\phi f = \pi_+(\phi f)$ where π_+ is the Riesz projection of $L_2(\mathbb{C}^p)$ onto $H_2(\mathbb{C}^p)$. H_ϕ will denote the Hankel operator $H_\phi : H_2(\mathbb{C}^p) \to H_2^-(\mathbb{C}^p)$ defined by $H_\phi f = \pi_-(\phi f)$, π_- being the Riesz projection of $L_2(\mathbb{C}^p)$ onto $H_2^-(\mathbb{C}^p)$. The convention we employ regarding a Hankel operator as acting from $H_2(\mathbb{C}^p)$ into $H_2^-(\mathbb{C}^p)$ is not in accordance with the one employed in systems theory, in which we act on $H_2^-(\mathbb{C}^p)$ into $H_2(\mathbb{C}^p)$: $H_\phi f = \pi_+(\phi f)$. It, however, conforms with the one employed by Adamjan-Arov-Krein and enables us to use their results without modifications, as well as to refer to them.

3. THE SCATTERING OPERATOR MODEL AND THE SCATTERING MATRIX ASSOCIATED WITH A STATIONARY STOCHASTIC PROCESS

Let H be a complex separable Hilbert space and let U be a unitary operator on H. A subspace D_+ is said to be outgoing for (U,H) if it satisfies

(i) $UD_+ \subset D_+$

(ii) $\bigwedge_{-\infty}^{\infty} U^n D_+ = \{0\}$ $\qquad (3.1)_+$

(iii) $\bigvee_{-\infty}^{\infty} U^n D_+ = H$.

A subspace D_- for which

(i) $U^* D_- \subset D_-$

(ii) $\bigwedge_{-\infty}^{\infty} U^n D_- = \{0\}$ $\qquad (3.1)_-$

(iii) $\bigvee_{-\infty}^{\infty} U^n D_- = H$

is said to be incoming for (U,H).

3.1 DEFINITION. A quadruple (U,H,D_+,D_-) satisfying (3.1) is said to be a scattering system.

We shall be interested in a scattering system arising in the following way. Let $(\Omega,\mathcal{A},\mathcal{P})$ be a fixed probability space and let

$$\{\underline{y}(n) : n\in \mathbb{Z}\} \quad , \quad \underline{y}(n) = \begin{pmatrix} y_1(n) \\ y_2(n) \\ \vdots \\ y_p(n) \end{pmatrix}$$

be a centered stationary process with $y_j(n)\in L_2(\Omega,\mathcal{A},\mathcal{P})$ $j=1,\ldots,p$. Let $f_{\underline{yy}}(\lambda) = (f_{kj}(\lambda))_{k,j=1}^{p}$, $\lambda\in[-\pi,\pi]$ be its spectral density satisfying

$$\frac{1}{2\pi}\int_{-\pi}^{\pi}\log\det f_{\underline{yy}}(\lambda)\, d\lambda > -\infty \tag{3.2}$$

i.e., the process is regular and of maximal rank. Let

$$H = H_{\underline{y}} = \bigvee_{n\in \mathbb{Z}} \{y_1(n),y_2(n),\ldots,y_p(n)\}\subset L_2(\Omega,\mathcal{A},\mathcal{P}),$$

be the space spanned by the process and let U be the unitary shift operator on H associated with the $\underline{y}$ process [18, p. 14]:

$$Uy_j(n) = y_j(n+1) \qquad j = 1,\ldots,p \ , \qquad n\in \mathbb{Z} \ .$$

We consider the past and future of $\{\underline{y}(n)\}_{-\infty}^{\infty}$ defined by

$$D_- \doteq H_{\underline{y}}^-(0) = \bigvee_{k\le 0} \{y_1(k),\ldots,y_p(k)\} \ ,$$

$$D_+ \doteq H_{\underline{y}}^+(0) = \bigvee_{k\ge 0} \{y_1(k),\ldots,y_p(k)\} \ .$$

By (3.2) it follows [18, Th. II.6.1]

$$\bigwedge_{-\infty}^{\infty} U^nD_- = \{0\} = \bigwedge_{-\infty}^{\infty} U^nD_+ \ .$$

We readily obtain that (U,H,D_+,D_-) is a scattering system.

3.2 THEOREM.(Translation Representation Theorem [14, Th. II.1.1]). Let (U,H,D_+) be outgoing. Then there exists a Hilbert space N_+ and a unitary map r_+ of H onto $l_2(-\infty,\infty;N_+)$ such that

$$\begin{aligned} &\text{(i)} \quad r_+[D_+] = l_2(0,\infty;N_+) \quad , \\ &\text{(ii)} \quad U_+ = r_+ U r_+^{-1} \end{aligned} \tag{3.3}$$

is the right shift operator on $l_2(-\infty,\infty;N_+)$. This representation is unique up to automorphisms of N_+.

PROOF.Standard (cf. [1, p. 77]). The proof will establish various quantities introduced later. By $(3.1)_+$-(ii) the operator $U|D_+$ is an isometry having no unitary part. By Wold's decomposition theorem [16, Th.I.1.1] we may write uniquely

$$D_+ = \sum_{n=0}^{\infty} \oplus\, U^n N_+ \quad , \quad N_+ = D_+ \ominus UD_+ \ . \tag{3.4}$$

Since for any $m > 0$

$$\begin{aligned} U^{-m} D_+ &= U^{-m}[(D_+ \ominus U^m D_+) \oplus U^m D_+] = U^{-m}\left[\left(\sum_{k=0}^{m-1} \oplus\, U^k N_+\right) \oplus U^m D_+\right] \\ &= \left(\sum_{k=-1}^{-m} \oplus\, U^k N_+\right) \oplus D_+ \quad , \end{aligned}$$

we obtain by $(3.1)_+$ - (iii) that

$$H = \sum_{-\infty}^{\infty} \oplus\, U^n N_+ \quad . \tag{3.5}$$

It follows that for arbitrary $h \in H$

$$h = \sum_{-\infty}^{\infty} \oplus\, U^n P_{N_+} U^{-n} h \quad , \quad \|h\|_H^2 = \sum_{-\infty}^{\infty} \|P_{N_+} U^{-n} h\|_H^2 \quad .$$

We readily conclude that the map

$$r_+ : H \to l_2(-\infty,\infty;N_+)$$

defined by

$$r_+h = \{P_{N_+} U^{-n}h\}_{n=-\infty}^{\infty} \tag{3.6}$$

is unitary. By (3.4) we obtain (i), and from (3.6) (ii) follows. By (3.5) U is a bilateral shift of multiplicity equal to dim N_+ and the uniqueness follows. ■

3.3 DEFINITION. The representation $(U_+, l_2(0,\infty;N_+), l_2(-\infty,\infty;N_+))$ is called an outgoing translation representation.

For (U,H,D_-) incoming we similarly obtain

$$D_- = \sum_{n=-\infty}^{0} \oplus\ U^n N_- \qquad N_- = D_- \ominus U^* D_- \quad , \tag{3.7}$$

and

$$H = \sum_{-\infty}^{\infty} \oplus\ U^n N_- \quad . \tag{3.8}$$

For the corresponding map r_- of H onto $l_2(-\infty,\infty;N_-)$ we define

$$r_-h = \{P_{N_-} U^{-(n+1)}h\}_{n=-\infty}^{\infty} \quad . \tag{3.9}$$

Thus

(i) $\quad r_-[D_-] = l_2(-\infty,-1;N_-) \quad ,$

(ii) $\quad U_- = r_- U r_-^{-1}$

is the right shift on $l_2(-\infty,\infty;N_-)$. The representation $(U_-, l_2(-\infty,-1;N_-), l_2(-\infty,\infty;N_-))$ is called an incoming translation representation.

Now let (U,H,D_+,D_-) be the scattering system associated with the regular maximal rank $\underline{y}$ process. The subspace $N_- = D_- \ominus U^* D_-$ $(N_+ = D_+ \ominus UD_+)$ is the forward (backward) innovation subspace at n = 0. Since for a scattering system (U,H,D_+,D_-) we have

$$\dim N_- = \text{multiplicity } U = \dim N_+ \quad ,$$

we can arrange the maps $r_\pm$ to be onto $l_2(-\infty,\infty;C^p)$.

3.4 DEFINITION ([1],[14]). The operator

$$S = r_- r_+^{-1} : l_2(-\infty,\infty;\mathcal{C}^p) \to l_2(-\infty,\infty;\mathcal{C}^p)$$

is called the abstract scattering operator.

Clearly S is unitary and commutes with the right shift on $l_2(-\infty,\infty;\mathcal{C}^p)$. It follows [9] that the unitary operator

$$\mathcal{F}S\mathcal{F}^{-1} : L_2(\mathcal{C}^p) \to L_2(\mathcal{C}^p)$$

is a Laurent operator $L_{\mathcal{S}}$, the operator of multiplication by $\mathcal{S} \in L_\infty(\mathcal{B}(\mathcal{C}^p))$ with

$$\mathcal{S}(e^{i\lambda})$$

a.e. a unitary map on $\mathcal{C}^p$.

3.5 DEFINITION ([1],[14]). $\mathcal{S}$ is called the scattering matrix.

It is clear from the translation representation theorem that $\mathcal{S}$ is determined to within right and left multiplication by unitary transformations on $\mathcal{C}^p$ (i.e., to within coincidence, see Definition 5.8).

We next compute the scattering matrix $\mathcal{S}$ for the $\underline{y}$ process. Let $\{\nu_1(0),\ldots,\nu_p(0)\}$ be an orthonormal basis for N_-, $\nu_j(n) = U^n\nu_j(0)$. By (3.8), the process $\{\underline{\nu}(n)\}_{-\infty}^{\infty}$ is a (centered) white noise process with covariance $R_{\underline{\nu}\underline{\nu}}(n) = \delta(n) I_{\mathcal{C}^p}$, constituting the forward innovation process for the $\underline{y}$ process. It is determined up to a choice of basis in N_-. By (3.7), we may write

$$\underline{y}(0) = \sum_{-\infty}^{\infty} A(k)\underline{\nu}(k) \qquad A(k) = (\alpha_{ij}(k))_{i,j=1}^{p} \qquad A(k) = [0]\ ,\ k > 0\ .$$

(Wold's representation cf. [18, p. 56]). It follows from (3.9)

$$r_- y_j(0) = \left\{ \sum_{m=1}^{p} \alpha_{jm}(k+1)\nu_m(0) \right\}_{k=-\infty}^{\infty} \quad . \tag{3.10}$$

Identifying N_- with $\mathcal{C}^p$ we readily obtain the representation

$$\underline{r}\_{y_j}(0) = \left\{ \begin{pmatrix} \alpha_{j1}(k+1) \\ \\ \alpha_{jp}(k+1) \end{pmatrix} \right\}_{k=-\infty}^{\infty}$$

Consider the function

$$\Lambda(z) = \sum_{-\infty}^{0} A'(k) z^k$$

Since

$$\sum_{k=-\infty}^{\infty} \sum_{i,j=1}^{p} |\alpha_{ij}(k)|^2 \leq \sum_{j=1}^{p} \|y_j(0)\|_H^2 < \infty ,$$

$\Lambda(z)$ is analytic in $|z| > 1$. For $\Lambda(z)$ we have [18, p. 57]

$$\frac{1}{2\pi} \Lambda^*(z)\Lambda(z) = f_{\underline{yy}}(\lambda) \qquad\qquad z = e^{i\lambda} \quad .$$

By the incoming properties

$$H_2^-(C^p) = \bigvee_{n<0} \{e^{in\lambda}\Lambda(e^{i\lambda})\underline{a} : \underline{a} \in C^p\}$$

i.e., Λ is conjugate outer [11, p. 121). Thus

$$(Fr\_y_1(0), \ldots, Fr\_y_p(0)) = \bar{\chi}\Lambda \ . \tag{3.11}$$

Since the translates (in $H_{\underline{y}}$) of $y_1(0),\ldots,y_p(0)$ and their linear combinations are dense in $H_{\underline{y}}$, $Fr\_$ is determined by the above expression.

We now consider the outgoing representation. Letting $\varepsilon_1(0),\ldots,\varepsilon_p(0)$ be an orthonormal basis in N_+, we similarly obtain

$$\underline{y}(0) = \sum_{-\infty}^{\infty} B(k)\underline{\varepsilon}(k) \qquad B(k) = (\beta_{ij}(k))_{i,j=1}^{p} \quad , \qquad B(k) = [0] \ , \qquad k < 0 \ .$$

This representation constitutes the representation of $\underline{y}(0)$ in terms of the backward innovation process $\{\underline{\varepsilon}(n)\}_{-\infty}^{\infty}$. We define

$$\Gamma(z) = \sum_{0}^{\infty} B'(k) z^k$$

which is analytic in $|z| < 1$. In a similar fashion we obtain by direct computation

$$\frac{1}{2\pi}\Gamma^*(z)\Gamma(z) = f_{\underline{y}\underline{y}}(\lambda) \qquad z = e^{i\lambda} ,$$

with Γ being outer. Also

$$(Fr_+ y_1(0), \ldots , Fr_+ y_p(0)) = \Gamma . \tag{3.12}$$

Combining (3.11) with (3.12), we obtain

$$S\Gamma = \bar{X}\Lambda$$

and thus

$$S = \bar{X}\Lambda\Gamma^{-1} .$$

We thus obtained

3.6 THEOREM. For a regular maximal rank process $\{\underline{y}(n)\}_{-\infty}^{\infty}$ we have

$$S = \bar{X}\Lambda\Gamma^{-1} ,$$

where S is determined up to left and right multiplication by constant unitary matrices.

For the case $p = 1$ we have

3.7 COROLLARY. For a regular process $\{y(n)\}_{-\infty}^{\infty}$

$$s = \bar{X}\,\frac{\bar{\Gamma}}{\Gamma} ,$$

and s is determined up to multiplication by a constant of unit modulus.

PROOF. The outer function $\bar{\Lambda}$ satisfies $|\bar{\Lambda}| = |\Gamma|$ on T and thus $\bar{\Lambda} = \gamma\Gamma$ a.e. where γ is a constant of unit modulus. ∎

3.8 REMARK. The scattering matrix S was defined by an outer and conjugate outer factors of the density $f_{\underline{y}\underline{y}}$. Since those are determined up to left multiplication by a constant unitary matrix, we may wish to make a canonical choice (which amounts to choosing specific orthonormal bases in

N_+, N_-) in the following fashion: For $\Gamma(0)$ we consider its polar decomposition $\Gamma(0) = KP$ (K unitary, $P > 0$) and define $\Gamma_1(z) = K^{-1}\Gamma(z)$. For Γ_1 we have $\Gamma_1(0) > 0$, and it is unique. Similarly for Λ. In this way, the density $f_{\underline{yy}}$ will have a unique S associated with it. From the viewpoint of seeing S as the phase function associated with $f_{\underline{yy}}$, this may be appealing. Note, however, that S measures the interaction between the past and future of the process $\underline{y}$ (see discussion following theorem 4.4) and uniqueness in the reverse direction (from S to $f_{\underline{yy}}$) is not possible.

4. THE INDUCED HANKEL AND TOEPLITZ OPERATORS

We call the unitary maps $F_- = Fr_-$, $F_+ = Fr_+$ - the incoming and outgoing spectral representations, respectively.

4.1 PROPOSITION. We have

(i) $$F_-[D_-] = H_2^-(C^p) \quad ,$$

(ii) $$F_+[D_+] = SH_2(C^p) \quad ,$$

(iii) $$F_-(Uh) = XF_-h \qquad h \in H_{\underline{y}} \quad .$$

PROOF. (i) and (iii) follow from the properties of the incoming translation representation. By the definition of S we obtain

$$F_-[D_+] = (Fr_-r_+^{-1}F^*)Fr_+[D_+] = SH_2(C^p) \quad .$$ ∎

4.2 LEMMA. The operator

$$P_-P_+ : D_+ \to D_- \qquad , \qquad P_\pm = P_{D_\pm}$$

is unitarily equivalent to the Hankel operator H_S, and the operator

$$P_{D_-^\perp} P_+ : D_+ \to D_-^\perp$$

is unitarily equivalent to the Toeplitz operator T_S.

PROOF. Straightforward.

By a theorem of Nehari [17] (a vector generalization of which was obtained by Sarason [20]), for a bounded Hankel operator H_ϕ, $\phi \in L_\infty$ there exists a function $\phi_\mu \in L_\infty$ such that $H_\phi = H_{\phi_\mu}$ and

$$\|H_\phi\| = \|\phi_\mu\|_\infty \quad . \tag{4.1}$$

Since always $\|H_\phi\| \leq \|\phi\|_\infty$, ϕ_μ is called a minifunction for H_ϕ [3]. The question regarding the uniqueness of the minifunction is of particular interest to us. In systems realization theory for a frequency response function $\phi \in H_\infty$ (same reasoning holds for the vector case), a central role is played by the Hankel operator $H_\phi : H_2^- \to H_2$, $H_\phi f = \pi_+(\phi f)$. Now observe that if $\phi_1, \phi_2 \in H_\infty$ are such that $H_{\phi_1} = H_{\phi_2}$ then $H_{\phi_1 - \phi_2} \equiv 0$, and the positive Fourier coefficients of $\phi_1 - \phi_2$ vanish. Thus $\phi_1 - \phi_2 =$ const. and H_ϕ determines (the analytic) ϕ up to an additive constant (recall in this regard that the composition of the reachability and observability maps determine the frequency response function up to an additive constant). However, in general $\|H_\phi\| < \|\phi\|_\infty$ and the uniqueness of ϕ inducing H_ϕ is guaranteed by analyticity (causality). In view of the central role played by S in realization of stationary sequences, the following theorem is of significance:

4.3 THEOREM. The Hankel operator H_S determines S uniquely. Indeed, S is its unique minifunction.

PROOF. From the simple identity

$$H_S^* H_S + T_S^* T_S = I_{H_2(\mathcal{C}^p)} \tag{4.2}$$

we note that $f \in \operatorname{Ker} T_S$ is an eigenvector of $H_S^* H_S$ corresponding to the eigenvalue $\|H_S\| = 1$. Since $S = \bar{X}\Lambda\Gamma^{-1}$ every column of Γ belongs to this kernel. Thus, the projection of the above eigenspace on the first coordinate in $l_2(0,\infty;\mathcal{C}^p)$ spans $\Gamma(0)$. Now observe that for $\Gamma(0)$ we have, because of its outer property in $H_2(\mathcal{B}(\mathcal{C}^p))$, (see e.g. [18, p. 76])

$$\log \frac{|\det\Gamma(0)|}{(2\pi)^{p/2}} = \frac{1}{4\pi}\int_{-\pi}^{\pi} \log \det f_{\underline{yy}}(\lambda)d\lambda > -\infty ,$$

so that $\Gamma(0)$ is of full rank. We conclude that the aforementioned projection is onto the first coordinate space. According to a result of Adamjan-Arov-Krein [2, Corollary 3.1] for a Hankel operator H_ϕ to have a unique minifunction, it is sufficient that the projection of the eigenspace of $H_\phi^* H_\phi$ corresponding to $\|H_\phi\|$ on the first coordinate space be onto. The result follows. ■

There is an alternative way to rephrase Theorem 4.3. If we consider the Laurent operator L_{ϕ_μ}, then (4.1) becomes

$$\|H_\phi\| = \|L_{\phi_\mu}\| .$$

Since

$$H_\phi = P_{H_2^-} L_{\phi_\mu} \mid H_2$$

one considers L_{ϕ_μ} a norm preserving lifting of H_ϕ. Thus, in systems theory, the uniqueness of the lifting is guaranteed by causality, the lifting being in general not norm preserving, while for stationary processes, the uniqueness is guaranteed by the lifting being norm preserving.

Viewing a linear time invariant system from the input-output point of view makes the frequency response function the sole accessible object containing all pertinent information about the system. As to the information contained in the scattering matrix, we have the following:

4.4 THEOREM. The scattering matrix S determines the density $f_{\underline{yy}}(\lambda)$ up to the form

$$K^* f_{\underline{yy}}(\lambda) K \tag{4.3}$$

where K is a constant pxp non-singular matrix, iff

$$\dim \operatorname{Ker} T_S = p . \tag{4.4}$$

PROOF. First note that for any representation of S

$$S = \bar{X}YX^{-1}$$

with the columns of X in $H_2(C^p)$ and those of $\bar{X}Y$ in $H_2^-(C^p)$, the columns of X belong to Ker T_S. Moreover (on T)

$$Y^*Y = (SX)^*SX = X^*X .$$

Assume (4.4) holds. It thus follows that

$$X(e^{i\lambda}) = \Gamma(e^{i\lambda})K \tag{4.5}$$

where K is a pxp full rank constant matrix. Thus,

$$\frac{1}{2\pi} X^*(z)X(z) = \frac{1}{2\pi} K^*\Gamma^*(z)\Gamma(z)K = K^* f_{\underline{yy}}(\lambda)K \qquad z = e^{i\lambda}$$

proving the 'if' part.

Now assume (4.4) not to hold, i.e., dim Ker $T_S > p$. We can thus find a pxp matrix $X(e^{i\lambda})$ of full rank a.e. λ such that the columns of X belong to Ker T_S and (4.5) does not hold. If we define

$$Y = \bar{X}SX$$

then the columns of $\bar{X}Y$ are in $H_2^-(C^p)$ and $S = \bar{X}YX^{-1}$ with $Y^*Y = X^*X$. The result follows. ■

That the scattering matrix determines $f_{\underline{yy}}$ up to the form (4.3) is a natural consequence of the scattering framework. Indeed, for an arbitrary non-singular K, the process $\underline{\xi}(n) = K^*\underline{y}(n)$ whose density equals

$$f_{\underline{\xi\xi}} = K^* f_{\underline{yy}} K$$

induces the same scattering system (U,H,D_+,D_-) as the $\underline{y}$ process.

We next characterize condition (4.4) on a process level.

4.5 PROPOSITION. We have

$$F_+^*[\text{Ker } T_S] = H_{\underline{y}}^-(0) \wedge H_{\underline{y}}^+(0) \quad .$$

PROOF. Let $0 \neq f \in \text{Ker } T_S$. From (4.2) it follows

$$H_S^* H_S f = f$$

i.e.,

$$\|H_S f\| = \|f\| \quad .$$

By Lemma 4.2, we obtain for $\xi = F_+^* f \in H_{\underline{y}}^+(0)$

$$\|P_-\xi\| = \|\xi\| \quad ,$$

and $\xi \in H_{\underline{y}}^-(0)$. Thus

$$F_+^*[\text{Ker } T_S] \subset H_{\underline{y}}^-(0) \wedge H_{\underline{y}}^+(0) \quad .$$

Now let $\xi \in H_{\underline{y}}^-(0) \wedge H_{\underline{y}}^+(0)$, from Lemma 4.2

$$H_S(F_+\xi) = F_-\xi \quad .$$

Let $f = F_+\xi \in H_2(C^p)$. We obtain

$$\|H_S f\| = \|F_-\xi\| = \|\xi\| = \|F_+\xi\| = \|f\|$$

and

$$H_S^* H_S f = f \quad .$$

Thus $f \in \text{Ker } T_S$ which implies

$$F_+[H_{\underline{y}}^-(0) \wedge H_{\underline{y}}^+(0)] \subset \text{Ker } T_S \quad .$$

The result follows. ∎

By the unitarity of F_+

$$\dim \text{Ker } T_S = \dim H_{\underline{y}}^-(0) \wedge H_{\underline{y}}^+(0) \quad ,$$

and since $\underline{y}$ is regular and of full rank we readily conclude

$$\dim H_{\underline{y}}^-(0) \wedge H_{\underline{y}}^+(0) = p \text{ iff } \dim H_{\underline{y}}^-(0) \wedge H_{\underline{y}}^+(1) = 0 \quad .$$

4.6 DEFINITION [6]. The process $\underline{y}$ is said to be completely non-deterministic if

$$H_{\underline{y}}^-(0) \wedge H_{\underline{y}}^+(1) = \{0\} \quad .$$

As is well-known (see e.g. [18, p. 73]) for a regular maximal rank process $\{\underline{y}(n)\}_{-\infty}^{\infty}$ no $\underline{y}_j(k)$ $k \geq 1$, $j=1,\dots,p$ can be predicted without error based on the past $H_{\underline{y}}^-(0)$. Being completely nondeterministic is more restrictive; indeed, no value in $H_{\underline{y}}^+(1)$ can be predicted without error based on $H_{\underline{y}}^-(0)$.

Summarizing, we can restate Theorem 4.4 in the following way:

4.7 THEOREM. The scattering matrix S determines $f_{\underline{yy}}$ up to the form (4.3) iff $\underline{y}$ is completely nondeterministic.

4.8 REMARK. It is of interest to observe that since for a completely nondeterministic process, the eigenvectors of $H_S^*H_S$ corresponding to $\|H_S\|$ are only the columns of Γ, the projection of this eigenspace on the first coordinate is not only onto, but also 1-1. In [2, Sec. 2] it is shown that for any Hankel operator $H : H_2(C^p) \to H_2^-(C^p)$ satisfying this condition, its unique minifunction is of the form

$$\rho S \qquad\qquad \rho = \|H\| \quad .$$

Thus up to a constant multiple $\rho > 0$ all minifunctions of such Hankel operators are in 1-1 correspondence with regular, full rank, completely nondeterministic processes.

4.9 EXAMPLE. Let $\{y(n)\}_{-\infty}^{\infty}$ have rational density

$$f_{yy}(\lambda) = \frac{|P(z)|^2}{|Q(z)|^2} \qquad\qquad z = e^{i\lambda} ,$$

where the polynomials P,Q have no zeros in $|z| < 1$ and are relatively prime. Since $f_{yy} \in L_1$, the polynomial Q has its zeros in $|z| > 1$. Write

$$P = P_1P_2$$

where P_1 of degree k has its zeros on T and P_2 in $|z| > 1$. For $P_1(z) = \prod_{j=1}^{k} (z-\alpha_j)$, $\{\alpha_j\}_{j=1}^{k} \subset T$ we have

$$\frac{\overline{P_1(e^{i\lambda})}}{P_1(e^{i\lambda})} = e^{-ik\lambda}(-1)^k \prod_{j=1}^{k} \bar{\alpha}_j \quad .$$

Thus

$$s = \gamma \bar{\chi}^{k+1} \frac{\bar{\psi}_e}{\psi_e} \qquad\qquad \psi_e = \frac{P_2}{Q} \;, \quad \gamma = (-1)^k \prod_{j=1}^{k} \bar{\alpha}_j \qquad (4.6)$$

where $|\gamma| = 1$ and ψ_e is outer. By a result of Adamjan-Arov-Krein [3, Th. 2.2] (4.6) is the general form of unimodular minifunctions and the dimension of the eigenspace corresponding to the singular value $1 = \|H_S\|$ is k+1. From (4.2) it readily follows that this dimension equals dim Ker T. Thus

$$\dim H_y^-(0) \wedge H_y^+(0) = k+1 \quad .$$

We conclude that a regular process with rational density is completely nondeterministic iff it has no zeros on T.

4.10 EXAMPLE. Let $\{y(n)\}_{-\infty}^{\infty}$ have density

$$0 < m \leq f_{yy}(\lambda) \leq M < \infty \qquad \text{a.e.} \quad .$$

It follows that the outer factor $\Gamma \in H_\infty$, and since Γ is outer $1/\Gamma \in H_\infty$. Now let $g \in \text{Ker } T_S$, for arbitrary $f \in H_2$

$$0 = (T_S g, f) = (\bar{\chi} \frac{\bar{\Gamma}}{\Gamma} g, f) = (\frac{g}{\Gamma}, \chi \Gamma f) \quad .$$

Since Γ is outer and $g/\Gamma \in H_2$, we conclude

$$\frac{g}{\Gamma} \in H_2 \ominus \chi H_2 = C \quad .$$

Thus,

$$\text{Ker } T_S = \{\lambda \Gamma : \lambda \in C\}$$

and dim Ker $T_S = 1$. The process y is, therefore, completely nondeterministic.

We now formulate a converse to Theorem 4.4. The question can be posed as follows: Under what conditions is a function $S \in L_\infty(\mathcal{B}(C^p))$ a scattering matrix of some full rank, p dimensional, completely nondeterministic process? On an abstract level we first observe that any $S \in L_\infty(\mathcal{B}(C^p))$ which is unitary valued a.e. on T is the scattering matrix of the canonical scattering system [1]

$$U = L_\chi \quad , \quad H = L_2(\mathcal{C}^p) \quad , \quad D_+ = SH_2(\mathcal{C}^p) \quad , \quad D_- = H_2^-(\mathcal{C}^p) \quad .$$

The above question amounts to characterizing all scattering systems (U,H,D_+,D_-) for which there exists a set $\{\xi_1,\ldots,\xi_p\}$ of linearly independent vectors such that

$$H = \overline{\text{span}\,\{U^n \xi_j : j=1,\ldots,p,\ n=0,\ \pm 1,\ \ldots\}}$$

$$D_+ = \overline{\text{span}\,\{U^n \xi_j : j=1,\ldots,p,\ n\geq 0\}} \tag{4.7}$$

$$D_- = \overline{\text{span}\,\{U^n \xi_j : j=1,\ldots,p,\ n\leq 0\}}$$

and such that any other linearly independent set satisfying (4.7) is of cardinality p. The corresponding process will be $\{\underline{\xi}(n)\}_{-\infty}^{\infty}$, $\xi_j(n) = U^n\xi$, $j=1,\ldots,p$, and the spectral density is obtained by

$$f_{\underline{\xi}\underline{\xi}}(\lambda) = \left(\frac{d(E_\lambda \xi_i,\xi_j)_H}{d\lambda}\right)_{i,j=1,\ldots,p} \quad , \qquad \{E_\lambda : \lambda \in [-\pi,\pi]\}$$

being the resolution of the identity for U. The answer is given in the following.

4.11 THEOREM. Let $S \in L(\mathcal{B}(\mathcal{C}^p))$ be such that

(i) $S(e^{i\lambda})$ is a.e. λ a unitary map on $\mathcal{C}^p$,

(ii) $\dim \operatorname{Ker} T_S = p$.

Then there exists a p-dimensional full rank completely nondeterministic process $\underline{y}$ whose scattering matrix is S.

PROOF. Let $\Gamma_1,\Gamma_2,\ldots,\Gamma_p$ span the kernel of T_S and define

$$\Gamma = [\Gamma_1|\Gamma_2| \ \ldots \ |\Gamma_p] \ .$$

Let

$$\Lambda = S\Gamma \quad ,$$

since $\Lambda_j = \pi_-(S\Gamma_j) + \pi_+(S\Gamma_j) = \pi_-(S\Gamma_j)$, j=1,...,p, the columns of $\Lambda = [\Lambda_1|\Lambda_2|\dots|\Lambda_p]$ are in $H_2^-(C^p)$ and by (i)

$$\Lambda^*(z)\Lambda(z) = \Gamma^*(z)\Gamma(z) \qquad\qquad z = e^{i\lambda} .$$

If we define

$$f_{\underline{yy}}(\lambda) = \frac{1}{2\pi}\Gamma^*(e^{i\lambda})\Gamma(e^{i\lambda})$$

the theorem follows provided we show that Γ is outer and $\chi\Lambda$ conjugate outer. Let $U = L_\chi$ and define:

$$\hat{D}_- = \bigvee_{n\leq -1} \{\chi^n\Lambda_1,\dots,\chi^n\Lambda_p\} \subset H_2^-(C^p) \quad ,$$

$$\hat{D}_+ = \bigvee_{n\geq 0} \{\chi^n S\Gamma_1,\dots,\chi^n S\Gamma_p\} \subset SH_2(C^p) \quad .$$

Let

$$\hat{H} = (\bigvee_{n\in Z} U^n\hat{D}_-) \vee (\bigvee_{n\in Z} U^n\hat{D}_+) \quad . \tag{4.8}$$

It is easily verified that $(3.1)_\pm$ - (i), (ii) holds for $(U,\hat{D}_\pm)$. In [1] Adamjan-Arov show [1, Th. 2.5] that a quadruple $(U,\hat{H},\hat{D}_+,\hat{D}_-)$ satisfying $(3.1)_\pm$ - (i), (ii) and (4.8) has a scattering matrix $\hat{S}$ which is unitary valued a.e. on T iff

$$\bigvee_{n\in Z} U^n\hat{D}_+ = \hat{H} = \bigvee_{n\in Z} U^n\hat{D}_-$$

and, moreover, from their generalized functional model [1, Th. 2.1] we need have

$$\hat{D}_- = H_2^-(C^p) \quad , \quad \hat{D}_+ = \hat{S}H_2(C^p) \quad .$$

A straightforward computation (mimicking the one in Section 3) gives

$$\hat{S} = S$$

and the result follows. ■

5. MARKOV PROCESSES, UNITARY DILATIONS AND FACTORIZATION OF THE SCATTERING MATRIX

In a Hilbert space setting, a centered stationary process $\{\underline{x}(n)\}_{-\infty}^{\infty}$ is said to be Markov if for all $n \geq s$

$$P_{H_{\underline{x}}^{-}(s)} x = P_{X(s)} x \qquad x \in H_{\underline{x}}^{+}(n) \quad , \tag{5.1}$$

where $X(s) = \text{span}\{x_j(s) : j=1,\ldots,m\}$. In our setting, all stationary processes will be generated by the shift U (on $H_{\underline{y}}$) associated with the $\underline{y}$ process. Thus, for a stationary process $\{\underline{x}(n)\}_{-\infty}^{\infty}$ (in $H_{\underline{y}}$) we will have $\underline{x}(n) = U^n \underline{x}(0)$. It readily follows from (5.1) that one can define the notion of a Markov subspace with respect to U in the following (see [15], [19]).

5.1 DEFINITION. A subspace $X \subset H_{\underline{y}}$ is said to be Markov with respect to U if for all $n \geq s$, $x \in X$

$$P_{\bigvee_{-\infty}^{s} U^m X} U^n x = P_{U^s X} U^n x \quad . \tag{5.2}$$

Thus X is a Markov subspace with respect to U iff the process $\{U^n X\}_{-\infty}^{\infty}$ has the (weak) Markov property. In what follows a Markov process $\{U^n X\}_{-\infty}^{\infty}$ will invariably arise in this fashion.

We shall be interested in Markov subspaces (with respect to U)[†] $X \subset H_{\underline{y}}$ for which

$$\{y_1(0),\ldots,y_p(0)\} \subset X \quad . \tag{5.3}$$

In this case X satisfies

$$H_{\underline{y}} = \bigvee_{-\infty}^{\infty} U^n X \quad ,$$

and we say that X (or $\{U^n X\}_{-\infty}^{\infty}$) is of full range. There is a direct relationship between Markov processes of full range and unitary dilations. Recall [16] that a unitary operator U on a Hilbert space H is said to be the minimal unitary (power) dilation of a contraction A on $X \subset H$ if

---

† We shall subsequently omit the phrase "with respect to U".

$$A^n = P_X U^n | X \qquad n \geq 0$$

and $H = \bigvee_{-\infty}^{\infty} U^n X$ (minimality).

5.2 PROPOSITION. $X \subset H_{\underline{y}}$ is a Markov subspace of full range iff U (on $H_{\underline{y}}$) is the minimal unitary (power) dilation of the state operator

$$A = P_X \; U|X : X \to X \quad .$$

PROOF. From (5.2), we obtain for $x, x' \in X$ and $m,n \geq 0$

$$(U^{-m}x, U^n x') = (U^{-m}x, P_X U^n x) \quad .$$

Thus, denoting $A(n) = P_X U^n | X$, we obtain

$$(x, A(m+n)x') = (x, U^{m+n}x') = (U^{-m}x, U^n x') = (U^{-m}x, P_X U^n x')$$

$$= (x, P_X U^m P_X U^n x') = (x, A(m)A(n)x') \quad .$$

We infer that $A(m+n) = A(m)A(n)$ and

$$A(n) = A^n(1) = A^n \quad .$$

Since X is of full range, we conclude that U in $H_{\underline{y}}$ is the minimal unitary dilation of A (in X). This proves the 'only if' part. The 'if' part follows by reversing the argument. ■

Much in the same way as for a regular process we make the following:

5.3 DEFINITION. A Markov process $\{U^n X\}_{-\infty}^{\infty} \subset H_{\underline{y}}$ is said to be regular if

$$\bigwedge_{n \leq 0} \bigvee_{k \leq n} U^k X = \{0\} = \bigwedge_{n \geq 0} \bigvee_{k \geq n} U^k X \quad .$$

We shall occasionally also refer to (the Markov subspace) X as being regular.

The notion of regularity for a Markov process is intimately related to the asymptotic stability of the state operator A (see Corollary 5.5).

The correspondence between Markov processes and scattering systems is established in the following:

5.4 THEOREM. Let $X \subset H_{\underline{y}}$ be a regular Markov subspace of full range. Then $H_{\underline{y}}$ decomposes and, moreover, uniquely into the orthogonal sum

$$H_{\underline{y}} = \mathcal{D}_- \oplus X \oplus \mathcal{D}_+ \tag{5.4}$$

where $(U, H_{\underline{y}}, \mathcal{D}_+, \mathcal{D}_-)$ is a scattering system.

PROOF. The proof employs the structure of the space of the minimal unitary dilation [16, Ch. II], see also [1, Sec. 3]. Define

$$\begin{aligned} \mathcal{D}_+ &= \Big(\bigvee_{n \geq 0} U^n X\Big) \ominus X \\ \mathcal{D}_- &= \Big(\bigvee_{n \leq 0} U^n X\Big) \ominus X \quad . \end{aligned} \tag{5.5}$$

We first show

$$U\mathcal{D}_+ \subset \mathcal{D}_+ \quad . \tag{5.6}$$

Note that

$$\mathcal{D}_+ = \bigvee_{n \geq 0} (I_{H_{\underline{y}}} - P_X) U^n X \quad .$$

For arbitrary $x, x' \in X$ and $n \geq 0$

$$\begin{aligned} (U(I_{H_{\underline{y}}} - P_X)U^n x, x') &= (U^{n+1}x, x') - (UP_X U^n x, x') = \\ &= (A^{n+1}x, x') - (UA^n x, x') = (A^{n+1}x, x') - (AA^n x, x') = 0 \quad . \end{aligned}$$

It follows that

$$U\mathcal{D}_+ \perp X \quad ,$$

and since $U\mathcal{D}_+ \subset \bigvee_{n \geq 0} U^n X$ we conclude (5.6). Similarly, we obtain $U^* \mathcal{D}_- \subset \mathcal{D}_-$.

To prove that $\mathcal{D}_- \perp \mathcal{D}_+$ we note that it suffices to prove

$$\mathcal{D}_+ \perp \bigvee_{m\leq 0} U^m X \ .$$

Since for arbitrary $x,x'\in X$, and $n,m \geq 0$

$$((I_{H_{\underline{y}}} - P_X)U^n x, U^{-m}x') = (U^n x, U^{-m}x') - (P_X U^n x, U^{-m}x') =$$

$$= (A^{m+n}x,x') - (A^m A^n x,x') = 0$$

the conclusion follows. To prove $\bigwedge_{-\infty}^{\infty} U^n\mathcal{D}_{\pm} = \{0\}$ observe that

$$\bigwedge_{-\infty}^{\infty} U^n X = H_{\underline{y}} \iff \bigwedge_{-\infty}^{\infty} U^n X^{\perp} = \{0\} \iff \bigwedge_{-\infty}^{\infty} U^n(\mathcal{D}_- \oplus \mathcal{D}_+) = \{0\}$$

$$\iff (\bigwedge_{\infty}^{\infty} U^n\mathcal{D}_-) \oplus (\bigwedge_{\infty}^{\infty} U^n\mathcal{D}_+) = \{0\} \ .$$

By regularity, we obtain that on the space $K_+ = \bigvee_{n\geq 0} U^n X$ the operator $U_+ = U|K_+$ is an isometry having no unitary part (Wold's decomposition [16, Th. I.1.1]). Moreover, from the structure of the space of the minimal isometric dilation (of A) [16, Section II.2], we obtain

$$K_+ = \sum_{n=0}^{\infty} \oplus U^{n+1}\mathcal{N}_- \qquad\qquad \mathcal{N}_- = \mathcal{D}_- \ominus U^*\mathcal{D}_- \ .$$

Since X is of full range, we have $H_{\underline{y}} = \sum_{-\infty}^{\infty} \oplus U^n\mathcal{N}_-$ and

$$\bigvee_{-\infty}^{\infty} U^n\mathcal{D}_- = H_{\underline{y}} \ . \tag{5.7$_-$}$$

In a similar way,

$$\bigvee_{-\infty}^{\infty} U^n\mathcal{D}_+ = H_{\underline{y}} \ . \tag{5.7$_+$}$$

We have therefore established that $(U,H_{\underline{y}},\mathcal{D}_+,\mathcal{D}_-)$ is a scattering system. To prove uniqueness, note that if

$$H_{\underline{y}} = \mathcal{D}_-' \oplus X \oplus \mathcal{D}_+'$$

is another decomposition, $UX \perp U\mathcal{D}'_- \supset \mathcal{D}'_-$ and, thus $UX \subset X \oplus \mathcal{D}'_+$. It follows $\mathcal{D}_+ \subset X \oplus \mathcal{D}'_+$, since $\mathcal{D}_+ \perp X$ we obtain

$$\mathcal{D}_+ \subset \mathcal{D}'_+ \quad .$$

Similarly,

$$\mathcal{D}_- \subset \mathcal{D}'_- \quad .$$

However, $\mathcal{D}_- \oplus \mathcal{D}_+ = \mathcal{D}'_- \oplus \mathcal{D}'_+$ and uniqueness follows. ■

5.5 COROLLARY. A full range Markov subspace is regular iff[†]

$$A^n \to 0 \quad , \quad A^{*n} \to 0 \qquad\qquad (n \to \infty)$$

strongly.

PROOF. From the proof of the last theorem it follows that X is regular iff (5.7) holds. Combining with [16, Th.II.1.1] gives the desired result. ■

5.6 DEFINITION. A scattering system (U,H,D_+,D_-) for which

$$\mathcal{D}_- \perp \mathcal{D}_+$$

is called a Lax-Phillips (L-P) scattering system.

Let $\{U^n X\}_{-\infty}^{\infty}$ be an arbitrary regular Markov process of full range, and $(U,H_{\underline{y}},\mathcal{D}_+,\mathcal{D}_-)_X$ its associated L-P scattering system. Let $\Theta_X(e^{i\lambda})$ be the corresponding scattering matrix. For the corresponding incoming spectral representation F_X^-, we obtain as in Proposition 4.1

$$F_X^-[\mathcal{D}_-] = H_2^-(C^p) \quad , \quad F_X^-[\mathcal{D}_+] = \Theta_X H_2(C^p) \quad .$$

Since $\mathcal{D}_+ \perp \mathcal{D}_-$

$$\Theta_X \in H_\infty(\mathcal{B}(C^p)) \quad .$$

To each regular full range Markov process there is thus associated an inner function Θ_X, which is the scattering matrix of the corresponding L-P

[†]For a finite dimensional X both convergences are equivalent.

system $(U,H_{\underline{y}},\mathcal{D}_+,\mathcal{D}_-)_X$. According to Proposition 5.2 and Theorem 5.4, characterizing those inner functions amounts to characterizing all those L-P systems $(U,H,\mathcal{D}_+,\mathcal{D}_-)$ for which U on H is the minimal unitary dilation of the contraction $A = P_X U|X$, $X = H \ominus (\mathcal{D}_- \oplus \mathcal{D}_+)$. By [14, Th. III.1.1], U on H is a dilation, and combining with [1, Th.3.3], we obtain

5.7 PROPOSITION. The scattering matrices Θ associated with regular full range Markov processes are precisely the inner functions $\Theta \in H_\infty(\mathcal{B}(\mathcal{C}^p))$ for which

$$\|\Theta(0)\| < 1 \quad . \tag{5.8}$$

An inner function $\Theta \in H_\infty(\mathcal{B}(\mathcal{C}^p))$ satisfying (5.8) is called purely contractive [16, p. 188]. For p = 1, this amounts to being non-trivial.

Recall that the scattering matrix was defined up to left and right multiplication by constant unitary matrices on $\mathcal{C}^p$. This follows from the arbitrariness in choice of orthonormal bases in the forward and backward innovation subspaces, and cannot be avoided. We make the following

5.8 DEFINITION [16, p. 132]. $\Theta \in H_\infty(\mathcal{B}(\mathcal{C}^p))$, and $\Theta_1 \in H_\infty(\mathcal{B}(\mathcal{C}^p))$ are said to coincide if for unitary maps τ_1,τ_2 on $\mathcal{C}^p$

$$\tau_2\Theta_1(z)\tau_1 = \Theta(z) \qquad |z| < 1 \quad .$$

The equivalence relations between Markov processes is obtained by

5.9 DEFINITION. We say that the Markov processes $\{U^n X\}_{-\infty}^{\infty}$, $\{U^n X_1\}_{-\infty}^{\infty}$ are equivalent if Θ_X,Θ_{X_1} coincide.

5.10 THEOREM. All equivalence classes of full range regular Markov processes are parametrized by the inner functions $\Theta \in H_\infty(\mathcal{B}(\mathcal{C}^p))$ such that $\|\Theta(0)\| < 1$.

PROOF. Immediate from Proposition 5.7.

5.11 PROPOSITION. Let X be a Markov subspace. Let $\{x_1(0),\ldots,x_k(0)\}$ $(k \leq \infty)$ be a complete orthonormal basis in X. We then have

$$\underline{x}(n+1) = [A]\underline{x}(n) + [B]\underline{w}(n) \qquad n \geq 0 \tag{5.9}$$

where

(i) $\{\underline{w}(n)\}_{-\infty}^{\infty}$ is a normalized[†] white noise vector process of dimension p.

(ii) $\{\underline{w}(n)\}_{n \geq 0}$ is orthogonal to $\underline{x}(0)$.

The above representation determines [A] (the matrix representation of the state operator A) and [B] up to unitary equivalence.

PROOF. Let $(U, H_{\underline{y}}, \mathcal{D}_+, \mathcal{D}_-)_X$ be the corresponding scattering system. Write $\mathcal{D}_+ = \sum_0^{\infty} \oplus U^n N_+$, N_+ the backward innovation subspace at t = 0 for $\mathcal{D}_+$, dim $N_+ = p$. Let the entries of $\underline{w}(0) = \begin{pmatrix} w_1(0) \\ \vdots \\ w_p(0) \end{pmatrix}$ be an orthonormal basis in N_+ with $\underline{w}(n) = U^n \underline{w}(0)$. Since $U^n N_+ \perp N_+$ $(n \neq 0)$ $\{\underline{w}(n)\}_{-\infty}^{\infty}$ is a normalized white noise process, and $X \perp \mathcal{D}_+$ implies that $\{\underline{w}(n)\}_{n \geq 0}$ is orthogonal to $\underline{x}(0)$. From (5.5)

$$UX \subset X \oplus N_+$$

and thus

$$Ux_j(0) = P_X U x_j(0) + P_{N_+} U x_j(0) = A x_j(0) + P_{N_+} U x_j(0) \quad . \tag{5.10}$$

Let

$$\alpha_{ij} = (A x_i(0), x_j(0))_{H_{\underline{y}}} \quad , \quad b_{ij} = (U x_i(0), w_j(0))_{H_{\underline{y}}} \quad ,$$

and define

$$[A] = (\alpha_{ij}) \quad , \quad [B] = (b_{ij}) \quad .$$

From (5.10)

$$\underline{x}(1) = [A]\underline{x}(0) + [B]\underline{w}(0) \quad ,$$

applying U to both sides gives

[†] Its covariance matrix equals $\delta(n) I_{\mathcal{C}^p}$.

$$\underline{x}(n+1) = [A]\underline{x}(n) + [B]\underline{w}(n) \quad .$$

Since the above representation is unique up to a choice of orthonormal bases in X, N_+ the theorem follows. ■

A regular Markov subspace $X \subset H_{\underline{y}}$ is said to be a representation for the process $\underline{y}$ if

$$\{y_1(0),\ldots,y_p(0)\} \subset X \quad .$$

Indeed, in this case we can write (notation as in Proposition 5.11)

$$\underline{y}(0) = [C]\underline{x}(0)$$

for some matrix [C], and applying U to both sides gives

$$\underline{y}(n) = [C]\underline{x}(n) \quad , \tag{5.11}$$

a dynamical representation for the regular process $\underline{y}$ in terms of the regular (and necessarily full range) Markov process $\underline{x}$.

While the correspondence

$$X \leftrightarrow (U, H_{\underline{y}}, \mathcal{D}_+, \mathcal{D}_-)$$

is 1-1, the same does not hold for Θ_X; two distinct Markov processes may have coinciding scattering matrices. However, by a fundamental result of Nagy-Foiaş [16, Th.VI.3.4], it then follows that the corresponding state operators are unitarily equivalent, and hence the Markov processes have an identical dynamical representation (5.9). Since in our context two processes having the same density are equivalent (indistinguishable), it readily follows that if X is a (regular) Markov subspace representing $\underline{y}$, then an equivalent Markov process to $\{U^n X\}_{-\infty}^{\infty}$ yields an identical representation (5.11) for a process equivalent to $\underline{y}$. Thus an equivalence class of Markov processes containing $\{U^n X\}_{-\infty}^{\infty}$ represents $\underline{y}$ and the corresponding equivalent process to $\underline{y}$. We shall subsequently refer to this equivalence class of Markov processes as representing $\underline{y}$.

We now consider the problem of finding all regular Markovian representations for $\underline{y}$ by translating the problem to the spectral representation in $L_2(\mathcal{C}^p)$. For $\Theta \in H_\infty(\mathcal{B}(\mathcal{C}^p))$ as in Theorem 5.10, we consider the orthogonal decomposition

$$L_2(\mathcal{C}^p) = \Theta^* H_2^-(\mathcal{C}^p) \oplus (H_2^-(\mathcal{C}^p) \ominus \Theta^* H_2^-(\mathcal{C}^p)) \oplus H_2(\mathcal{C}^p) \tag{5.12}$$

inducing the L-P scattering system

$$\left(\chi, L_2(\mathcal{C}^p) \quad , \quad H_2(\mathcal{C}^p) \quad , \quad \Theta^* H_2^-(\mathcal{C}^p)\right) \quad ,$$

which corresponds to the full range regular Markov subspace (with respect to χ)

$$X_\Theta = H_2^-(\mathcal{C}^p) \ominus \Theta^* H_2^-(\mathcal{C}^p) \quad .$$

Let $X \subset H_{\underline{y}}$ be a regular Markov subspace representing $\underline{y}$ for which Θ_X coincides with Θ, $(U, H_{\underline{y}}, \mathcal{D}_+, \mathcal{D}_-)_X$ its L-P scattering system, and F_X^+ the corresponding outgoing spectral representation. Since $\{y_1(0), \dots, y_p(0)\} \subset X$ it follows

$$F_X^+[H_{\underline{y}}^-(0)] \subset F_X^+\left[\bigvee_{n \le 0} U^n X\right] = F_X^+[\mathcal{D}_- \oplus X] = H_2^-(\mathcal{C}^p) \quad . \tag{5.13}$$

Thus, $F_X^+[H_{\underline{y}}(0)]$ is a full range left shift invariant subspace of $H_2^-(\mathcal{C}^p)$. Let V be the corresponding inner function obtained from the Beurling-Lax theorem, i.e.,

$$F_X^+[H_{\underline{y}}^-(0)] = V^* H_2^-(\mathcal{C}^p) \quad .$$

By the unitarity of F_X^+ we readily conclude

$$F_X^+(\underline{y}(0)) = \bar{\chi} V^* \Lambda \quad . \tag{5.14}$$

Thus

$$\{y_1(0), \dots, y_p(0)\} \subset X$$

translates under F_X^+ to the equivalent condition

$$\bar{\chi}V^*\Lambda \in H_2^-(\mathcal{C}^p) \ominus \Theta^* H_2^-(\mathcal{C}^p) \quad . \tag{5.15}$$

Conversely, if Θ, V are inner functions for which (5.15) holds, the mapping

$$\underline{y}(0) \longmapsto \bar{\chi}V^*\Lambda$$

induces in a natural fashion a spectral representation $\mathcal{F}_{\Theta,V}$ for which

$$X = \mathcal{F}_{\Theta,V}^{-1}\,[H_2^-(\mathcal{C}^p) \ominus \Theta^* H_2^-(\mathcal{C}^p)]$$

is a Markovian representation for $\underline{y}$, and its corresponding L-P scattering system has as its scattering matrix Θ_X coinciding with Θ. Thus, finding all regular Markovian representations for $\underline{y}$ reduces to the following:

Find all inner functions Θ_1 such that[†]

$$\bar{\chi}V^*\Lambda \in H_2^-(\mathcal{C}^p) \ominus \Theta_1^* H_2^-(\mathcal{C}^p) \tag{5.16}$$

for some inner function V.

5.12 THEOREM. All regular Markovian representations $\underline{y}$ are parametrized by precisely those inner functions Θ_1 for which

$$V^*S = \Theta_1^*\Theta_2 \qquad \Theta_2 \in H_\infty(\mathcal{B}(\mathcal{C}^p)) \quad ^{\dagger\dagger} \tag{5.17}$$

for some inner function V.

PROOF. (5.16) holds iff $\Theta_1\bar{\chi}\Lambda \in H_2(\mathcal{C}^p)$ iff $\Theta_1 V^* S\Gamma \in H_2(\mathcal{C}^p)$. Since Γ is outer the latter holds iff $\Theta_1 V^* S \in H_\infty(\mathcal{B}(\mathcal{C}^p))$, i.e., iff (5.17) holds. ∎

If $X \subset H_{\underline{y}}^-(0)$, then $\mathcal{D}_+ \oplus X = \bigvee_{n\leq 0} U^n X \subset H_{\underline{y}}^-(0)$ and combining with (5.13) we obtain that V is a constant unitary matrix. We thus have

5.13 COROLLARY. All regular Markov subspaces $X \subset H_{\underline{y}}^-(0)$ representing $\underline{y}$ are parametrized by those and only those inner functions Θ_1 for which

$$S = \Theta_1^*\Theta_2 \qquad \Theta_2 \in H_\infty(\mathcal{B}(\mathcal{C}^p)) \quad . \tag{5.18}$$

---

† Since Λ is conjugate outer the corresponding Markov subspace will automatically be of full range.

†† Θ_2 is necessarily inner since S is unitary valued on T.

The possibility of writing the scattering matrix S in the form (5.18) has an interpretation on a process level. By the Beurling-Lax theorem, (5.18) holds iff (the invariant subspace for the left shift):

$$H_2^-(C^p) \ominus (\text{range } H_S) \quad \text{is of full range (for } X\text{).} \tag{5.19}$$

By Lemma 4.2 this is equivalent to

$$H_{\underline{y}}^-(0) \ominus P_{H_{\underline{y}}^-(0)} H_{\underline{y}}^+(0) \quad \text{is of full range (for U).} \tag{5.20}$$

An $L_\infty(B(C^p))$ function satisfying (5.19) is called strictly non-cyclic [10]. A process $\underline{y}$ satisfying (5.20) (i.e., having a strictly non-cyclic scattering matrix) is called strictly non-cyclic.

Recall that two inner functions $U,V \in H_\infty(B(C^p))$ are left coprime if for no non-trivial [i.e., for no unitary matrix in $B(C^p)$] inner function W we have

$$U = WU_1 \quad , \quad V_1 = WV_1$$

with U_1, V_1 inner. Similarly for U,V being right coprime. Let

$$S = Q_1^* Q_2$$

be the (left) coprime factorization for S. For the inner function Q_1 we have by the Beurling-Lax theorem

$$\overline{\text{range } H_S} = H_2^-(C^p) \ominus Q_1^* H_2^-(C^p) \quad .$$

If $S = U_1^* U_2$ is some other factorization, then we necessarily have for some (non-trivial) inner function W

$$U_1 = WQ_1 \quad , \quad U_2 = WQ_2 \quad .$$

It follows that for the (regular full range) Markov subspaces X_{Q_1}, X_{U_1} we have

$$X_{Q_1} \subset X_{U_1} \quad .$$

Thus, among all those regular Markov subspaces $X \subset H_{\underline{y}}^{-}(0)$ representing $\underline{y}$, X_{Q_1} is the smallest (setwise). Let us call a regular Markov subspace $X \subset H_{\underline{y}}$ representing $\underline{y}$ minimal if no proper subspace of $X^{\dagger}$ is a Markovian representation of $\underline{y}$. We have thus established the following:

5.14 PROPOSITION. Let $\underline{y}$ be strictly non-cyclic, and

$$S = Q_1^* Q_2$$

the (left) coprime factorization of S. Then (the equivalence class of) all regular minimal Markov subspaces $X \subset H_{\underline{y}}^{-}(0)$ representing $\underline{y}$ are parametrized by Q_1. This class is represented in $L_2(C^p)$ by the subspace

$$X_{Q_1} = H_2^-(C^p) \ominus Q_1^* H_2^-(C^p) = \overline{\text{range } H_S} \quad .$$

In a similar fashion, one obtains that all factorizations

$$S = U_2 U_1^*$$

parametrize via the inner function U_1 all regular Markov subspaces $X \subset H_{\underline{y}}^{+}$ representing $\underline{y}$. The minimal one is obtained by the (right) coprime factorization $S = P_2 P_1^*$ and its representation (under the incoming spectral representation) in $L_2(C^p)$ is $X_{P_1} = H_2(C^p) \ominus P_1 H_2(C^p)$.

As for the general case, when X is not necessarily contained in $H_{\underline{y}}^{\pm}(0)$, we start with some preliminaries. For a purely contractive inner function Θ, we denote by A_Θ the representation of the state operator A under the incoming spectral representation F_X^- : $A_\Theta = F_X^- A$. A_Θ (the model operator, c.f. [16]) acts on the model space according to:

$$(A_\Theta^* f)(z) = \frac{1}{z}\,[f(z) - f(0)] \qquad f \in H_2(C^p) \ominus \Theta H_2(C^p) \quad .$$

Let M_1, M_2 be left coprime and N_1, N_2 right coprime inner functions. We further assume that M_1, N_1 are purely contractive and, moreover,

---

†Which is necessarily regular.

$$M_1^* M_2 = N_2 N_1^* \quad .$$

By [16, Ths. VI.3.1,3.6] and [10, Th.II.14-11] it follows that the contractions A_{M_1}, A_{N_1} are quasi-similar, i.e., there exists an injective bounded linear transformation having dense range, from the model space of A_{M_1} into that of A_{N_1} intertwining A_{M_1}, A_{N_1} (cf. [16, p. 70]). It follows [16, p. 370] that A_{M_1}, A_{N_1} have the same Jordan operator and in particular

$$\det M_1 = \det N_1 \quad .$$

5.15 THEOREM. Let

$$S = Q_1^* Q_2 = P_2 P_1^*$$

be respectively the left, right coprime factorization of S. Then (the equivalence classes of) all minimal regular Markov subspaces representing $\underline{y}$ are parametrized by those and only those inner functions Θ_1 such that

$$V^* S = \Theta_1^* \Theta_2 \tag{5.21}$$

where Θ_1, Θ_2 are left coprime and V is an arbitrary left divisor of P_2. Moreover, we have

$$\det \Theta_1 = \det Q_1 \quad .$$

PROOF. First assume that V, P_2 are left coprime. It follows (c.f. [10, Lemma III.5-8])

$$\det \Theta_1 = \det P_1 \det V$$

and since $\det P_1 = \det Q_1$ we have

$$\det \Theta_1 = \det Q_1 \det V \quad . \tag{5.22}$$

However,

$$V^* S = V^* Q_1^* Q_2 = (Q_1 V)^* Q_2 \quad ,$$

combining with (5.22) we conclude

$$Q_1 V = \Theta_1 .$$

It follows

$$V^* X_{Q_1} = V^* H_2^-(C^p) \ominus V^* Q_1^* H_2^-(C^p) \subset X_{\Theta_1}$$

and, moreover, by (5.14) we infer that X_{Θ_1} properly contains a regular Markov subspace representing $\underline{y}$. Thus X_{Θ_1} is not minimal. On the other hand, if C is the greatest left inner factor for V, P_2, i.e., $V = CV_1$, $P_2 = CP_3$ then

$$V^* S = V_1^* P_3 P_1^* .$$

Let M_1, M_2 be left coprime with $P_3 P_1^* = M_1^* M_2$. It follows

$$V^* S = V_1^* P_3 P_1^* = (M_1 V_1)^* M_2$$

and, moreover, $M_1 V_1 = \Theta_1$. Since $C^* S = M_1^* M_2$ by Theorem 5.12 X_{M_1} is a regular Markov subspace representing $\underline{y}$

$$F^+_{M_1,C}(\underline{y}(0)) \in X_{M_1} .$$

Since $M_1 V_1 = \Theta_1$ we have

$$F^+_{\Theta_1,V}(\underline{y}(0)) = V_1^* F^+_{M_1,C}(y(0)) \in V_1^* X_{M_1} \subset X_{\Theta_1} .$$

We have thus established that the minimal class is a subclass of all those Markov subspaces parametrized by the Θ_1's for which V is a left divisor of P_2.

Now assume that V is a left divisor of P_2

$$P_2 = VP_3 .$$

Thus $V^* S = P_3 P_1^*$ where P_1, P_3 are right coprime. It follows

$$\det \Theta_1 = \det P_1 = \det Q_1 .$$

If X_{Θ_1} is not minimal, there exists a regular Markov subspace properly contained in X_{Θ_1} and containing $F^+_{\Theta_1,V}(\underline{y}(0))$. Thus, for some (purely contractive) inner function R_1 we have

$$\Theta_1 = WR_1 \qquad\qquad W \text{ non-trivial} \quad . \tag{5.23}$$

Since $F^+_{\Theta_1,V}(y(0)) \in X_{R_1}$ it follows from Theorem 5.12

$$V^* S = R_1^* R_2 \quad .$$

Since (5.21) is the coprime factorization of $V^* S$, we have that det θ_1 divides det R_1. This contradicts (5.23) and completes the proof. ■

Note that the general degree theory for strictly non-cyclic functions [10, Ch.III.5], upon which we have drawn considerably, arises naturally in our context. All regular Markovian subspaces $X \subset H_{\underline{y}}$ representing $\underline{y}$ (which are in the equivalence class) parametrized by the (purely contractive) inner function $\Theta \in H_\infty(\mathcal{B}(\mathcal{C}^p))$ will be of degree

$$d(\Theta) = \det \Theta \quad ,$$

an inner function in H_∞ (its scalar multiple, see [16, p. 216]). From the proof of the last theorem, we infer that for any (equivalence class of) regular Markov subspaces representing $\underline{y}$ parametrized by the (purely contractive) inner function Θ, we have

$$d(Q_1) \text{ divides } d(\Theta) \quad .$$

Thus, the degree of the minimal equivalence class is the lowest, in the sense that $d(Q_1)$ is the weakest (i.e., it is an inner divisor) among the degrees of all other classes of regular Markovian subspaces representing $\underline{y}$.

5.16 COROLLARY. Let X,X' be minimal regular Markov subspaces representing $\underline{y}$ and A,A' their corresponding state operators:

$$A = P_X U|X \quad , \quad A' = P_{X'} U|X' \ .$$

Then A,A' are quasi-similar.

PROOF. Let Θ_1 be the scattering matrix for X, and V the corresponding left divisor of $P_2 : P_2 = VP_3$. Thus

$$V^* S = P_3 P_1^* = \Theta_1^* \Theta_2$$

where P_3, P_1 are right coprime and Θ_1, Θ_2 left coprime. Since A_{P_1}, A_{Θ_1} are quasi-similar and A, A_{Θ_1} are unitarily equivalent, we have that A and A_{P_1} are quasi-similar. Since quasi-similarity is transitive, the result follows. ∎

5.17 COROLLARY. For a p = 1 dimensional process y the equivalence class of minimal Markovian representations of y is parametrized by the inner function q_1 for which

$$s = \bar{q}_1 q_2$$

is a coprime factorization.

REFERENCES

1. Adamjan, V.M. and Arov, D.Z. Amer. Math. Soc. Transl. 95: 17-129, 1970.
2. Adamjan, V.M., Arov, D.Z., and Krein, M.G. Amer. Math. Soc. Transl. 3: 133-156, 1978.
3. Adamjan, V.M., Arov, D.Z., and Krein, M.G. Funktsional'nyl Analiz i Ego Prilozheniya 2: 1-18, 1968.
4. Adamjan, V.M., Arov, D.Z., and Krein, M.G. Mat. Sb. 86: 34-75, 1971; Math. U.S.S.R. Sb.: 31-73, 1971.
5. Avniel, Y. Relization and Approximation of Stationary Stochastic Processes. Report LIDS-TH-1440, Laboratory for Information and Decision Systems, MIT, Cambridge, MA, Feb. 1985.
6. Bloomfield, P.B., Jewell, N.P., and Hayashi, E. Pacific J. of Math 107: 307-317, 1983.
7. Douglas, R.G. Banach Algebra Techniques in Operator Theory. Academic Press, New York, 1972.
8. Foiaş, C. and Frazho, E. Acta Sci. Math. 45: 165-175, 1983.
9. Fourès, Y. and Segal, I.E. Trans. Am. Math. Soc. 78: 385-407, 1955.
10. Fuhrmann, P.A. Linear Systems and Operatorsin Hilbert Space, McGraw-Hill, New York, 1981.
11. Helson, H., Lectures on Invariant Subspaces. Academic Press, New York, 1964.
12. Helton, J.W. J. Funct. Anal. 16: 15-38, 1974.
13. Hoffman, K. Banach Spaces of Analytic Functions. Prentice-Hall, Englewood Cliffs, 1962.
14. Lax, P.D. and Phillips, R.S. Scattering Theory. Academic Press, New York, 1967.

15. Lindquist, A. and Picci, G. Stochastic Systems: The Mathematics of Filtering and Identification and Applications, (Eds. M. Hazewinkel and J.C. Willems), Reidel, 1981, pp. 169-204.
16. Sz-Nagy, B. and Foiaş, C. Harmonic Analysis of Operators on Hilbert Space. North-Holland, Amsterdam, 1970.
17. Nehari, Z. Annals of Math. 65: 153-162, 1957.
18. Rozanov, Y.A. Stationary Random Processes. Holden-Day, San Francisco, 1963.
19. Ruckebusch, G. Ann. Inst. Henri Poincaré 16: 225-297, 1980.
20. Sarason, D.E. Trans. Amer. Math. Soc. 127: 179-203, 1967.

8

REALIZATION AND REDUCTION OF S.I.S.O. NONMINIMUM PHASE STOCHASTIC SYSTEMS

JITENDRA K. TUGNAIT

Exxon Production Research Company, P.O. Box 2189, Houston, TX 77252-2189

ABSTRACT

We consider the order reduction problem for single-input/ single-output (SISO) nonminimum phase linear stochastic systems. First, the realization of SISO state-space models for stationary processes generated by time-invariant finite-order, possibly nonminimum phase, linear systems is considered. It is shown that in order to obtain a "correct" phase realization, it is necessary to consider some higher-order statistics of the process in addition to the usual second-order statistics. A model in the controllable canonical form is obtained by utilizing a partial set of second- and fourth- order cumulant functions of the process. Next, model approximation by order reduction is defined in this framework. An example is presented to illustrate the proposed method.

INTRODUCTION

Approximation of high-order linear systems by lower-order models for the purposes of analysis, design and simulation is of considerable interest in several applications involving signal processing, estimation/filtering, and control. The order reduction problem has received considerable attention, both for stochastic systems [1]-[7] as well as deterministic systems [8]-[10]. This chapter is concerned with approximation of single-input/ single-output (SISO) discrete stochastic systems.

Current approaches to stochastic model reduction ([1]-[7] and references therein) may be classified into two categories: realization approach and transformation approach [3]. In the realization approach [2],[3],[5], [6] it is desired to obtain a low-order (approximate) stochastic realization, given some statistics of the original model output; the original model is not necessarily available. In the transformation approach [7] the original model is assumed to be given. This distinction is quite important in signal estimation and identification problems where only noisy outout data may be available. In this case the statistics of the noisy output can be calculated without explicitly knowing the original model (by appealing to ergodicity and/or by exploiting multiple data records as in [2]).

Current approaches to stochastic model reduction (as well as to stochastic realization) are restrictive in that they choose to exploit only the second-order statistics of the process. The main drawback of using only the second-order statistics is the fact that the the autocorrelation function is phase-insensitive, i.e., it is a function of only the magnitude of the system transfer function. Thus, essentially one is choosing a minimum-phase model for the stochastic process. In certain applications such as signal reconstruction/ deconvolution [11],[15],[13], [19], the signal/system phase cannot be simply assumed to be minimum phase since an incorrect phase assumption can have a deleterious effect on the final result. The system phase is also important in control system analysis and design [12] where neglect of nonminimum phase zeros can result in poor performance.

This chapter is concerned with the realization approach. We choose a fixed-order model to approximate some statistical properties of the original system output. That is, we view the problem as that of approximating a stochastic process by another. Our proposed measure of approximation is sensitive to both the power spectrum and the phase of the original system/process.

The chapter is organized as follows. First we define the class of models of interest in this chapter. Then we briefly review some definitions and facts concerning the higher order statistics of a stochastic process. Next we propose a new approximation criterion for the realization of SISO state-space models which are not necessarily minimum-phase. The criterion relies on simultaneously approximating the autocorrelation function and the fourth-order cumulant function of the original process; the latter is a function of the system phase also. Model approximation by order reduction is defined in this framework. The computational details are discussed in some detail. Finally, a numerical example is considered to illustrate the proposed approach. Some mathematical details are provided in the Appendix.

MODEL ASSUMPTIONS

Let $\{y(t)\}$ be discrete-time, scalar, stationary stochastic process that is generated by the following p-th order state-space model

$$x(t+1) = Ax(t) + bw(t) \tag{1}$$

$$y(t) = cx(t) + v(t), \quad t = 1,2,3.... \tag{2}$$

where $x(t)$ is $p \times 1$, and $w(t)$, $y(t)$ and $v(t)$ are scalars. The constant matrices A, b, and c are of appropriate dimensions. The processes $\{w(t)\}$ and $\{v(t)\}$ are zero-mean, mutually independent, i.i.d. (independent and identically distributed) sequences with variances q and r, respectively.

We impose the following restrictions on the model (1) and (2).

(C1) The constant matrix A has all its eigenvalues inside the unit circle.

(C2) The driving noise $\{w(t)\}$ is non-Gaussian such that the fourth-order cumulant γ_4 of the random variable $w(t)$ is nonzero where

$$\gamma_4 = E\{w^4(t)\} - 3[E\{w^2(t)\}]^2$$

(C3) The measurement noise $\{v(t)\}$ is Gaussian.

The model (1)-(2) along with the restrictions (C1)-(C3) defines the original stochastic process $\{y(t)\}$. It is desired to determine a reduced-order state-space model of dimension $m < p$ whose stationary output is a good approximation to the stochastic process $\{y(t)\}$. Let the reduced-order model be given by

$$\bar{x}(t+1) = \bar{A}\bar{x}(t) + \bar{b}\bar{w}(t) \tag{3}$$

$$\bar{y}(t) = \bar{c}\,\bar{x}(t) + \bar{v}(t) \tag{4}$$

where $\bar{x}(t)$ is $m\times 1$ and $\bar{w}(t)$, $\bar{y}(t)$ and $\bar{v}(t)$ are scalars. As in the original model, the processes $\{\bar{w}(t)\}$ and $\{\bar{v}(t)\}$ are assumed to be zero-mean, mutually independent i.i.d. sequences with variances $\bar{q}$ and $\bar{r}$, respectively. It is also assumed that the conditions (C1)-(C3) also hold for the reduced-order model (3)-(4), i.e., $\bar{A}$ is exponentially stable, the fourth-order cumulant $\bar{\gamma}_4$ of $\bar{w}(t)$ is nonzero, and $\bar{v}(t)$ is Gaussian.

Since an infinite number of triples $(\bar{A},\bar{b},\bar{c})$ lead to identical input-output properties, without loss of any generality we will restrict the triple $(\bar{A},\bar{b},\bar{c})$ to controllable canonical form [14, p. 343] as follows

$$\bar{A}(\theta) = \begin{bmatrix} 0 & 1 & \cdots & 0 \\ 0 & 0 & \cdots & 0 \\ \cdot & \cdot & & \cdot \\ \cdot & \cdot & & \cdot \\ 0 & 0 & \cdots & 1 \\ -a_m(\theta) & -a_{m-1}(\theta) & \cdots & -a_1(\theta) \end{bmatrix}$$

$$\bar{b} = [0 \;\; 0 \;\; \cdot \;\; \cdot \;\; 0 \;\; 1]^T$$

$$\bar{c}(\theta) = [h_{m-1}(\theta)\; h_{m-2}(\theta) \cdot\cdot\; h_0(\theta)]^T, \qquad h_0(\theta) = 1 \tag{5}$$

where

$$\theta = (a_1,a_2,\cdots,a_m,h_1,h_2,\cdots,h_{m-1},\bar{q},\bar{r},\bar{\gamma}_4)$$

is the vector of free parameters of the reduced-order model and we have $a_i(\theta) = a_i$, $h_i(\theta) = h_i$, etc. Let Θ denote the parameter set consistent with the model assumptions. This implies that

$$\Theta \subset R^{2m+2} \cap D_S$$

where R is the real line and

$D_S = \{\theta$: eigenvalues of $\bar{A}(\theta)$ lie inside the unit circle $\}$.

The problem then is to choose θ so that the stochastic process $\{\bar{y}(t)\}$ is "close" to $\{y(t)\}$.

Notice that the original and the reduced models are not restricted to be minimum phase. In the next section we briefly review the necessary background needed to handle the phase problem.

Remark 1. We note that for random variables with symmetric distributions the third-order cumulant is zero. The fourth-order cumulant γ_4 is nonzero for a large class of random variables including Bernoulli-Gaussian variate [11], sub- and super- Gaussian variates of [13], and uniformly distributed random variables. Our approach goes through so long as $\gamma_k \neq 0$ for some $k>2$ which is true for any non-Gaussian random variable with all bounded moments. □

PRELIMINARIES

In this section we briefly review some definitions and facts concerning the cumulant functions of a scalar stochastic process. Another objective is to introduce the notation used in the sequel. Finally, Lemma 1 provides the motivation and the justification for the proposed approach to realization and to approximate realization of nonminimum phase stochastic systems.

Given a real random variable x, its cumulants are defined by the following identity in ν [26, pp. 69-74]

$$\phi(\nu) = E\{\exp(j\nu x)\} = \exp[\sum_{k=0}^{\infty} \gamma_k (j\nu)^k / k!] \tag{6}$$

where $\phi(\nu)$ is the characteristic function of x, $j=\sqrt{-1}$, γ_k is the k-th cumulant of x, and γ_k exists if k-th and lower central moments of x exist. For real random k-vector $\{y(t_1),y(t_2),\cdots,y(t_k)\}$ with $E\{|y(t_i)|^k\} < \infty$ $(i=1,2,\cdots,k)$, we define $\mathrm{cum}_k(y(t_1),y(t_2),\cdots,y(t_k))$, the joint cumulant of $\{y(t_1),y(t_2),\cdots,y(t_k)\}$, to be the coefficient of the product $\nu_1\nu_2\cdots\nu_k$ in the Taylor series expansion of the cumulant generating function

$$\ln E\{\exp[j\sum_{i=1}^{k} \nu_i y(t_i)]\} \tag{7}$$

Thus for $k=2$, $\mathrm{cum}_2(y(t_1),y(t_2))$ is the usual covariance function.

For a stationary time-series $\{y(t)\}$ we have for any τ

$$\begin{aligned} &\mathrm{cum}_k(y(t),y(t+t_1),\cdots,y(t+t_{k-1})) \\ &\quad = \mathrm{cum}_k(y(t+\tau),y(t+\tau+t_1),\cdots,y(t+\tau+t_{k-1})) \\ &\quad =: C_k(t_1,t_2,\cdots,t_{k-1}) \end{aligned} \tag{8}$$

For a zero-mean stationary time-series $\{y(t)\}$ it can be shown [16]-[18] that

$$C_4(t_1,t_2,t_3) = R_4(t_1,t_2,t_3) - R_2(t_1)R_2(t_3-t_2)$$

$$-R_2(t_2)R_2(t_3-t_1) - R_2(t_3)R_2(t_2-t_1) \tag{9}$$

where

$$R_4(t_1,t_2,t_3) = E\{y(t)y(t+t_1)y(t+t_2)y(t+t_3)\}$$

$$R_2(\tau) = E\{y(t)y(t+\tau)\}.$$

Furthermore, for a Gaussian random variable, k-th order cumulant γ_k vanishes for $k > 2$. For a Gaussian time series, the (joint) k-th order cumulant function vanishes for $k > 2$. Therefore, since $\overline{v}(t)$ in (4) is Gaussian, it follows that

$$C_4(t_1,t_2,t_3) = C_{s4}(t_1,t_2,t_3)$$
$$:= \mathrm{cum}_4(y_s(t),y_s(t+t_1),y_s(t+t_2),y_s(t+t_3)) \tag{10}$$

where we define $y_s(t) = \overline{c}\,\overline{x}(t)$.

Take 3-dimensional two-sided z-transform of the fourth-order cumulant function $C_4(t_1,t_2,t_3)$ to obtain

$$S_4(z_1,z_2,z_3) = \sum_{t_1=-\infty}^{\infty} \sum_{t_2=-\infty}^{\infty} \sum_{t_3=-\infty}^{\infty} C_4(t_1,t_2,t_3)\, z_1^{-t_1} z_2^{-t_2} z_3^{-t_3} \tag{11}$$

For the model (3)-(4) for which conditions (C1)-(C3) hold (see also (10)), it can be shown that [16]-[18] that

$$S_4(z_1,z_2,z_3) = \gamma_4 H(z_1)H(z_2)H(z_3)H([z_1z_2z_3]^{-1}) \tag{12}$$

where $H(z)$ is the transfer function of the noise-free model (3)-(4) and is given by

$$H(z) = \frac{\sum_{i=1}^{m} h_{i-1}(\theta)z^{-i}}{1+\sum_{i=1}^{m} a_i(\theta)z^{-i}} \tag{13}$$

Finally, with $z_i = \exp(j\omega_i)$, $(i=1,2,3)$, $j = \sqrt{-1}$, $S_4(z_1,z_2,z_3)$ is called the **trispectrum** of the series $\{\overline{y}(t)\}$.

Lemma 1 is a slight modification of [15, Lemma 1]. It shows that to estimate the system phase, we need a higher-order spectrum (equivalently cumulant function) in addition to the usual power spectrum.

Lemma 1 Consider $y_s(t) = \overline{c}\,\overline{x}(t)$ as specified by (3)-(4). Assume that $H(1) \neq 0$ and that the conditions (C1) and (C2) hold. Then the transfer function $H(z)$ can be determined for $z = \exp(j\omega)$, $-\pi \leqslant \omega < \pi$, from the power spectral density and the trispectrum of $\{y_s(t)\}$.

Proof: Here we will make frequent references to [15, Lemma 1]. Let

$$m(\omega) = |H(\exp(j\omega))| = \text{magnitude of the system transfer function}$$

$$h(\omega) = \arg\{H(\exp(j\omega))H(1)/|H(1)|\}$$

$$= \text{transfer function phase upto } \pm\pi$$

Let $h'(\omega)$ denote the first derivative of $h(\omega)$ w.r.t. ω. Following [15, Lemma 1] it can be shown that

$$h'(0) - h'(\omega) = \lim_{\Delta \to 0}\left(\frac{1}{2\Delta}\right)\{h(\omega)+2h(\Delta)-h(\omega+2\Delta)\} \tag{14}$$

where the terms inside the braces in (14) can be determined for all ω for which $m(\omega)\neq 0$. Now with $\alpha = h'(0)$, we have

$$h(\omega) = \int_0^{\omega}[h'(u)-h'(0)]du + \alpha\omega =: h_1(\omega)+\alpha\omega \tag{15}$$

which can be computed from (14) if $[0,\omega)$ contains no ω for which $m(\omega) = 0$. Similarly, we have $(\omega_2 > \omega_1)$

$$h(\omega_2)-h(\omega_1) = \int_{\omega_1}^{\omega_2}[h'(u)-h'(0)]du + \alpha(\omega_2-\omega_1)$$

$$= h_1(\omega_2) - h_1(\omega_1) + \alpha(\omega_2-\omega_1) \tag{16}$$

for any (ω_1,ω_2) which contains no ω for which $m(\omega) = 0$. For the system model (13) there can be only finitely many zeros on the unit circle $|z|=1$. Therefore, by continuity of $h(\omega)$ in ω, we can piece together the finitely many segments given by (15) and (16) to obtain $h(\omega)$ for all ω $(-\pi \leqslant \omega < \pi)$:

$$h(\omega) = h_1(\omega) + \alpha\omega \tag{17}$$

Thus we have

$$H(\exp(j\omega)) = [H(1)/|H(1)|]m(\omega)\exp(jh(\omega)) \tag{18}$$

where $m(\omega)$ is computed from the power spectral density and $h(\omega)$ from (17). Now (18) determines the transfer function $H(\exp(j\omega))$ upto a sign and a linear phase $\alpha\omega$. However, the assumption $h_0=1$ in (5) implies that the phase of $H(\exp(j\omega))$ can have no term linear in ω. Moreover the restriction $h_0=1$ also resolves the sign ambiguity. Thus, restriction to the transfer function of the type (13) (impicitly) resolves both the linear phase and the sign ambiguity problems. □

STOCHASTIC REALIZATION OF NONMINIMUM PHASE SYSTEMS

Following the usual approach to stochastic realization and to model reduction via approximate stochastic realization, we assume that the original model (1)-(2) is not necessary available; only certain statistics of $\{y(t)\}$ are known. As noted earlier, we need more than the second-order statistics to resolve the phase problem. Therefore, we assume that we are given the sequences $\{R_{2y}(\tau), 0 \geqslant \tau \geqslant -L_1\}$ and $\{C_{4y}(t_1,t_2,t_3), 0 \geqslant t_3 \geqslant t_2 \geqslant t_1 \geqslant -L_2\}$ where $L_1>0$, $L_2>0$, and

$$R_{2y}(\tau) = E\{y(t)y(t-\tau)\} \tag{19}$$

$$C_{4y}(t_1,t_2,t_3) = \text{4th-order cumulant function of } \{y(t)\} \tag{20}$$

We propose to select the parameter vector θ, given the model order m, to minimize the following scalar cost function

$$\min_{\theta \in \Theta} J(\theta) \tag{21}$$

where

$$J(\theta) = J_1(\theta) + \lambda J_2(\theta) \tag{22}$$

$$J_1(\theta) = 0.5 \sum_{\tau=-L_1}^{0} [R_{2y}(\tau) - R_2(\tau|\theta)]^2 \tag{23}$$

$$J_2(\theta) = 0.5 \sum_{t_1=-L_2}^{0} \sum_{t_2=t_1}^{0} \sum_{t_3=t_2}^{0} [C_{4y}(t_1,t_2,t_3) - C_4(t_1,t_2,t_3|\theta)]^2 \tag{24}$$

$$R_2(\tau|\theta) = E\{\bar{y}(t)\bar{y}(t-\tau)|\theta\} \tag{25}$$

$$C_4(t_1,t_2,t_3|\theta) = \text{4th-order cumulant function of } \{\bar{y}(t)\} \text{ parametrized by } \theta \tag{26}$$

$$\lambda = \frac{\sum_{\tau=-L_1}^{0} [R_{2y}(\tau)]^2}{\sum_{t_1=-L_2}^{0} \sum_{t_2=t_1}^{0} \sum_{t_3=t_2}^{0} [C_{4y}(t_1,t_2,t_3)]^2} \tag{27}$$

The choice of the scalar λ as specified by (27) is motivated by the desire to equally penalize errors in matching the original correlation and 4th-order cumulant functions, respectively.

The motivation for proposing (21) as the criterion for realization and approximate realization rests on Lemma 1 which suggests that one has to use higher-order statistics, in addition to the power spectrum of the original process, to capture its phase properties. We now show that if m=p, then the criterion (21) has the global minimum at $\theta = \theta_0$, the true parameter vector, under certain restrictions on the maximum lags L_1 and L_2. This result later serves as the main motivation for the proposed order-reduction approach.

Let $H(z\,|\,\theta)$ denote the transfer function of the reduced order model (3)-(5) with $\bar{v}(t)\equiv 0$. That is, it is given by (13) where we now indicate the dependence on θ explicitly.

Theorem. Consider the reduced-order model (3)-(5). Suppose that the following conditions hold for some $\theta_i \in \Theta$, (i=1,2).

(C4) $R_2(\tau|\theta_1) = R_2(\tau|\theta_2), \quad 0 \geqslant \tau \geqslant -2m$

(C5) $C_4(t_1,t_2,t_3|\theta_1) = C_4(t_1,t_2,t_3|\theta_2), \quad 0 \geqslant t_3 \geqslant t_2 \geqslant t_1 \geqslant -2m$

(C6) $H(z|\theta_i)$ has no pole-zero cancellations for at least one $i \in \{1,2\}$ and $H(1|\theta_i) \neq 0$, (i=1,2).

Then (C4) hold for $-\infty < \tau < \infty$ and (C5) holds for $-\infty < t_i < \infty$, (i=1,2,3).

Corollary 1. The hypotheses of the Theorem also imply that $\theta_1 = \theta_2$.

Corollary 2. Let m=p, and θ_0 be such that $R_2(\tau|\theta_0) \equiv R_{2y}(\tau)$ and $C_4(t_1,t_2,t_3)|\theta_0) \equiv C_{4y}(t_1,t_2,t_3)$. If $L_i \geqslant 2p$ (i=1,2) and (C6) holds for $\theta_i = \theta_0$, then $J(\theta)=0$ implies that $\theta = \theta_0$. That is, the true system yields the global minimum of $J(\theta)$.

Proofs of the above results are given in the Appendix.

Remark 2. If $H(z|\theta)$ has any pole-zero cancellations then the statistics of $\{\bar{y}(t)\}$ (parametrized by θ) remain unaffected by that pole-zero pair. This pair can be chosen arbitrarily (within the system stability restriction) leading to $\theta_1 \neq \theta_2$, in general, but to identical output statistics. □

Remark 3. Since the parametrization for the model (3)-(5) for m>p is contained in that for m=p, it also follows that all global minima of $J(\theta)$ for m>p also lead to $J(\theta) = 0$. In view of Remark 2 any global minimum for m>p will lead to a non-minimal triple $(\bar{A},\bar{b},\bar{c})$. □

Remark 4. Let m be the smallest positive integer such that for some $\theta \in \Theta$ and for all $L_i \geqslant 2m$ (i=1,2) we have $J(\theta)=0$. Then, by Corollaries 1 and 2 and Remark 2, we have m=p=true order, if the original order is minimal. This suggests a method for determining the true system order p from the process statistics. For example, if we are given the functions R_{2y} and C_{4y} up to maximum lags $\bar{L}_i$ (i=1,2), then we only need to check if $J(\theta) = 0$ for some $\theta \in \Theta$ and for $L_i = 2m$ and $L_i = \bar{L}_i$ (i=1,2), respectively. □

APPROXIMATE STOCHASTIC REALIZATION

The results of the previous section show that the proposed criterion yields the true system (more precisely, its canonical equivalent) under certain conditions if m=p. This suggests that use of the criterion (21) is also appropriate for reduced-order realization. Note, however, that now the choice $L_i = 2m$ (i=1,2) is not so well-motivated. In general, one should take L_1 and L_2 as large as possible within the computational constraints and availability of the output statistics. In certain cases it may be possible to take $L_i = 2m$.

To see this, consider (19)-(27). Suppose that $\{y(t)\}$ can also be generated exactly via a reduced order model of order $m<p$. Then we have

$$R_{2y}(\tau) = R_2(\tau|\theta_0)$$

$$C_{4y}(t_1,t_2,t_3) = C_4(t_1,t_2,t_3|\theta_0) \qquad \text{for every } t_i,\ (i=1,2,3)$$

for some $\theta_0 \in \Theta$. Now let $\theta_1 = \theta_0$ in the Theorem. If we can find a θ_2 such that (C4)-(C6) hold, then clearly $J(\theta_2) = J(\theta_0) = 0$. Indeed, by the Cor. 1, we have $\theta_2 = \theta_0$. Heuristically, then, if a reduced order model of order m is capable of approximating the second-order and the fourth-order statistics of the original model output "closely enough," then $L_i \geqslant 2m$ $(i=1,2)$ may lead to a "good" approximate model in the sense of (21).

Stability. It is required that $\bar{A}(\theta)$ satisfy (C1). This constraint is imposed during optimization of $J(\theta)$ by confining the search to the set D_S defined earlier. This set can also be defined as

$$D_S = \{\theta : \text{roots of } z^m(1+\sum_{i=1}^{m} a_i(\theta)z^{-i}) \text{ satisfy } |z|<1\}.$$

The required constraint is checked by computing "reflection coefficients" for the denominator polynomial of $H(z|\theta)$ by using the (fast) inverse-Levinson algorithm [21]. The Schur-Cohn criterion implies that the required stability condition holds if and only if the magnitude of the reflection coefficients is less than one. Thus, during the iterative gradient-type optimization, the step-size must be such that the stability constraint is not violated.

A Performance Index. We now propose a performance index which can be used to compare the goodness of approximation due to various models obtained by using different values of the system order m, maximum lags L_1 and L_2, etc. Let $\bar{L}_1$ and $\bar{L}_2$ denote some large values such that $L_1 \leqslant \bar{L}_1$ and $L_2 \leqslant \bar{L}_2$ for all possible values of L_1 and L_2 used for model realization. Let $\bar{\lambda}$ denote the value obtained from (27) with $L_1=\bar{L}_1$ and $L_2=\bar{L}_2$. Let θ denote the parameter vector corresponding to a selected model where θ implicitly inculdes the model order m. Then we define a performance index η as follows

$$\eta = 1-\frac{J(\theta;\bar{L}_1,\bar{L}_2,\bar{\lambda})}{J(\theta=0;\bar{L}_1,\bar{L}_2,\bar{\lambda})}$$

where $J(\theta;\bar{L}_1,\bar{L}_2,\bar{\lambda})$ is given by (22) with $L_1=\bar{L}_1$, $L_2=\bar{L}_2$ and $\lambda=\bar{\lambda}$, and $J(\theta=0;\bar{L}_1,\bar{L}_2,\bar{\lambda})$ is used simply for normalization. Clearly, $\eta \leqslant 1$. When $m \geqslant p$ and θ minimizes (21), we have $\eta=1$. Thus, the value of η should be close to one for a good approximate model.

The above measure allows us to compare models obtained as a result of using different values of lags L_1 and L_2 and/or different values of model order m. Since a selected model must be able to generate all of the given second-order and the fourth-order statistics with high fidelity, we should take $\overline{L}_1$ and $\overline{L}_2$ to be as large as possible to judge the effectiveness of the selected model.

Finally, we note that for all "reasonable" problems, we do not expect $J(\theta=0;\overline{L}_1,\overline{L}_2,\lambda)$ to be smaller than $J(\theta;\overline{L}_1,\overline{L}_2,\lambda)$ because the former corresponds to the approximate model $\overline{y}(t)\equiv 0$. Therefore, in such cases, we also have $\eta > 0$.

Later we illustrate the use of η by numerical examples.

THE OPTIMIZATION PROBLEM

We now address the problem of minimization of $J(\theta)$. Any standard gradient-type algorithm may be used to accomplish this. We have tried the software package NL2SOL [24]. It requires computation of $J(\theta)$ and its gradients w.r.t. each element of θ. This in turn requires computation of $R_2(\tau|\theta)$ $(0\geqslant\tau\geqslant -L_1)$, $C_4(t_1,t_2,t_3|\theta)$ $(0\geqslant t_3\geqslant t_2\geqslant t_1\geqslant -L_2)$, and their gradients w.r.t. each element of θ. In this section we discuss how to compute the above functions and their gradients. Note that the gradient-type schemes achieve only a local minimum. Since, in general, many minima are possible for m<p, one may have to try several different "initial guesses" to try to obtain a global minimum.

Computation of Autocorrelation Function and Its Gradients

It is well known [20] that

$$R_2(\tau|\theta) = -\sum_{i=1}^{m} a_i(\theta)R_2(\tau-i|\theta) \quad \text{for } \tau\geqslant m+1 \tag{28}$$

With θ_l denoting the l-th element of the vector θ, we have from (28)

$$\frac{\partial R_2(\tau|\theta)}{\partial\theta_l} = -\sum_{i=1}^{m}\frac{\partial a_i(\theta)}{\partial\theta_l}R_2(\tau-i|\theta) - \sum_{i=1}^{m} a_i(\theta)\frac{\partial R_2(\tau-i|\theta)}{\partial\theta_l} \tag{29}$$

for $\tau\geqslant m+1$.

It remains to calculate $R_2(\tau|\theta)$ and its gradients for $0\leqslant\tau\leqslant m$. With $P(\theta) = \lim_{t\to\infty} E\{\overline{x}(t)\overline{x}^T(t)|\theta\}$, it is well known [14,p.70] that $P(\theta)$ is the unique solution of the matrix Lyapunov equation

$$P(\theta) - \overline{A}(\theta)P(\theta)\overline{A}^T(\theta) = \overline{q}(\theta)\overline{b}\,\overline{b}^T \tag{30}$$

if $\overline{A}(\theta)$ is exponentially stable. Moreover, $P(\theta)$ is symmetric, Toeplitz, and nonsingular [21]. From [21], it can be shown that

$$P^{-1}(\theta) = \overline{q}^{-1}(\theta)[D_1(\theta)D_1^T(\theta) - D_2(\theta)D_2^T(\theta)] \tag{31}$$

where

$$D_1(\theta) = \begin{bmatrix} 1 & 0 & 0 & \cdots & 0 \\ a_1(\theta) & 1 & 0 & \cdots & 0 \\ a_2(\theta) & a_1(\theta) & 1 & \cdots & 0 \\ \cdot & \cdot & \cdot & & \cdot \\ \cdot & \cdot & \cdot & & \cdot \\ \cdot & \cdot & \cdot & & \cdot \\ a_{m-1}(\theta) & a_{m-2}(\theta) & a_{m-3}(\theta) & a_{m-3}(\theta) & 1 \end{bmatrix} \tag{32}$$

$$D_2(\theta) = \begin{bmatrix} a_m(\theta) & 0 & 0 & \cdots & 0 \\ a_{m-1}(\theta) & a_m(\theta) & 0 & \cdots & 0 \\ \cdot & \cdot & \cdot & & \cdot \\ \cdot & \cdot & \cdot & & \cdot \\ \cdot & \cdot & \cdot & & \cdot \\ a_2(\theta) & a_3(\theta) & a_4(\theta) & \cdots & 0 \\ a_1(\theta) & a_2(\theta) & a_3(\theta) & \cdots & a_m(\theta) \end{bmatrix} \tag{33}$$

While (31) can be directly inverted to yield $P(\theta)$, a more efficient solution can be obtained by the use of the inverse Levinson algorithm [21]. Note that since $P(\theta)$ is symmetric and Toeplitz, its first column (or last row) completely determines the rest of the matrix.

From (3), (4), and [14, p.72], we have ($\tau \geqslant 0$)

$$R_2(\tau|\theta) = \overline{c}(\theta)P(\theta)(\overline{A}^T(\theta))^\tau\, \overline{c}^T(\theta) + \overline{r}(\theta)\delta(\tau) \tag{34}$$

where $\delta(\tau)$ is the Kronecker's delta. Thus, (31) and (34) may be used to calculate $R_2(\tau|\theta)$ for $0 \leqslant \tau \leqslant m$.

In order to calculate the gradients of $R_2(\tau|\theta)$, ($0 \leqslant \tau \leqslant m$), w.r.t. each component of θ, we need the first derivatives of $P(\theta)$. Here we use (31)-(34) and the fact that

$$\frac{\partial P(\theta)}{\partial \theta_l} = -P(\theta)\frac{\partial P^{-1}(\theta)}{\partial \theta_l}P(\theta) \tag{35}$$

Finally, we note that $R_2(\tau|\theta) = R_2(-\tau|\theta)$ so that one gets the desired quantities for negative lags from those for the positive lags.

Computation of Fourth-Order Cumulant Function and Its Gradients

We first recall some useful results. From [18, p.211], given an i.i.d. sequence $\{w(t)\}$, its k-th order cumulant function, $\mathrm{cum}_k(w(t_1),w(t_2),\cdots,w(t_k))$, is given by

$$\mathrm{cum}_k(w(t_1),w(t_2),\cdots,w(t_k)) = \begin{cases} \gamma_k & \text{if } t_1=t_2=\cdots=t_k \\ 0 & \text{otherwise} \end{cases} \tag{36}$$

where γ_k denotes the k-th order cumulant of the random variable $w(t)$. Let

$$z(t) = \sum_{i_t \in I_t} f(i_t,t)u(i_t,t)$$

where $t=1,2,\cdots,k$; given t, I_t is an index set, $u(.,.)$ and $z(.)$ are random variables, and $f(.,.)$ are constants. Then [16, p.1357]

$$\mathrm{cum}_k(z(1),z(2),\cdots,z(k))$$
$$= \sum_{i_1 \in I_1} \sum_{i_2 \in I_2} \cdots \sum_{i_k \in I_k} f(i_1,1)f(i_2,2)\cdots f(i_k,k)$$
$$\times \mathrm{cum}_k(u(i_1,1),u(i_2,2),\cdots,u(i_k,k)) \tag{37}$$

where all the cumulants are assumed to exist.

We now turn to computation of the fourth-order cumulant functions for the model (3)-(5). This model can also be expressed in the following form

$$y_s(t) = -\sum_{i=1}^{m} a_i(\theta)y_s(t-i) + \sum_{i=0}^{m-1} h_i(\theta)\overline{w}(t-i-1) \tag{38}$$

$$y_s(t) = \overline{c}(\theta)\overline{x}(t) \tag{39}$$

$$\overline{y}(t) = y_s(t) + \overline{v}(t)$$

Since $\{\overline{v}(t)\}$ is a Gaussian process by assumption, we have from (4),(26), (38), and (39)

$$C_4(t_1,t_2,t_3|\theta) = \mathrm{cum}_4(y_s(t),y_s(t+t_1),y_s(t+t_2)y_s(t+t_3)|\theta) \tag{40}$$

Now consider $y_s(t)$ defined via (38), together with $y_s(t-t_1)$, $y_s(t-t_2)$, and $y_s(t-t_3)$, and use the result (37) to obtain

$$C_4(-t_1,-t_2,-t_3|\theta) = -\sum_{i=1}^{m} a_i(\theta)C_4(i-t_1,i-t_2,i-t_3|\theta) + d(t_1,t_2,t_3|\theta) \tag{41}$$

where $-\infty<t_i<\infty$, $(i=1,2,3)$, and

$$d(t_1,t_2,t_3|\theta) = \sum_{i=0}^{m} h_i(\theta)\mathrm{cum}_4(\overline{w}(t-i-1),y_s(t-t_1),y_s(t-t_2),y_s(t-t_3)|\theta) \tag{42}$$

Now let $\{g(i,\theta), i\geq 0\}$ denote the impulse response function of the causal system (38) so that we have

$$y_s(t-\tau) = \sum_{i=0}^{\infty} g(i,\theta)\overline{w}(t-\tau-i) \tag{43}$$

where the impulse response satisfies the recursion

$$g(i,\theta) = -\sum_{j=1}^{m} a_j(\theta)g(i-j,\theta) + \sum_{j=0}^{m-1} h_j(\theta)u(i-j-1) \tag{44}$$

with $u(i) = \delta(i)$ and $g(i,\theta) = 0$ for $i<0$. From (37) and (43) it follows that for $t_i \geq 0$,

$(i=1,2,3)$,

$$\mathrm{cum}_4(\overline{w}(t-i+1),y_s(t-t_1),y_s(t-t_2),y(t-t_3)|\theta)$$

$$= \sum_{i_1=0}^{\infty}\sum_{i_2=0}^{\infty}\sum_{i_3=0}^{\infty} g(i_1,\theta)g(i_2,\theta)g(i_3,\theta)$$

$$\times \mathrm{cum}_4(\overline{w}(t-i-1),\overline{w}(t-t_1-i_1),\overline{w}(t-t_2-i_2),\overline{w}(t-t_3-i_3)|\theta)$$

$$= \begin{cases} 0, & \text{if } \max(t_1,t_2,t_3) > m-1 \text{ and } 0\leqslant i+1\leqslant m-1 \\ \overline{\gamma}_4(\theta)g(i-t_1+1,\theta)g(i-t_2+1,\theta)g(i-t_3+1,\theta), & \text{otherwise} \end{cases} \quad (45)$$

For $t_i\geqslant 0$ $(i=1,2,3)$ and $t_j\leqslant m$ for at least one $j\in\{1,2,3\}$, it follows from (42) and (45) that

$$d(t_1,t_2,t_3|\theta) = 0, \quad \text{if } \lim = \max(t_1,t_2,t_3)\geqslant m$$

$$= \sum_{i=\lim}^{m-1} h_i(\theta)\overline{\gamma}_4(\theta)g(i+1-t_1,\theta)g(i+1-t_2,\theta)g(i+1-t_3,\theta), \quad (46)$$

$$\text{if } \lim\leqslant m-1$$

Thus, from (41) and (46), we have the recursion

$$C_4(-t_1,-t_2,-t_3|\theta) = -\sum_{i=1}^{m} a_i(\theta)C_4(i-t_1,i-t_2,i-t_3|\theta) \quad (47)$$

for $t_i\geqslant 0$ $(i=1,2,3)$ and $t_j\geqslant m$ for at least one $j\in\{1,2,3\}$.

To calculate the derivatives of the cumulants w.r.t. θ_l, $1\leqslant l\leqslant 2p+2$, simply differentiate both sides of (47).

Next we calculate $C_4(-t_1,-t_2,-t_3|\theta)$ and its gradients for $0\leqslant t_i\leqslant m-1$, $(i=1,2,3)$. Unfortunately, here we do not yet have any fast method. We give a brute-force technique. It is easy to show [19] that for $-\infty<t_i<\infty$ $(i=1,2,3)$

$$C_4(-t_1,-t_2,-t_3|\theta) = \overline{\gamma}_4(\theta)\sum_{i=0}^{\infty} g(i,\theta)g(i-t_1,\theta)g(i-t_2,\theta)g(i-t_3,\theta) \quad (48)$$

To calculate the required derivatives differentiate both sides of (44) and (48) w.r.t. θ_l, $1\leqslant l\leqslant 2p+2$. Efficient computation of the derivatives of (44) can be carried out following a "trick" given in [22]; we omit the details. Also, by the exponential stability of $\overline{A}(\theta)$, $g(i,\theta)\approx 0$ for $i\geqslant$ some positive integer F. Therefore, in practice, the upper limit in the summation of (48) may be replaced with F. This aspect is discussed further in the following subsection.

Finally, the above procedure provides values of $C_4(t_1,t_2,t_3|\theta)$ and its derivatives for $t_i\leqslant 0$, $(i=1,2,3)$. To obtain the fourth-order cumulants and their derivatives for the other values of the lags we note the identities

$$C_4(t_1,t_2,t_3|\theta) = C_4(-t_3,t_1-t_3,t_2-t_3|\theta)$$

$$= C_4(-t_2, t_1-t_2, t_3-t_2|\theta)$$

$$= C_4(-t_1, t_2-t_1, t_3-t_1|\theta) \tag{49}$$

for $-\infty<t_i<\infty$, (i=1,2,3). Thus given the fourth-order cumulants and their derivatives for all nonpositive lag triples, the cumulants and their derivatives for arbitrary lags can be obtained through (49). Also, note the identities

$$C_4(t_1,t_2,t_3|\theta) = C_4(t_2,t_1,t_3|\theta) = C_4(t_2,t_3,t_1|\theta)$$

$$= C_4(t_3,t_1,t_2|\theta) = C_4(t_3,t_2,t_1|\theta)$$

$$= C_4(t_1,t_3,t_2|\theta) \tag{50}$$

Constrained Optimization: Convergence

Minimization of $J(\theta)$ is performed under a stability constraint on $\overline{A}(\theta)$ as discussed earlier, with $\theta \in \Theta$, the parameter set. Moreover, we claimed that the upper limit in the summation of (48) may be replaced with a large integer F. In this subsection we investigate the convergence of the search procedure under the above conditions.

Define a compact set Θ_C as

$$\Theta_C \subset \Theta \subset R^{2m+2} \cap D_S$$

Note that both the sets R^{2m+2} and D_S are open. (The set Θ is not necessarily open.) More specifically, take

$$\Theta_C = \{\theta : |h_i(\theta)| \leqslant M < \infty \ (1 \leqslant i \leqslant m-1),\ 0 \leqslant \overline{q}(\theta) \leqslant M,$$

$$0 \leqslant \overline{r}(\theta) \leqslant M,\ |\overline{\gamma}_4(\theta)| \leqslant M, \text{ and roots of}$$

$$z^m(1+\sum_{i=1}^{m} a_i(\theta) z^{-i}) \text{ satisfy } |z| \leqslant 1-\delta \text{ for some } \delta>0\} \tag{51}$$

Note that as $M \to \infty$ and $\delta \to 0$, we have $\Theta_C \to R^{2m+2} \cap D_S \cap \{\overline{q} \geqslant 0, \overline{r} \geqslant 0\}$. Finally, let $\overline{J}(\theta, F)$ denote the function $J(\theta)$ when the latter is computed with the upper limit in the summation of (48) replaced with a fixed F.

Proposition 1. $J(\theta)$ as well as $\overline{J}(\theta, F)$ has a minimum in Θ_C.

Proposition 2. $\lim_{F\to\infty} \overline{J}(\theta, F) = J(\theta)$ uniformly in θ for $\theta \in \Theta_C$.

Proposition 3. Let $\theta' \in \Theta_C$ denote a minimum of $J(\theta)$. Then as $F \to \infty$, a minimum of $\overline{J}(\theta, F)$ converges to θ'.

Proofs of these propositions may be found in [25].

Thus, in practice, for large enough F (the choice of which can be made independent of $\theta \in \Theta_C$), a local minimum of $\overline{J}(\theta, F)$ will be arbitrarily close to a local minimum of $J(\theta)$.

Remark 5. It is interesting to note that $J(\theta)$ can not have a minimum on the unit circle unless the triple $(\bar{A}(\theta), \bar{b}(\theta), \bar{c}(\theta))$ is not minimal, for it would lead to an infinite $R(0|\theta)$, hence to an infinite $J(\theta)$. □

EXAMPLE

We now illustrate the proposed approach by a numerical example. The original model is given by a fifth-order Markov model (see (1)-(2)) with

$$A = \begin{bmatrix} 0 & 1 & 0 & 0 & 0 \\ 0 & 0 & 1 & 0 & 0 \\ 0 & 0 & 0 & 1 & 0 \\ 0 & 0 & 0 & 0 & 1 \\ 0.364 & -2.032 & 4.743 & -5.790 & 3.700 \end{bmatrix}, \quad b = \begin{bmatrix} 0 \\ 0 \\ 0 \\ 0 \\ 1 \end{bmatrix}$$

$$c = [0.375 \ \ -1.975 \ \ 3.750 \ \ -3.100 \ \ 1.000]$$

$$q = 1.0, \quad r = 0.2, \quad \gamma_4 = 1.0$$

The pole-zero distribution of this system is

poles : $0.8 \pm j0.4$, $0.7 \pm j0.4$, 0.7

zeros : $1.0 \pm j0.5$, 0.5, 0.6

For comparison via the performance index η we will use $\bar{L}_1 = 14$, and $\bar{L}_2 = 70$ for all the results in this section.

We first seek a 4th-order approximation to the given system. We begin with the smallest values of the maximum lags suggested by the discussion in the section on approximate stochastic realization, namely, take $L_1 = L_2 = 8$. The resulting system is as follows.

m=4: Criterion (21), $L_1 = 8$, $L_2 = 8$, minimum cost $= 0.001193$, $\eta = 0.9995$

$$\bar{A} = \begin{bmatrix} 0 & 1 & 0 & 0 \\ 0 & 0 & 1 & 0 \\ 0 & 0 & 0 & 1 \\ -0.4953 & 2.1040 & -3.6427 & 2.9874 \end{bmatrix}, \quad \bar{b} = \begin{bmatrix} 0 \\ 0 \\ 0 \\ 1 \end{bmatrix}$$

$$\bar{c} = [-0.4792 \ \ 2.0165 \ \ -2.3832 \ \ 1.0]$$

$$\bar{q} = 0.9920, \quad \bar{r} = 0.1942, \quad \bar{\gamma}_4 = 0.9943$$

poles : $0.7910 \pm j0.4191$, $0.7027 \pm j0.3526$

zeros : $0.9999 \pm j0.5000$, 0.3834

Since the performance index turns out to be very close to 1, there is no point in increasing the maximum lags. The impulse response of the approximate model obtained above is compared in Fig. 1 to the original impulse response. It is seen that the two are virtually identical.

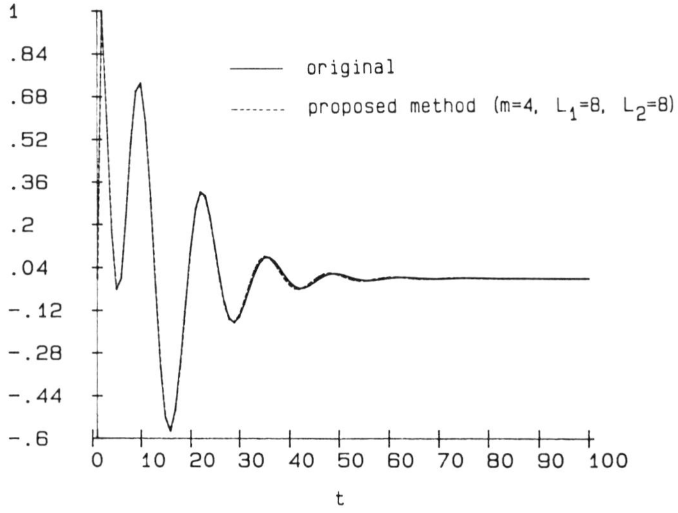

Fig.1. Impulse responses of the original model and the reduced order model of order m=4

m=3: Criterion (21), L_1=6, L_2=6, minimum cost = 1.3298, η =0.7373

$$\bar{A} = \begin{bmatrix} 0 & 1 & 0 \\ 0 & 0 & 1 \\ 0.6946 & -2.1665 & 2.4031 \end{bmatrix}, \quad \bar{b} = \begin{bmatrix} 0 \\ 0 \\ 1 \end{bmatrix}$$

$$\bar{c} = [1.2956 \quad -2.0313 \quad 1.0]$$

$$\bar{q} = 0.9330, \quad \bar{r} = 0.4221, \quad \bar{\gamma}_4 = 1.1831$$

poles : 0.8274±j0.4936, 0.7483

zeros : 1.0156±j0.5139

Since the performance index η is now not as close to one as that for m=4, we increase the maximum lags.

m=3: Criterion (21), L_1=60, L_2=10, minimum cost = 6.9412, η =0.9130

$$\bar{A} = \begin{bmatrix} 0 & 1 & 0 \\ 0 & 0 & 1 \\ 0.5755 & -1.9585 & 2.3016 \end{bmatrix}, \quad \bar{b} = \begin{bmatrix} 0 \\ 0 \\ 1 \end{bmatrix}$$

$\bar{c} = [1.3414 \;\; -2.0654 \;\; 1.0]$

$\bar{q} = 1.0600, \;\; \bar{r} = 0.6903, \;\; \bar{\gamma}_4 = 1.3254$

poles : $0.8251 \pm j0.4502, \; 0.6515$

zeros : $1.0327 \pm j0.5243$

Notice that increasing the maximum lags has improved the performance index as expected, leading to better fidelity to the given statistics. Fig. 2 compares the impulse responses due to the approximations to the original impulse response. It is clear from the figure that we do have better approximation for larger maximum lags.

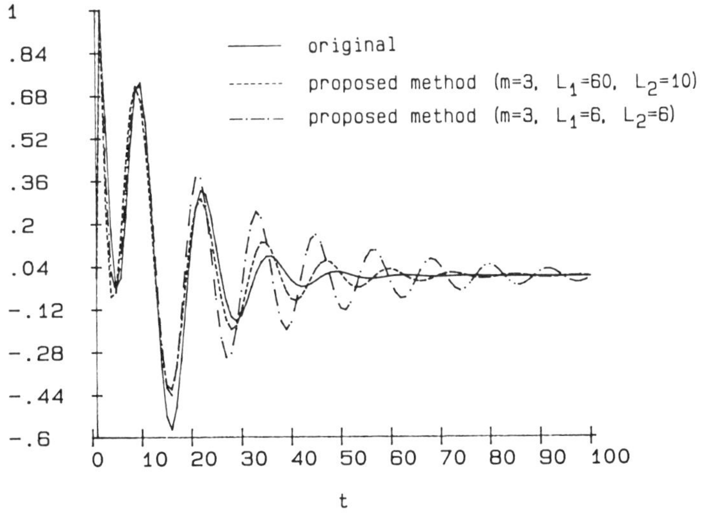

Fig.2. Impulse responses of the original and the reduced order models of order m=3

Finally, we also sought approximate models based on the correlation function only (i.e, in (22) take $\lambda=0$). This provides a minimum-phase approximate model.

m=3: Criterion (21) with $\lambda=0$, $L_1=60$, minimum cost ≈ 0

$$\bar{A} = \begin{bmatrix} 0 & 1 & 0 \\ 0 & 0 & 1 \\ 0.4550 & -1.6300 & 2.1000 \end{bmatrix}, \quad \bar{b} = \begin{bmatrix} 0 \\ 0 \\ 1 \end{bmatrix}$$

$\bar{c} = [0.3000 \;\; 1.1000 \;\; 1.00]$

$\bar{q} = 1.5625, \;\; \bar{r} = 0.2000$

poles : 0.7±j0.4, 0.7

zeros : 0.5, 0.6

Almost identical results were obtained with L_1=6. Figure 3 compares the impulse response of the above approximate model to that of the original system and of the approximate model obtained by our new method. It is seen that the model based on the second-order statistics alone provides a very poor approximation to the true system response in the presence of nonminimum phase zeros. The all-pass pole-zero pairs (0.8+j0.4,1-j0.5) and (0.8-j0.4,1+j0.5) in the original model have no effect on the output autocorrelation function. Therefore, a third-order fit based only on autocorrelations provides a "perfect" approximation by simply deleting the all-pass pairs.

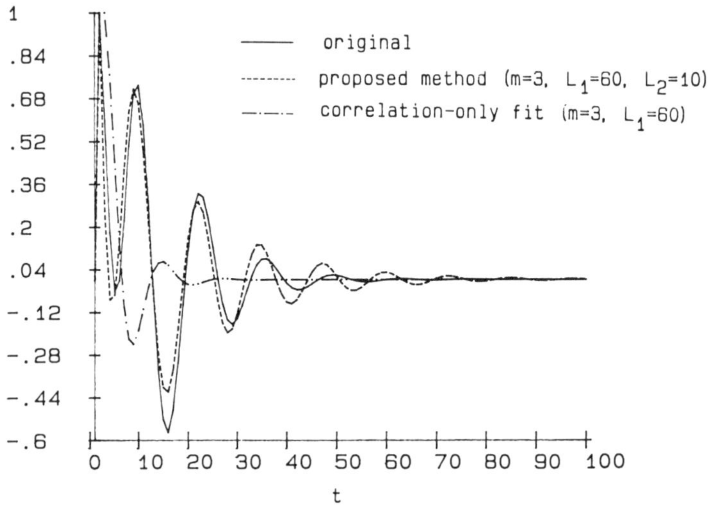

Fig.3. Impulse responses of the original and the reduced order models of order m=3

CONCLUDING REMARKS

A new realization approach to the problem of order reduction of SISO nonminimum phase discrete stochastic systems was proposed. It was shown that to capture the system phase characteristics in the reduced order models, one must also consider higher order statistics of the system output in addition to the usual autocorrelation function. An approximation criterion was proposed and analyzed. The algorithmic aspects of the order reduction problem were discussed in some detail. Finally, an example was presented to illustrate the proposed method.

APPENDIX

In the Appendix we will provide proofs of the Theorem and the Corollaries. We first need several preliminary results.

Consider the numerator polynomial of $H(z|\theta)$ defined in (13), namely

$$z^m\left[\sum_{i=1}^{m} h_{i-1}(\theta)z^{-i}\right] \tag{A1}$$

It has $m-1$ roots. Suppose that it has n_r real roots and n_c complex roots so that $n_r+n_c=m-1$. Since the coefficients of the polynomial (A1) are real, the complex roots must occur in conjugate pairs so that n_c is even. The p.s.d. (power spectral density) of the observations $\{\bar{y}(t)\}$ of the reduced-order model (3)-(5) is given by

$$S_2(z|\theta) = \bar{q}(\theta)H(z|\theta)H(z^{-1}|\theta) + \bar{r}(\theta) \tag{A2}$$

Therefore, there are at most 2^l $(l=n_r+n_c/2)$ possible combinations of the zeros of $H(z|\theta)$ which yield, up to a constant, the same p.s.d. (A2). A zero located at f or at f^{-1} yields the same product term within a constant

$$(z-f)(z^{-1}-f) = f^2(z-f^{-1})(z^{-1}-f^{-1}) \tag{A3}$$

If there are any zeros on the unit circle, then the number of possible combinations of zeros yielding the same p.s.d. will be less than 2^l.

Let $D(\theta_1)$ denote the set of all combinations of the parameter vector θ derived from θ_1 by considering all possible locations of the zeros of $H(z|\theta_1)$ as discussed above. Therefore, the elements of $D(\theta_1)$ differ in the coefficients h_i's and $\bar{q}$ where, whenever h_i's are changed, $\bar{q}$ may have to be changed too to maintain identical p.s.d.; (see (A3)). Thus, by definition,

$$D(\theta_1) = \{\theta : S_2(z|\theta) = S_2(z|\theta_1),\ \theta\in\Theta\} \tag{A4}$$

Note that $D(\theta_1)$ is a finite set, given $\theta_1\in\Theta$.

Lemma 2. Let $S_2(z|\theta_1) = S_2(z|\theta_2)$ for any $\theta_i\in\Theta$, $(i=1,2)$. If $H(z|\theta_i)$ has no pole-zero cancellations for i=1 and 2, then $\bar{r}(\theta_1) = \bar{r}(\theta_2)$.

Proof. The product $H(z|\theta_i)H(z^{-1}|\theta_i)$, $(i=1,2)$, vanishes at $z=0$. The desired result then follows by using (A2). □

Lemma 3. Given $\theta_1\in\Theta$ such that $H(z|\theta_1)$ has no pole-zero cancellations, if $C_4(t_1,t_2,t_3|\theta) = C_4(t_1,t_2,t_3|\theta_1)$ for $0\geq t_i\geq -2m$, $(i=1,2,3)$, then $a_j(\theta) = a_j(\theta_1)$, $(j=1,2,\dots,m)$.

Proof. Given $C_4(t_1,t_2,t_3|\theta)$ for $0\geq t_i\geq -2m$, $(i=1,2,3)$, we can calculate it for all triples (t_1,t_2,t_3) such that $-m\leq t_i\leq m$, $(i=1,2,3)$, by the use of the identities (49). Therefore, the hypothesis of the lemma implies that

$$C_4(t_1,t_2,t_3|\theta) = C_4(t_1,t_2,t_3|\theta_1), \quad |t_i| \leqslant m \ (i=1,2,3) \tag{A5}$$

Now consider the recursion (47). It may be rewritten as

$$f(i|\theta) = -\sum_{j=1}^{m} a_j(\theta) f(i-j|\theta), \quad i \geqslant 0 \tag{A6}$$

where

$$f(i|\theta) = C_4(t_1-i, t_2-i, t_3-i|\theta)$$

$$t_l < 0 \quad (l=1,2,3)$$

$$t_k \leqslant -m \text{ for at least one } k \in \{1,2,3\}.$$

With (t_1,t_2,t_3) fixed, (A6) defines a recursion in i. By (A6) we have

$$\underline{f}(\theta) = -F(\theta)\underline{a}(\theta) \tag{A7}$$

where

$$\underline{f}(\theta) = [f(m+1|\theta), f(m+2|\theta), \cdots, f(2m|\theta)]^T$$

$$\underline{a}(\theta) = [a_m(\theta), a_{m-1}(\theta), \cdots, a_1(\theta)]^T$$

$$F(\theta) = \begin{bmatrix} f(1|\theta) & f(2|\theta) & \cdots & f(m|\theta) \\ f(2|\theta) & f(3|\theta) & \cdots & f(m+1|\theta) \\ \cdot & \cdot & & \cdot \\ \cdot & \cdot & & \cdot \\ \cdot & \cdot & & \cdot \\ f(m|\theta) & f(m+1|\theta) & \cdots & f(2m-1|\theta) \end{bmatrix}$$

By the assumption (C6), $H(z|\theta)$ has no pole-zero cancellations. Therefore, there exists a triple (t_1,t_2,t_3) with $t_i \leqslant 0$ $(i=1,2,3)$ and $t_l \leqslant -m$ for at least one $l \in \{1,2,3\}$ such that the recursion (A6) holds with $a_m(\theta_1) \neq 0$ and $\theta = \theta_1$. (The proof of this claim is by contradiction. If it is not true, then we can calculate $C_4(t_1,t_2,t_3|\theta)$ for all triples (t_1,t_2,t_3) via (47) and (49) with m replaced with $m' < m$ in (47), leading to a transfer function $H(z|\theta)$ (see Lemma 1) that has a denominator polynomial of degree $m' < m$. This contradicts (C6).) Therefore, by [23, Theorem 7, p.244], $F(\theta_1)$ is invertible and we have

$$\underline{a}(\theta_1) = -F^{-1}(\theta_1)\underline{f}(\theta_1) \tag{A8}$$

Now by the hypothesis of the lemma, $\underline{f}(\theta) = \underline{f}(\theta_1)$ and $F(\theta) = F(\theta_1)$. Therefore, by (A7) and (A8), we have

$$\underline{a}(\theta) = -F^{-1}(\theta_1)\underline{f}(\theta_1) = \underline{a}(\theta_1)$$

which yields the desired result. □

Lemma 4. Given $\theta_1 \in \Theta$, suppose that $a_j(\theta) = a_j(\theta_1)$, $(j=1,2,...,m)$, for some $\theta \in \Theta$. If $C_4(t_1,t_2,t_3|\theta) = C_4(t_1,t_2,t_3|\theta_1)$, for $0 \geq t_i \geq -2m$, $(i=1,2,3)$, then the equality holds for all triples (t_1,t_2,t_3), i.e., for $-\infty < t_i < \infty$, $(i=1,2,3)$.

Proof. As shown in the proof of Lemma 3, (A5) holds. Then, by the recursion (47) and the hypothesis of the lemma, $C_4(t_1,t_2,t_3|\theta) = C_4(t_1,t_2,t_3|\theta_1)$ for all negative triples. Finally, use (49) to obtain the desired result. □

We now turn to the proof of the Theorem.

Proof of Theorem. Suppose that (C4) holds. Let $\theta_i' \in D(\theta_i)$, $(i=1,2)$, such that $H(z|\theta_i')$ is minimum phase, where $D(\theta_i)$ has been defined in (A4). Then (C4) also holds with θ_i replaced with θ_i', $(i=1,2)$. This fact implies that [20] $R_2(\tau|\theta_1') = R_2(\tau|\theta_2')$ for $-\infty < \tau < \infty$, hence $R_2(\tau|\theta_1) = R_2(\tau|\theta_2)$ for $-\infty < \tau < \infty$. The other conclusion of the Theorem follows easily from Lemmas 3 and 4 and identities (50). □

Proof of Cor. 1. The conclusions of the Theorem together with Lemma 2 imply that the power spectra of $\{y_s(t)\}$ parametrized by θ_1 and θ_2, respectively, are identical. The same holds for the trispectra of $\{y_s(t)\}$. Now use Lemma 1 to conclude that $H(z|\theta_1') = H(z|\theta_2')$, which implies that $a_i(\theta_1) = a_i(\theta_2)$ for $1 \leq i \leq m$, and $h_i(\theta_1) = h_1(\theta_2)$ for $1 \leq i \leq m-1$. Also, from Lemma 2 we have $\bar{r}(\theta_1) = \bar{r}(\theta_2)$ and hence from (A2), $\bar{q}(\theta_1) = \bar{q}(\theta_2)$. Finally, it can be shown that we also have $\bar{\gamma}_4(\theta_1) = \bar{\gamma}_4(\theta_2)$; see, e.g., the z-domain expressions for the tripsectrum. Thus, we have $\theta_1 = \theta_2$. □

Proof of Cor. 2. It follows trivially from the Theorem and Cor.1. □

REFERENCES

[1] Baram, Y. and Be'eri, Y., **IEEE Trans. Automatic Control,** vol. AC-26, pp. 379-390, April 1981.

[2] Baram, Y., **IEEE Trans. Automatic Control,** vol. AC-26, pp. 1225-1231, Dec. 1981.

[3] Desai, U.B. and Pal, D., in **Proc. 21st IEEE Conf. Decision and Control,** pp. 1105-1112, Orlando, FL, Dec. 1982.

[4] Larimore, W.E., in **Proc. 1983 American Control Conf.,** pp. 445-451, San Francisco, CA, June 1983.

[5] Arun, K.S., Bhaskar Rao, D.V., and Kung, S.Y., in **Proc. 22nd IEEE Conf. Decision and Control,** pp. 1353-1355, San Antonio, TX, Dec. 1983.

[6] Vaccaro, R.J., in **Proc. 1984 American Control Conference,** pp. 393-396, San Diego, CA, June 1984.

[7] Wagie, D.A. and Skelton, R.E., in **Proc. 23rd IEEE Conf. Decision and Control,** pp. 133-138, Las Vegas, NV, Dec. 1984.

[8] Genesio, R. and Milanese, R., **IEEE Trans. Automatic Control,** vol. AC-21, pp. 118-122, Feb. 1976.

[9] Hickin, J. and Sinha, N.K., **IEEE Trans. Automatic Control,** vol. AC-25, pp.1121-1127, Dec. 1980.

[10] Moore, B.C., **IEEE Trans. Automatic Control,** vol. AC-26, pp. 17-32, Feb. 1981.

[11] Mendel, J.M., **Optimal Seismic Deconvolution.** New York: Academic Press, 1983.

[12] Freudenberg, J.S. and Looze, D.P., in **Proc. 22nd IEEE Conf. Decision and Control,** San Antonio, TX, pp. 625-630, Dec. 1983.

[13] Benveniste, A., Goursat, M. and Ruget, G., **IEEE Trans. Automatic Control,** vol. AC-25, pp. 385-398, June 1980.

[14] Anderson, B.D.O. and Moore, J.B., **Optimal Filtering.** Englewood Cliffs, N.J.: Prentice-Hall, 1979.

[15] Lii, K.S. and Rosenblatt, M., **Annals of Statistics,** vol. 10, pp. 1195-1208, 1982.

[16] Brillinger, D.R., **Annals Math. Statistics,** vol. 36, pp. 1351-1374, 1965.

[17] Brillinger, D.R. and Rosenblatt, M., in **Spectral Analysis of Time Series,** B.Harris (ed.), New York: Wiley, 1967, pp. 153-188.

[18] Brillinger, D.R. and Rosenblatt, M., in **Spectral Analysis of Time Series,** B.Harris (ed.), New York: Wiley, 1967, pp. 189-232.

[19] Tugnait, J.K., in **Proc. 23rd IEEE Conf. Decision and Control,** pp. 342-347, Las Vegas, NV, Dec. 1984. (Also, to appear in **Automatica,** 1986.)

[20] Mehra, R.K., **IEEE Trans. Automatic Control,** vol. AC-16, pp. 12-21, Feb. 1971.

[21] Bitmead, R.R., **IEEE Trans. Automatic Control,** vol. AC-26, pp. 11-1294, Dec. 1981.

[22] Astrom, K.J., **Automatica,** vol. 16, pp. 551-574, 1980.

[23] Gantmacher, F.R., **Applications of the Theory of Matrices.** New York: Interscience, 1959.

[24] Dennis, J.E., Jr. and Schnabel, R.B., **Numerical Methods for Unconstrained Optimization and Nonlinear Equations.** Englewood Cliff, NJ: Prentice-Hall, 1983, (Sec. 10.3).

[25] Tugnait, J.K., **IEEE Trans. Automatic Control,** to appear.

[26] Kendall, M.G. and Stuart, A., **The Advanced Theory of Statistics, Vol. I.** New York: Macmillan, 4th Edition, 1977.

9

ON STOCHASTIC BILINEAR SYSTEMS

Arthur E. Frazho

School of Aeronautics and Astronautics, Purdue University, West Lafayette, IN 47907

ABSTRACT

This paper uses noncommutative dilation theory to develop a theory of stochastic bilinear systems. A certain stochastic bilinear realization and partial realization problem is solved. A new Caratheodory interpolation problem is presented and solved.

1. INTRODUCTION

In this paper we use and extend the noncommutative dilation theory in [8, 18, 19] to solve a certain stochastic bilinear realization problem and a stochastic partial bilinear realization problem. We obtain a minimal stochastic bilinear realization for an infinite "covariance" sequence and present a recursive algorithm for obtaining a "maximal entropy" bilinear realization for a finite "covariance" sequence. Motivated by bilinear systems, we present and solve a new Caratheodory interpolation problem. In particular it is shown that a bilinear system is uniquely determined by and uniquely determines a sequence of Schur contractions or reflection operators. The paper begins by using the dilation theory in [35] to review the connection between positive definite functions, the Caratheodory interpolation problem, Schur contractions and linear stochastic realization theory presented in [1,3-7,9-12,16,22-28,30,32,33] and elsewhere. Then we extend these results and techniques to the bilinear setting.

Throughout we follow the standard notation for Hilbert space in [21] [35]. The orthogonal projection onto a subspace H is denoted by P_H. An operator T is a *contraction* if $||T|| \leq 1$. If T is a contraction then D_T is the positive square root of $I - T^*T$ and $\mathcal{D}_T$ is the closed range of D_T. Let T be a contraction on X. Recall that an isometry U on $K(\supseteq X)$ is the *minimal isometric dilation* of T if

$$T^n = P_X U^n | X \qquad \text{(for all } n \geq 0) \tag{1.1}$$

and

$$K = \bigvee_{n \geq 0} U^n X \,. \tag{1.2}$$

It is easy to verify that the operator U on K defined by

$$U = \begin{bmatrix} T & 0 & 0 & \cdots \\ D_T & 0 & 0 & \cdots \\ 0 & I & 0 & \cdots \\ 0 & 0 & I & \cdots \\ \cdot & \cdot & \cdot & \cdots \\ \cdot & \cdot & \cdot & \cdots \\ \cdot & \cdot & \cdot & \cdots \end{bmatrix}$$

$$K = H \oplus D_T \oplus D_T \oplus D_T \oplus D_T \oplus \cdots \tag{1.3}$$

is a minimal isometric dilation of T. The minimal isometric dilation of T is unique up to an isomorphism, see [35] for further results.

We begin by reviewing some of the results in [5-7,9-12,20,22-27,32,33] on the positive definite Caratheodory interpolation problem. Then we extend these results in Section 6 to the bilinear setting. Recall that a sequence of operators $\{R_n\}_1^\infty$ on H is *positive definite if*

$$\sum_{0,0}^{\infty,\infty} (R_{i-j}h_i, h_j) \geq 0 \tag{1.4}$$

for all sequences $\{h_i\}_0^\infty$ in H with finite support. Throughout it is always assumed that $R_n^* = R_{-n}$ for all n and $R_o = I$, the identity. By converting (1.4) to a matrix it is easy to show that $\{R_n\}_0^\infty$ is positive definite if and only if the Toeplitz matrix

$$\hat{R}_m = \begin{bmatrix} I & R_1 & \cdots & R_{m-1} \\ R_{-1} & I & \cdots & R_{m-2} \\ R_{-2} & R_{-1} & \cdots & R_{m-3} \\ \cdot & \cdot & \cdots & \cdot \\ \cdot & \cdot & \cdots & \cdot \\ \cdot & \cdot & \cdots & \cdot \\ R_{1-m} & R_{2-m} & \cdots & I \end{bmatrix} \tag{1.5}$$

is positive (≥ 0) for all $m \geq 1$. The following is the Naimark dilation theorem [35], which plays a central role in our approach.

THEOREM 1.1 *A sequence of operators* $\{R_n\}_1^\infty$ *on* H *is positive definite if and only if there exists an isometry* U *on* $K(\supseteq H)$ *such that*

$$R_n = P_H U^n \big| H \qquad \text{(for } n \geq 0\text{)}. \tag{1.6}$$

Furthermore if

$$K = \bigvee_{n \geq 0} U^n H \tag{1.7}$$

then U *is unique up to an isomorphism.*

By *unique up to an isomorphism,* we mean that if U' on K' is an isometry satisfying (1.6) and (1.7) where U' replaces U and K' replaces K, then there exists a unitary operator W mapping K onto K' such that $U'W = WU$ and $W|H = I$. The isometry U on K satisfying (1.6) and (1.7) is called the *minimal isometric dilation* of $\{R_n\}_1^\infty$.

REMARK 1.2 The sequence of operators $\{R_n\}_1^\infty$ on H is positive definite if and only if there exists a contraction T on $X(\supseteq H)$ such that

$$R_n = P_H T^n | H \qquad \text{(for all } n \geq 0) . \tag{1.8}$$

To see this assume that (1.8) holds. Let U be the minimal isometric dilation of T. Equations (1.1) and (1.8) give (1.6). By Theorem 1.1 the sequence $\{R_n\}_1^\infty$ is positive definite. Theorem 1.1 also proves the other half by choosing U = T.

2. THE CHOICE SEQUENCE FOR $\{R_n\}_o^\infty$.

In this section we obtain a choice sequence or the Schur contractions for $\{R_n\}_1^\infty$. The results in this section and the following section are not new. They are also contained in [5-7,9-12,20,22-27,32,33]. We review the linear setting in this section and the next to see exactly how these ideas can be extended to bilinear systems. Throughout t denotes the transpose

$$[A, B]^t = \begin{bmatrix} A \\ B \end{bmatrix} .$$

Notice that the transpose does not operate on A and B. Finally recall that any contraction Γ mapping H into H' uniquely determines the unitary operator

$$\begin{bmatrix} \Gamma & D_{\Gamma^*} \\ D_\Gamma & -\Gamma^* \end{bmatrix} \tag{2.1}$$

mapping $H \oplus D_{\Gamma^*}$ onto $H' \oplus D_\Gamma$. We call the operator in (2.1) is called the *Halmos extension* of Γ, see [21].

To begin let U on K be the minimal isometric dilation of $\{R_n\}_1^\infty$ on H. The minimality condition (1.7) implies that U has a matrix representation of the form:

$$U = \begin{bmatrix} \Gamma_1 & * & * & \cdots \\ B & * & * & \cdots \\ 0 & * & * & \cdots \\ 0 & 0 & * & \cdots \\ 0 & 0 & 0 & \cdots \\ \cdot & \cdot & \cdot & \cdots \\ \cdot & \cdot & \cdot & \cdots \\ \cdot & \cdot & \cdot & \cdots \end{bmatrix} \tag{2.2}$$

$$K = H \oplus D_1 \oplus D_2 \oplus D_3 \oplus \cdots$$

where

$$D_m = H_{m+1} \ominus H_m \text{ and } \qquad (m \geq 1) \tag{2.3}$$

$$H_m = \bigvee_{n=0}^{m-1} U^n H = H \oplus D_1 \oplus D_2 \oplus \cdots \oplus D_{m-1} \,.$$

Since U is an isometry the columns must be isometric and orthogonal. The first column in (2.2) is easy to obtain. Equation (1.6) gives $R_1 = P_H U|\,H = \Gamma_1$. This and the isometric property shows that

$$\|Bh\|^2 = \|h\|^2 - \|\Gamma_1 h\|^2 = \|D_1 h\|^2 \qquad (h \in H)$$

where $D_1 = D_{\Gamma_1}$. So there exists a unitary operator W mapping $\overline{BH}$ onto $D_1 = D_{\Gamma_1}$ such that $WB = D_1$. Without loss of generality we can choose $W = I$. This follows because U is uniquely determined up to an isomorphism by Theorem 1.1. Therefore the first column of U is

$$[\Gamma_1, D_1, 0, 0, 0, \ldots]^t$$

Furthermore $H_2 = H \oplus D_1$ where $D_1 = D_{\Gamma_1}$.

Now let $D_{1^*} = D_{\Gamma_1^*}$ and $D_{1^*} = D_{\Gamma_1^*}$. Using the fact that the Halmos extension of Γ_1 is unitary we have

$$\begin{bmatrix} H \\ D_1 \end{bmatrix} \ominus \begin{bmatrix} \Gamma_1 \\ D_1 \end{bmatrix} H = \begin{bmatrix} D_{1^*} \\ -\Gamma_1^* \end{bmatrix} D_{1^*} \,. \tag{2.4}$$

Notice that the operator $[D_{1^*}, -\Gamma_1^*]^t$ is an isometry. Since the second column in (2.2) is orthogonal to the first column and isometric, (2.4) and the minimality condition (1.7) shows that the second column can be identified with

$$[D_{1^*}\Gamma_2, -\Gamma_1^*\Gamma_2, D_2, 0, 0, 0, \ldots]^t$$

where Γ_2 is a *uniquely determined* contraction mapping D_1 into D_{1^*}. (As expected

$D_n = D_{\Gamma_n}$ and $D_{n*} = D_{\Gamma_n^*}$ and $\mathcal{D}_n = \mathcal{D}_{\Gamma_n}$ and $\mathcal{D}_{n*} = \mathcal{D}_{\Gamma_n^*}$ for all n.) Clearly $H_2 = H \oplus \mathcal{D}_1 \oplus \mathcal{D}_2$. Continuing in this fashion, i.e., using (1.7) and the fact that all the columns of U must be isometric and orthogonal shows that U can be identified with a matrix of the form:

$$U = \begin{bmatrix} \Gamma_1 & D_{1*}\Gamma_2 & D_{1*}D_{2*}\Gamma_3 & D_{1*}D_{2*}D_{3*}\Gamma_4 & \cdots \\ D_1 & -\Gamma_1^*\Gamma_2 & -\Gamma_1^*D_{2*}\Gamma_3 & -\Gamma_1^*D_{2*}D_{3*}\Gamma_4 & \cdots \\ 0 & D_2 & -\Gamma_2^*\Gamma_3 & -\Gamma_2^*D_{3*}\Gamma_4 & \cdots \\ 0 & 0 & D_3 & -\Gamma_3^*\Gamma_4 & \cdots \\ 0 & 0 & 0 & D_4 & \cdots \\ 0 & 0 & 0 & 0 & \cdots \\ 0 & 0 & 0 & 0 & \cdots \\ \cdot & \cdot & \cdot & \cdot & \cdots \\ \cdot & \cdot & \cdot & \cdot & \cdots \\ \cdot & \cdot & \cdot & \cdot & \cdots \end{bmatrix} \tag{2.5}$$

$$\mathcal{K} = H \oplus \mathcal{D}_1 \oplus \mathcal{D}_2 \oplus \mathcal{D}_3 \oplus \cdots$$

Here $\Gamma_1 = R_1$ and Γ_{n+1} is a uniquely determined contraction mapping $\mathcal{D}_n$ into $\mathcal{D}_{n*}$ for $n \geq 1$.

The isometry U on $\mathcal{K}$ in (2.5) can be put in a simple form. To see this let U_n be the isometry defined by

$$U_n = I \oplus I \oplus \cdots \oplus I \oplus \begin{bmatrix} \Gamma_n & D_{n*} \\ D_n & -\Gamma_n^* \end{bmatrix} \oplus I \oplus I \oplus \cdots \tag{2.6}$$

where Γ_n is the contraction appearing in the n-th position and I is the identity operator on the appropriate space. A simple calculation shows that U in (2.5) is given by

$$U = \prod_{n=1}^{\infty} U_n = U_1U_2U_3U_4 \cdots \tag{2.7}$$

Clearly the minimal isometric dilation U of $\{R_n\}_1^\infty$ and the sequence of contractions $\{\Gamma_n\}_1^\infty$ uniquely determine each other. Following [5] we call a set of operators $\{\Gamma_n\}_1^\infty$ a choice sequence if Γ_1 is a contraction on H and Γ_{n+1} is a contraction mapping $\mathcal{D}_n$ into $\mathcal{D}_{n*}$ for $n \geq 1$. Combining the above with the uniqueness of U yields the following well known result [5,6,9-12,20,24].

THEOREM 2.1 *There is a one to one correspondence between the set of all positive definite functions* $\{R_n\}_1^\infty$ *on* H *and the set of all choice sequences* $\{\Gamma_n\}_1^\infty$. *In this correspondence the minimal isometric dilation* U *on* K *for* $\{R_n\}_1^\infty$ *can be identified with the isometry in* (2.5) *or equivalently* (2.6) (2.7).

3. POSITIVE EXTENSIONS.

In this section we review the Caratheodory interpolation problem. In particular, it is shown that $\{R_n\}_1^m$ and the choice sequence $\{\Gamma_n\}_1^m$ uniquely determine each other.

A sequence of operators $\{R_n\}_1^m$ on H has a *positive extension* if there exists a sequence of operators $\{R_n\}_{m+1}^\infty$ on H such that $\{R_n\}_1^\infty$ is positive definite. Recall that the *Caratheodory interpolation problem* is to find all positive extensions (if any) of $\{R_n\}_1^m$.

Let $\{R_n\}_1^\infty$ be positive definite and $\{\Gamma_n\}_1^\infty$ be its choice sequence. One can use the Levinson algorithm [22,23,27,29] or the layer peeling algorithm [24,25,26] to recursively compute the choice sequence $\{\Gamma_n\}_1^\infty$ from $\{R_n\}_1^\infty$ and visa versa. Here we will follow the ideas in [5,11,20] to recursively obtain the choice sequence $\{\Gamma_n\}_1^\infty$ from $\{R_n\}_1^\infty$ and solve the Caratheodory interpolation problem. Then these results will be extended to bilinear systems in Sec. 7.

To begin let U on K be the minimal isometric dilation of $\{R_n\}_1^\infty$ in (2.5) or (2.7). Let T_m on H_m be the mxm matrix contained in the upper mxm left hand corner of U in (2.5), that is,

$$T_m = P_{H_m}U|H_m = P_{H_m}U_1U_2 \cdots U_m|H_m$$

$$H_m = H \oplus D_1 \oplus D_2 \oplus \cdots \oplus D_{m-1}\,. \tag{3.1}$$

This (1.6) and the special form of U in (2.5) yields

$$R_m = P_H T_{m-1}^m|H + D_{1*} \cdots D_{m-1*}\Gamma_m D_{m-1} \cdots D_1 \qquad (m \geq 2)\,. \tag{3.2}$$

(Recall that $R_1 = \Gamma_1$). The second term consisting of T_{m-1} is determined by $\{\Gamma_n\}_1^{m-1}$. The operator Γ_m only appears in the last term. So (3.2) shows that $\{R_n\}_1^m$ and $\{\Gamma_n\}_1^m$ uniquely determine each other. Equation (3.2) also gives the following recursive algorithm [5,11,20].

PROCEDURE 3.1 Equation (3.2) can be used to recursively calculate the choice sequence $\{\Gamma_n\}_1^m$ given $\{R_n\}_1^m$. This procedure starts by setting $\Gamma_1 = R_1$. Now assume that we have computed $\{\Gamma_n\}_1^{m-1}$. Using $\{\Gamma_n\}_1^{m-1}$ we recursively compute $T_{m-1}^m|H$. this and R_m recursively yields Γ_m by inverting the self adjoint operators D_{i*} and D_i in (3.2).

REMARK 3.2 If at any step in this procedure Γ_m is not a contraction then the procedure stops. In this case the sequence $\{R_n\}_1^m$ does not have a positive definite extension, by Theorem 2.1. On the other hand if $\{\Gamma_n\}_1^m$ computed by (3.2) is a choice sequence, then the set of all positive definite extensions $\{R_n\}_1^\infty$ of $\{R_n\}_1^m$ is the set of all positive definite functions generated by the choice sequence $\{\Gamma_n\}_1^\infty$, where $\{\Gamma_n\}_{m+1}^\infty$ is an arbitrary choice sequence such that Γ_{m+1} is a contraction mapping D_m into D_{m*}. In particular there exists a unique positive definite extension of $\{R_n\}_1^m$ if and only if $\{\Gamma_n\}_1^m$ is a choice sequence and $D_m = \{0\}$ or $D_{m*} = \{0\}$.

REMARK 3.3. Assume that $\{R_n\}_1^m$ on H admits a positive definite extension and $\{\Gamma_n\}_1^m$ is its choice sequence. Obviously all positive definite extensions of $\{R_n\}_1^m$ have the same first m contractions $\{\Gamma_n\}_1^m$ in its choice sequence. So all positive definite extensions determine the same matrix T_m. Since T_m is contained in the upper left hand corner of U Equation (1.6) gives

$$R_i = P_H T_m^i | H \qquad (0 \le i \le m) . \tag{3.3}$$

Consider the sequence $\{R_n^o\}_1^\infty$ on H generated by

$$R_n^o = P_H T_m^n | H \qquad (n \ge 0) . \tag{3.4}$$

Remark 1.2 and (3.3) show that $\{R_n^o\}_1^\infty$ is a positive definite extension of $\{R_n\}_1^m$. We claim that $\{R_n^o\}_1^\infty$ is *the maximal entropy extension of* $\{R_n\}_1^m$, that is, the choice sequence for $\{R_n^o\}_1^\infty$ is

$$\{\Gamma_1, \Gamma_2, ..., \Gamma_m, 0, 0, 0, \cdots\} . \tag{3.5}$$

To prove our claim let U_o be the isometry in (2.5) generated by the choice sequence in (3.5). Using $U = U_o$ in (2.5) with $\{\Gamma_n\}_{m+1}^\infty$ zero shows that H_m is invariant for U_o^*. So U_o is an isometric dilation of T_m. Since H is cyclic for T_m and T_m is contained in the upper left hand corner of U_o we see that H is cyclic for U_o. Thus U_o is the minimal isometric dilation of T_m. This (3.4) and the dilation property in (1.1) gives

$$R_n^o = P_H T_m^n | H = P_H U_o^n | H \qquad (n \ge 0) . \tag{3.6}$$

Therefore $\{R_n^o\}_1^\infty$ is the maximally entropy extension of $\{R_n\}_1^m$.

Finally it is noted that if $\{R_n\}_1^m$ admits a unique positive definite extension ($D_m = \{0\}$ or $D_{m*} = \{0\}$) then this unique extension is given by (3.4) where T_m is computed recursively by Procedure 3.1.

REMARK 3.4 Assume that $\{R_n\}_1^m$ on H admits a positive definite extension and $\{\Gamma_n\}_1^m$ is its choice sequence. Consider the matrix L_m mapping $\bigoplus_o^{m-1} H$ into H_m defined by

$$L_m = [I|H, T_m|H, T_m^2|H, ..., T_m^{m-1}|H] \tag{3.7}$$

We claim that L_m is an upper triangular factorization of the Toeplitz matrix $\hat{R}_m$, defined in (1.5) that is, $L_m^* L_m = \hat{R}_m$.

To prove this result let U on K in (2.5) be a minimal isometric dilation $\{R_n\}_1^\infty$ where $\{R_n\}_1^\infty$ is a positive definite extension of $\{R_n\}_1^m$. Since T_m is contained in the upper left hand corner of U we have

$$T_m^i | H = U^i | H \qquad (\text{for } 0 \leq i \leq m-1) . \tag{3.8}$$

This and the form of U in (2.5) shows that L_m is upper triangular. To prove that $L_m^* L_m = \hat{R}_m$ let $\{h_n\}_o^{m-1}$ be a sequence in H. Then (1.6) (3.7) and (3.8) gives:

$$(\hat{R}_m(\bigoplus_o^{m-1} h_n), (\bigoplus_o^{m-1} h_n)) = \sum_{o,o}^{m-1,m-1} (R_{n-i}h_n, h_i) =$$

$$\| \sum_o^{m-1} U^n h_n \|^2 = \| L_m (\bigoplus_o^{m-1} h_n) \|^2 .$$

Since $\{h_n\}_o^{m-1}$ is arbitrary this yields $L_m^* L_m = \hat{R}_m$ and verifies our claim.

Finally it is emphasized that one can obtain the upper triangular factorization L_m of $\hat{R}_m$ in (3.7) recursively, by Procedure 3.1. These factorizations are also discussed in [25] by using a layer peeling algorithm and scattering techniques.

We complete this section with the following well known fact. A proof is given only because a similar result and proof will be extended to bilinear systems.

THEOREM 3.5 *A sequence of operators* $\{R_n\}_1^m$ *on* H *admits a positive extension if and only if the Toeplitz matrix* $\hat{R}_{m+1}$ *in* (1.5) *is positive* (≥ 0).

PROOF Assume $\hat{R}_m$ is positive. To show that $\{R_n\}_1^m$ admits a positive extension we construct a contraction A on $X(\supseteq H)$ such that

$$R_n = P_H A^n | H \qquad (0 \leq n \leq m) . \tag{3.9}$$

Then by Remark 1.2 the sequence $\{P_H A^n | H\}_1^\infty$ is a positive extension of $\{R_n\}_1^m$. To this end let F be the quotient space defined by $(\bigoplus_o^m H)/\ker \hat{R}_m$ with the inner product

$$(f, g)_F = (\hat{R}_{m+1} f, g) = \sum_{i=0,j=0}^{m,m} (R_{i-j} f_i, g_j) \tag{3.10}$$

where $f = \bigoplus_o^m f_i$ and $g = \bigoplus_o^m g_i$ are in F. Consider the operator A on X defined by

$$A(f_o \oplus f_1 \oplus \cdots \oplus f_{m-1}) = \tag{3.11}$$

$$P_X(0 \oplus f_o \oplus f_1 \oplus \cdots \oplus f_{m-1}) \quad (\bigoplus_o^{m-1} f_i \in X)$$

where X is the subspace of F identified with $X = \bigoplus_o^{m-1} H$. By virtue of (3.10) (3.11) we have

$$(A^n h, h')_F = (R_n h, h') \qquad (\text{for } 0 \leq n \leq m)$$

where h and h′ are in H. (Note H is identified with $H \oplus \{0\} \oplus \{0\} \oplus \cdots \oplus \{0\}$.) Hence (3.9) holds. Finally A is a contraction. This follows from (1.5) (3.10) and

$$\|A(f_o \oplus f_1 \oplus \cdots \oplus f_{m-1})\|_X^2 = \|P_X(0 \oplus f_o \oplus f_1 \oplus \cdots \oplus f_{m-1})\|_F^2 \leq$$

$$\|0 \oplus f_o \oplus f_1 \oplus \cdots \oplus f_{m-1}\|_F^2 = \|f_o \oplus f_1 \oplus \cdots \oplus f_{m-1} \oplus 0\|_X^2 .$$

Therefore $\{R_n\}_o^m$ admits a positive extension. The other half of the proof is obvious and omitted.

We conclude this section by noting that the positive definite extension $\{P_H A^n | H\}_1^\infty$ of $\{R_n\}_1^m$ in the previous proof is precisely the maximal entropy extension of $\{R_n\}_1^m$. The proof is left to the reader.

4. RANDOM PROCESSES

In this section we sketch how the above theory can be used to obtain state space realizations for a stationary Gaussian process. Then we generalize this setup and introduce a stochastic bilinear system which motivates a new Caratheodory interpolation problem. Our solution to this problem will follow the ideas of Sections 2 and 3 and lead to new results in bilinear systems theory.

To begin let y(n) be a stationary Gaussian random process with values in the Hilbert space H. A *state space realization* of y(n) is a system of the form:

$$x(n+1) = Tx(n) + Bu(n) \tag{4.1}$$

$$y(n) = Cx(n)$$

where T on X, B mapping U into X and C mapping X into H are bounded operators, and u(n) is a stationary Gaussian random process with values in U. By definition $E(u(n)u^*(m)) = I$ if $n = m$ and $E(u(n)u^*(m)) = 0$ if $n \neq m$, where E is the expectation. It is well known that the (steady) state covariance $E(x(n)x^*(n)) = Q$ satisfies the following Lyapunov equation:

$$Q = TQT^* + BB^* \qquad (0 < \delta I \leq Q < \infty) . \tag{4.2}$$

The system in (4.1) and (4.2) is denoted by {T, B, C, Q}. It is always assumed that the state covariance Q satisfies $0 < \delta I \leq Q < \infty$ where $\delta > 0$ is a scalar. The output covariance for y(n) in (4.1) is

$$R_n = E(y(n)y^*(0)) = CT^n QC^* \qquad (n \geq 0) . \tag{4.3}$$

(Without loss of generality it is always assumed that $E(y(0)y^*(0)) = R_o = I$.) Since

$\{R_n\}_1^\infty$ is generated by a stochastic process it is positive definite. Finally we say that $\{T, B, C, Q\}$ is ***minimal*** or equivalently ***stochastically controllable and observable*** if QC^*H is cyclic for T and C^*H is cyclic for T^*, that is,

$$X = \bigvee_{n\geq 0} T^n QC^* H \text{ and } X = \bigvee_{n\geq 0} T^{*n} C^* H \qquad (4.4)$$

respectively. Minimal realizations guarantee that the state space is of the lowest possible dimension, they are canonical and possess many important properties, see [1,16,22,28,30] for further details.

A system of operators $\{T, B, C, Q\}$ satisfying (4.1) to (4.3) is called ***a stochastic realization*** of $\{R_n\}_1^\infty$. This motivates the following ***stochastic realization problem:*** Given a positive definite sequence find a (minimal) stochastic realization $\{T, B, C, Q\}$ of $\{R_n\}_1^\infty$. One solution to this problem is given by the following which is the ***positive real lemma*** [3,4]. Our proof is based on dilation theory.

LEMMA 4.1 *A sequence of operators* $\{R_n\}_1^\infty$ *on H is positive definite if and only if there exists a minimal stochastic realization of* $\{R_n\}_1^\infty$.

PROOF Assume that $\{R_n\}_1^\infty$ is positive definite. Let U on K be its minimal isometric dilation. Consider the operator T on X defined by

$$T^* = U^*|X \text{ where } X = \bigvee_{n\geq 0} U^{*n} H . \qquad (4.5)$$

Let $C = P_H|X$ and $B = D_{T^*}$ be the positive square root of $(I-TT^*)$. Obviously $Q = I$ satisfies the Lyapunov equation in (4.2). We claim that $\{T, B, C, I\}$ is a minimal stochastic realization of $\{R_n\}$. The definition of T in (4.5) gives

$$CT^n QC^* = P_H U^n | H = R_n \qquad (n \geq 0) . \qquad (4.6)$$

So $\{T, B, C, I\}$ is a stochastic realization of $\{R_n\}_1^\infty$. The definition of T in (4.5) with $C^*H = H$ also shows that C^*H is cyclic for T^*. Using $C^*H = H$, the minimality of U (see (1.7)) and $P_X U = TP_X$ we have

$$\bigvee_{n\geq 0} T^n C^* H = \bigvee_{n\geq 0} P_X U^n H = P_X K = X . \qquad (4.7)$$

Therefore C^*H is cyclic for T and $\{T, B, C, I\}$ is a minimal stochastic realization of $\{R_n\}_1^\infty$.

Assume that $\{T, B, C, Q\}$ is a minimal stochastic realization of $\{R_n\}_1^\infty$. Without loss of generality we can assume that $Q = I$. The stochastic realization can always be converted to $Q = I$ by means of a similarly transform. In this case (4.2) shows that T is a contraction. Let U be the minimal isometric dilation of T and $\{h_n\}_o^\infty$ a sequence in H with finite support. By (4.3) with $Q = I$ and (1.1):

$$\sum_{n\geq 0, m\geq 0} (R_{n-m}h_n, h_m) =$$

$$\sum_{n\geq m} (T^{n-m}C^*h_n, C^*h_m) + \sum_{m<n} (T^{*m-n}C^*h_n, C^*h_m) = \tag{4.8}$$

$$\sum_{n\geq m} (U^{n-m}C^*h_n, h_m) + \sum_{m<n} (C^*h_n, U^{m-n}C^*h_m) =$$

$$||\sum U^n C^* h_n||^2 \geq 0 \, .$$

Therefore $\{R_n\}_1^\infty$ in (4.3) is positive definite. This completes the proof.

For further details on linear stochastic realization theory, see [1,16,22,28,30].

In many applications one is not given the entire covariance sequence. This leads to the following *partial stochastic realization problem:* Assume that $\{R_n\}_1^m$ admits a positive extension find a stochastic system $\{T, B, C, Q\}$ such that

$$R_n = CT^nQC^* \qquad (\text{for } 0 \leq n \leq m) \, . \tag{4.9}$$

Remark 3.3 and (3.3) provide a recursive solution to this problem, that is $\{T_m, D_{T_m^*}, P_H | H_m, I\}$ is a stochastic realization of $\{R_n\}_1^m$. (One can show that $D_{T_m^*}$ can be identified with D_{m*}). In fact this provides a maximal entropy realization of $\{R_n\}_1^m$

A stochastic bilinear system is a system of the form

$$x(n+1) = Tx(n) + Nx(n)v(n) + Bu(n) \tag{4.10}$$

$$y(n) = Cx(n)$$

where T, N are operators on X, B maps U into X and C maps X into H. Here u(n) and v(n) are independent mean zero Gaussian white noise processes with values in U and C^1 respectively. Obviously (4.10) reduces to (4.1) if N = 0. As before we assume that (4.10) is a stationary representation of y(n) and the (steady) state covariance $E(x(n)x^*(n)) = Q$. In this case Q satisfies the following Lyapunov equation:

$$Q = TQT^* + NQN^* + BB^* \qquad (0 < \delta I \leq Q < \infty) \, . \tag{4.11}$$

The stochastic bilinear system in (4.10) and (4.11) is denoted by $\{T, N, B, C, Q\}$. As before it is always assumed that $0 < \delta I \leq Q < \infty$.

Using the fact that u(n) and v(n) are mean zero and independent with (4.10) we have

$$R(z) = CTQC^* = E(y(1)y^*(0)),$$

$$R(\lambda) = CNQC^* = E(y(1)y^*(0)v(0)),$$

$$R(z^2) = CT^2QC^* = E(y(2)y^*(0)),$$

$$R(z\lambda) = CTNQC^* = E(y(2)y^*(0)v(0)),$$

$$R(\lambda z) = CNTQC^* = E(y(2)y^*(0)v(1)),$$

$$R(\lambda^2) = CN^2QC^* = E(y(2)y^*(0)v(0)v(1)),...,$$

$$R(z^4) = CT^4QC^* = E(y(4)y^*(0)),...,$$

$$R(z^3\lambda) = CT^3NQC^* = E(y(4)y^*(0)v(0)),...,$$

$$R(z^2\lambda z) = CT^2NTQC^* = E(y(4)y^*(0)v(1)),...,$$

$$R(z\lambda z\lambda^2 z) = CTNTN^2TQC^* =$$

$$E(y(6)y^*(0)v(1)v(2)v(4)),...$$

Notice also that N appears in the j-th position on the left of QC^* if v(j-1) appears in the expectation. It is always assumed that $R(1) = CC^*$ and $CC^* = I$ the identity.

The idea is that {T, N, B, C, Q} uniquely determines a sequence of operators $\{R(\pi_n)\}_1^\infty$ on *H* satisfying

$$R(\pi_n) = C\pi_n(T, N)QC^* \qquad (n \geq 0 \text{ and } \pi_n \in P_n) \tag{4.12}$$

Here $\pi_n = \pi_n(z, \lambda)$ is a product of length n of two noncommutative variables z and λ. For example

$$\pi_{10}(z, \lambda) = z^3\lambda^2 z\lambda^4 \neq z^4\lambda^6$$

$$\pi_{10}(T, N) = T^3N^2TN^4 .$$

The operator $\pi_n(T, N)$ is obtained by substituting T for z and N for λ. By definition $\pi_o = 1$ and $\pi_o(T, N) = I$. The set of all products of length n of the noncommutative variables z and λ is denoted by P_n. By a slight abuse of notation we set

$$\{R(\pi_n)\}_1^j = \{R(\pi_n):1 \leq n \leq j \text{ and } \pi_n \in P_n\} .$$

Finally we set $R(\pi_o) = I$. A theory of deterministic bilinear realizations using noncommutative variables is given in [14,15]. Here we use noncommutative variables for notational purposes only. One could just as easily use the regular transform theory in [17,31]. Both methods have their advantages and disadvantages [17,31].

We call {T, N, B, C, Q} a ***stochastic bilinear realization*** of $\{R(\pi_n)\}_1^\infty$ if {T, N, B, C, Q} satisfies (4.11) and (4.12). This system is ***minimal or stochastically controllable and observable*** if the space QC^*H and C^*H is cyclic for the pair of operators {T, N} and $\{T^*, N^*\}$ on X respectively. To be precise

$$X = \vee\{\pi_n(T, N)QC^*H{:}n \geq 0 \text{ and } \pi_n \in P_n\} \tag{4.13}$$

$$X = \vee\{\pi_n(T^*, N^*)C^*H{:}\ n \geq 0 \text{ and } \pi_n \in P_n\}$$

, respectively. Consulting the deterministic theory in [17,31] we see that (4.13) implies that if we view (4.10) as a deterministic bilinear realization of $\{R(\pi_n)\}_1^\infty$ in (4.12) where $B = QC^*$, then the state space X is minimal. In other words if (4.13) holds the state space X of a stochastic bilinear realization of $\{R_n\}_1^\infty$ is of the lowest possible dimension over the class of all stochastic bilinear realizations of $\{R_n\}_1^\infty$. Furthermore if $\{R_n\}_1^\infty$ admits a finite dimensional stochastic realization ($\dim X < \infty$), then the operators C, T and N are unique up to a similarity transform. So minimal stochastic bilinear realizations are "canonical".

The above discussion leads to the following stochastic bilinear realization problems: (i) Given a sequence of operators $\{R(\pi_n)\}_1^\infty$ find (if any) a minimal stochastic bilinear realization {T, N, B, C, Q} of $\{R(\pi_n)\}_1^\infty$. (ii) Given a finite sequence of operators $\{R(\pi_n)\}_1^m$ find a recursive procedure to obtain a stochastic bilinear system {T, N, B, C, Q} such that (4.12) holds for $0 \leq n \leq m$. In the following sections we shall mimic our previous theory to solve these problems and a new Caratheodory interpolation problem.

5. MINIMAL ISOMETRIC DILATIONS FOR TWO OPERATORS

In this section we develop a dilation theory for positive definite functions generated by stochastic bilinear systems. To begin let T and N be two operators on X such that [T, N] is a contraction, or equivalently,

$$TT^* + NN^* \leq I\,. \tag{5.1}$$

Following [8,18,19] we call {U, V} an ***isometric dilation*** of {T, N} if {U, V} are two isometries on $K \supseteq X$ with orthogonal range, or equivalently

$$UU^* + VV^* \leq I \tag{5.2}$$

such that X is invariant for both U^* and V^* and

$$T^* = U^*|X \text{ and } N^* = V^*|X\,. \tag{5.3}$$

The pair {U, V} is a *minimal* isometric dilation of {T, N} if {U, V} satisfy the additional condition:

$$K = \bigvee\{\pi_n(U, V)X : n \geq 0 \text{ and } \pi_n \in P_n\} . \tag{5.4}$$

As expected all minimal isometric dilations of {T, N} are isomorphic. To be precise if $\{U', V'\}$ on $K'(\supseteq X)$ is another minimal isometric dilation of {T, N}, then there exists a unitary operatory W mapping K onto K′ such that $WU = U'W$ and $WV = V'W$ and $W|H = I$. The proof of this is similar to the identical result in dilation theory for one operator [35] and is omitted.

To construct a minimal isometric dilation for {T, N} we need the following shift operators on $l_2(G)$:

$$S_o(h_o, h_1, h_2, \ldots) = (0, h_o, 0, h_1, 0, h_2, 0, \ldots) \tag{5.5}$$

$$S_e(h_o, h_1, h_2, \ldots) = (0, 0, h_o, 0, h_1, 0, h_2, 0, \ldots)$$

Clearly S_o and S_e are two isometries on $l_2(G)$ with orthogonal range. They place the nonzero entries in the odd or even positions. Now let D be the closed range of $D = D_{[T,N]}$. Condition (5.1) guarantees that [T, N] is a contraction and D is well defined. A simple calculation verifies that {U, V} on K, defined below is the minimal isometric dilation of {T, N}:

$$U(h \oplus f) = Th \oplus (DE_1 h + S_o f) \tag{5.6}$$

$$V(h \oplus f) = Nh \oplus (DE_2 h + S_e f) \qquad (h \oplus f \in H \oplus l_2(D))$$

$$K = H \oplus D \oplus D \oplus D \oplus D \oplus \cdots$$

Here E_1 and E_2 is the operator that embeds H into $H \oplus \{0\}$ and H into $\{0\} \oplus H$ respectively.

A pair of isometries {U, V} on K is an ***isometric dilation*** of $\{R(\pi_n)\}_1^\infty$ if U and V have orthogonal range and

$$R(\pi_n) = P_H \pi_n(U, V)|H \qquad (n \geq 0 \text{ and } \pi_n \in P_n) . \tag{5.7}$$

The pair {U, V} is a ***minimal isometric dilation*** of $\{R(\pi_n)\}_1^\infty$ if in addition to (5.7):

$$K = \bigvee\{\pi_n(U, V)H : n \geq 0 \text{ and } \pi_n \in P_n\} . \tag{5.8}$$

As expected two minimal isometric dilations of $\{R(\pi_n)\}_1^\infty$ are unique up to an isomorphism: If $\{U', V'\}$ on K' is another minimal isometric dilation of $\{R(\pi_n)\}_1^\infty$, then there is a unitary operator W mapping K onto K' such that $WU = U'W$ and $WV = V'W$ and $W|H = I$. The proof of this is similar to the proof of Theorem 1.1 or equivalently Theorem 7.1 pg. 25 in [35] and is omitted. Finally it is noted that if $\{R(\pi_n)\}_1^\infty$ admits an isometric dilation {U, V} it also admits a (unique) minimal isometric dilation. The minimal isometric dilation is obtained by restricting U and V to the subspace generated by H in the left hand side of (5.8).

Some further notation is needed. Recall that z and λ are noncommutative variables and $\pi_n = \pi_n(z, \lambda)$ is in P_n. For $j \geq i$ we define $\pi_i^* \pi_j$ by

$$\pi_i^* \pi_j = \pi_k \text{ if } \pi_j = \pi_i \pi_k \qquad (j \geq i) \tag{5.9}$$

$$= \phi \quad \text{otherwise} \qquad (j \geq i).$$

This says that $\pi_i^* \pi_j \neq \phi$ if and only if π_i is a left divisor of π_j. Note $\pi_o^* \pi_j = \pi_j$. If U and V are two isometries with orthogonal range then

$$\pi_i^*(U, V)\pi_j(U, V) = \pi_k(U, V) \text{ if } \pi_j = \pi_i \pi_k \qquad (j \geq i) \tag{5.10}$$

$$= 0 \qquad \text{otherwise} \qquad (j \geq i).$$

where $\pi_i^*(U, V)$ is the adjoint of $\pi_i(U, V)$. Let $\{R(\pi_n)\}_1^\infty$ be a sequence of operators on H. We set $R(\phi) = 0$ and

$$R(\pi_j^* \pi_i) = R^*(\pi_i^* \pi_j) \qquad (\text{if } j \geq i)\ . \tag{5.11}$$

Finally we say that $\{R(\pi_n)\}_1^\infty$ is *positive definite* if

$$\sum_{o,o}^{\infty,\infty} \sum_{\pi_i \epsilon P_i, \pi_j \epsilon P_j} (R(\pi_i^* \pi_j)h(\pi_j), h(\pi_i)) \geq 0 \tag{5.12}$$

where $\{h(\pi_j)\}_o^\infty$ is any sequence in H with finite support. Using this we present the following "generalization" of the Naimark dilation theorem, Theorem 1.1 or equivalently Theorem 7.1 pg. 25 in [35].

THEOREM 5.1. *The sequence of operators* $\{R(\pi_n)\}_1^\infty$ *on H is positive definite if and only if* $\{R(\pi_n)\}_1^\infty$ *admits a minimal isometric dilation. In this case the minimal isometric dilation of* $\{R(\pi_n)\}_1^\infty$ *is unique up to an isomorphism.*

PROOF. The proof of this result is similar to the proof of Theorem 7.1 pg. 25 in [35]. For completeness it is given. Assume that $\{U, V\}$ on K is the minimal isometric dilation of $\{R(\pi_n)\}_1^\infty$ and $\{h(\pi_n)\}_o^\infty$ is a sequence of finite support in H. Using (5.7), (5.10) and (5.11) we have:

$$0 \leq \|\sum_o^\infty \sum_{\pi_j \epsilon P_j} \pi_j(U, V)h(\pi_j)\|^2 =$$

$$\sum_{o,o}^{\infty,\infty} \sum_{\pi_i \epsilon P_i, \pi_j \epsilon P_j} (\pi_i^*(U, V)\pi_j(U, V)h(\pi_j), h(\pi_i)) =$$

$$\sum_{o,o}^{\infty,\infty} \sum_{\pi_i \epsilon P_i, \pi_j \epsilon P_j} (R(\pi_i^* \pi_j)h(\pi_j), h(\pi_i))\ . \tag{5.13}$$

Hence $\{R(\pi_n)\}_1^\infty$ is positive definite.

Assume that $\{R(\pi_n)\}_1^\infty$ is positive definite. Let K be the (quotient) Hilbert space generated by sequences in H with finite support of the form $\{h(\pi_n)\}_o^\infty$. The inner product on K is defined by

$$(\{h(\pi_j)\}_o^\infty, \{h(\pi_i)\}_o^\infty)_K = \sum_{o,o}^{\infty,\infty} (R(\pi_i^* \pi_j)h(\pi_j), h(\pi_i)) . \tag{5.14}$$

Equation (5.12) guarantees that this inner product is positive. We identify the zero element in K with the set of all sequences f such that $\|f\|_K = 0$. This allows us to turn K into a Hilbert space by factoring out the sequences with zero norm. Note f = g in K if and only if f - g = 0 in K. (For a further discussion see the proof of Theorem 7.1 pg. 25 in [35] where the same technique is used.)

Consider the shift operators U and V on K generated by

$$Uh(\pi_n) = h(z\pi_n) \text{ and } V(h(\pi_n)) = h(\lambda\pi_n) . \tag{5.15}$$

Using (5.9) (5.11) (5.14) and $R(\phi) = 0$ it is easy to verify that U and V are two isometries on K with orthogonal range. We embed H in K by setting $h = h(\pi_o)$ where $h \in H$. For h, h′ in H we have

$$(\pi_n(U, V)h, h')_K = (h(\pi_n), h'(\pi_o))_K = (R(\pi_n)h, h') . \tag{5.16}$$

Thus (5.7) holds. Obviously H is cyclic for the pair {U, V}, that is, (5.8) holds. So {U, V} is a minimal isometric dilation of $\{R(\pi_n)\}_1^\infty$. The proof of uniqueness is similar to the proof of uniqueness in Theorem 7.1 pg. 25 in [35] and is omitted. This completes the proof.

One can also obtain Theorem 5.1 directly from the classical Naimark dilation theorem (Theorem 7.1 pg. 25 in [35]) where G is the group determined by the two noncommutative variables z and λ. The orthogonality of U and V will be a consequence of $R(\phi) = 0$ and (5.9). The classical Naimark dilation theorem also yields a minimal unitary dilation of $\{R(\pi_n)\}_1^\infty$. Using this unitary dilation one can extend the results in [16] to stochastic bilinear system with finite dimensional state spaces. Rather than elaborate on this let us state the following bilinear positive real lemma.

LEMMA 5.2 *A sequence of operators* $\{R(\pi_n)\}_1^\infty$ *on* H *is positive definite if and only if there exists a (minimal) stochastic bilinear realization of* $\{R(\pi_n)\}_1^\infty$.

PROOF. The proof is similar to Lemma 4.1. Assume that $\{R(\pi_n)\}_1^\infty$ is positive definite and {U, V} on K is its minimal isometric dilation. Consider the operators T and N on X defined by

$$T^* = U^*|X \text{ and } N^* = V^*|X \tag{5.17}$$

$$X = \bigvee\{\pi_n(U^*, V^*): n \geq 0 \text{ and } \pi_n \in P_n\} .$$

Set $C = P_H|X$ and B equal to the positive square root of $(I - TT^* - NN^*)$. (Since

U and V have orthogonal range this implies that [T, N] is a contraction and B is well defined.) Obviously Q = I satisfies the Lyapunov equation in (4.11). Equation (5.7) and (5.17) give

$$C\pi_n(T, N)QC^* = P_H \pi_n(U, V)|H = R(\pi_n) \quad (n \geq 0 \text{ and } \pi_n \in P_n). \tag{5.18}$$

So {T, N, B, C, I} is a stochastic bilinear realization of $\{R(\pi_n)\}_1^\infty$. The definition of X in (5.17) with $C^*H = H$ guarantees that the second equation in (4.13) holds. The first equation in (4.13) follows from the minimality of {U, V} (see (5.8)) and

$$TP_X = P_X U \text{ and } NP_X = P_X V. \tag{5.19}$$

Therefore {T, N, B, C, I} is a minimal stochastic bilinear realization of $\{R(\pi_n)\}_1^\infty$.

Assume that {T, N, B, C, Q} is a stochastic bilinear realization of $\{R(\pi_n)\}_1^\infty$. Without loss of generality assume that Q = I. (One can always convert to this case by means of a similarity transform.) The Lyapunov equation (4.11) with Q = I shows that [T, N] is a contraction, or equivalently (5.1) holds. Let {U, V} on K be the minimal isometric dilation of {T, N} and $\{h(\pi_n)\}_o^\infty$ a sequence in H with finite support. Using Q = I with (4.12), (5.3) (5.9), (5.10), (5.11) and $R(\phi) = 0$ we have

$$\sum_{o,o}^{\infty,\infty} \sum_{\pi_i \epsilon P_i, \pi_j \epsilon P_j} (R(\pi_i^* \pi_j)h(\pi_j), h(\pi_i)) =$$

$$\sum_{\substack{\pi_j = \pi_i \pi_k \\ j \geq i}} (\pi_k(T, N)C^*h(\pi_j), C^*h(\pi_i)) +$$

$$\sum_{\substack{\pi_i = \pi_j \pi_k \\ i > j}} (\pi_k^*(T, N)C^*h(\pi_j), C^*h(\pi_i)) = \tag{5.20}$$

$$\sum_{o,o}^{\infty,\infty} \sum_{\pi_i \epsilon P_i, \pi_j \epsilon P_j} (\pi_i^*(U, V)\pi_j(U, V)C^*h(\pi_j), C^*h(\pi_i)) =$$

$$\|\sum_{o}^{\infty} \pi_j(U, V)h(\pi_j)\|^2 \geq 0 .$$

Thus $\{R(\pi_n)\}_1^\infty$ is positive definite and the proof is complete.

Lemma 5.2 with the deterministic realization theory in [17,31] can be used to construct a minimal stochastic bilinear realization of $\{R(\pi_n)\}_1^\infty$. For an alternate approach to stochastic bilinear systems see [13].

6. CHOICE SEQUENCES REVISITED

In this section we generalize the results in Section 2 and show that a sequence is positive definite if and only if it uniquely determines a choice sequence. In our new setting a sequence of operators $\{[\Gamma_n, \Omega_n]\}_1^\infty$ is a choice sequence if $[\Gamma_1, \Omega_1]$ is a contraction mapping $H \oplus H$ into H and $[\Gamma_{n+1}, \Omega_{n+1}]$ is a contraction mapping $D_n \oplus D_n$ into D_{n*} for all $n \geq 1$. Here

$$D_n = D_{[\Gamma_n, \Omega_n]} \text{ and } D_{n*} = D_{[\Gamma_n, \Omega_n]^*} \qquad (n \geq 1)\,. \tag{6.1}$$

The closed range of D_n and D_{n*} is denoted by D_n and D_{n*}, respectively.

Let $\{[\Gamma_n, \Omega_n]\}_1^\infty$ be a choice sequence. By virtue of (6.1) and the Halmos extension (2.1) the operators U_n and V_n defined below are isometries for all $n \geq 1$:

$$U_n = I \oplus I \oplus \cdots \oplus I \oplus \begin{bmatrix} \Gamma_n & D_{n*} \\ D_n E_1 & -[\Gamma_n, \Omega_n]^* \end{bmatrix} \oplus I \oplus I \cdots \tag{6.2}$$

$$V_n = I \oplus I \oplus \cdots \oplus I \oplus \begin{bmatrix} \Omega_n & D_{n*} \\ D_n E_2 & -[\Gamma_n \Omega_n]^* \end{bmatrix} \oplus I \oplus I \oplus \cdots$$

where Γ_n and Ω_n appear in the n-th position. The operators E_1 and E_2 embed a space G into $G \oplus \{0\}$ and $\{0\} \oplus G$, respectively. Now consider the isometries U and V on K defined by

$$U = \mathop{\pi}_{i=1}^{\infty} U_n = U_1 U_2 U_3 U_4 \cdots$$

$$V = \mathop{\pi}_{n=1}^{\infty} V_n = V_1 V_2 V_3 V_4 \cdots \tag{6.3}$$

$$K = H \oplus D_1 \oplus D_2 \oplus D_3 \oplus D_4 \oplus \cdots$$

The matrix form for U and V is

$$U = \begin{bmatrix} \Gamma_1 & D_{1*}\Gamma_2 & D_{1*}D_{2*}\Gamma_3 & \cdots \\ D_1E_1 & -[\Gamma_1, \Omega_1]^*\Gamma_2 & -[\Gamma_1, \Omega_1]^*D_{2*}\Gamma_3 & \cdots \\ 0 & D_2E_1 & -[\Gamma_2, \Omega_2]^*\Gamma_3 & \cdots \\ 0 & 0 & D_3E_1 & \cdots \\ 0 & 0 & 0 & \cdots \\ \cdot & \cdot & \cdot & \cdots \\ \cdot & \cdot & \cdot & \cdots \\ \cdot & \cdot & \cdot & \cdots \end{bmatrix} \tag{6.4}$$

$$V = \begin{bmatrix} \Omega_1 & D_{1*}\Omega_2 & D_{1*}D_{2*}\Omega_3 & \cdots \\ D_1E_2 & -[\Gamma_1, \Omega_1]^*\Omega_2 & -[\Gamma_1, \Omega_1]^*D_{2*}\Omega_3 & \cdots \\ 0 & D_2E_2 & -[\Gamma_2, \Omega_2]^*\Omega_3 & \cdots \\ 0 & 0 & D_3E_2 & \cdots \\ 0 & 0 & 0 & \cdots \\ \cdot & \cdot & \cdot & \cdots \\ \cdot & \cdot & \cdot & \cdots \\ \cdot & \cdot & \cdot & \cdots \end{bmatrix}$$

By using the matrix forms of U and V one can easily verify that U, V satisfy the minimality condition in (5.8). Furthermore the range of U is orthogonal to the range of V. To see this first notice that the range of the following isometric operators are orthogonal

$$\begin{bmatrix} \Gamma_n \\ D_nE_1 \end{bmatrix}, \begin{bmatrix} \Omega_n \\ D_nE_2 \end{bmatrix} \begin{bmatrix} D_{n*} \\ -[\Gamma_n, \Omega_n]^* \end{bmatrix} \tag{6.5}$$

This and the matrix form of U and V shows that all the columns of U and V are isometric and orthogonal. In particular the range of U is orthogonal to the range of V. Therefore the choice sequence $\{[\Gamma_n, \Omega_n]\}_1^\infty$ uniquely determines {U, V} and its positive definite function $\{R(\pi_n)\}_1^\infty$ by (5.7).

The other way is also true any positive definite function $\{R(\pi_n)\}_1^\infty$ uniquely determines a choice sequence $\{[\Gamma_n, \Omega_n]\}_1^\infty$. In other words if {U, V} on K is the minimal isometric dilation of $\{R(\pi_n)\}_1^\infty$ then U, V can be identified with the operators in (6.3), (6.4). The proof of this follows ideas similar to Section 2 and is only sketched. First we define the spaces H_m and D_m by

$$H_m = UH_{m-1} \vee VH_{m-1} \qquad (m > 1 \text{ and } H_1 = H) \tag{6.6}$$

$$D_m = H_{m+1} \ominus H_m \,.$$

The minimality condition (5.8) gives

$$K = H_\infty \text{ and } K = H \oplus D_1 \oplus D_2 \oplus D_3 \oplus \cdots \tag{6.7}$$

Therefore U and V admit a matrix representation of the form (2.2) (where V replaces U and Ω_1 replaces Γ_1). Let Γ_1 and Ω_1 be the contractions on H defined by

$$P_H U|H = R(z) = \Gamma_1 \text{ and } P_H V|H = R(\lambda) = \Omega_1 \,. \tag{6.8}$$

Since U and V have orthogonal range $[\Gamma_1, \Omega_1]$ is a contraction. For h and g in H we have

$$||P_{D_1}(Uh + Vg)||^2 = ||h \oplus g||^2 - ||P_H(Uh + Vg)||^2 =$$

$$||h \oplus g||^2 - ||\Gamma_1 h + \Omega_1 g||^2 = ||D_1(h \oplus g)||^2 .$$

This implies that there exists a unitary operator W mapping $H_2 \ominus H_1$ onto the closed range of D_1 such that

$$WP_{D_1}(Uh + Vg) = D_1(h \oplus g) .$$

Since {U,V} is unique up to an isomorphism without loss of generality we choose W = I. Then $H_2 \ominus H_1 = D_1$ the closed range of D_1. This gives the first column of U and V in (6.4). Because U and V have orthogonal range, all the columns of U and V must be isometric and orthogonal. So using this with the orthogonality of the operators in (6.5) and following the arguments in Section 2 shows that U and V can be identified with the operators U and V in (6.4) or (6.3). Summing up

THEOREM 6.1. *There is a one to one correspondence between the set of all positive definite functions* $\{R_n(\pi_n)\}_1^\infty$ *and the set of all choice sequences* $\{[\Gamma_n, \Omega_n]\}_1^\infty$. *In this correspondence the minimal isometric dilation* {U, V} *on* K *of* $\{R(\pi_n)\}_1^\infty$ *is given by* (6.3) *and* (6.4).

REMARK 6.2. Let T, N be two operators on X satisfying (5.1) where $X = H$. Consider the sequence $\{R_n(\pi_n)\}_1^\infty$ on H defined by

$$R(\pi_n) = \pi_n(T, N) \qquad (n \geq 0 \text{ and } \pi_n \in P_n) . \tag{6.9}$$

Let B be the positive square root of $(I - TT^* - NN^*)$. Obviously {T, N, B, I, I} is a minimal stochastic bilinear realization of $\{R(\pi_n)\}_1^\infty$ in (6.9). In this case $\{R(\pi_n)\}_1^\infty$ and {T, N} admit the same minimal isometric dilation {U, V}. The above theorem and the uniqueness of minimal isometric dilations show that U and V can be identified with the operators in (6.4) and (5.6). Therefore the choice sequence for $\{R(\pi_n)\}_1^\infty$ in (6.9) is $[\Gamma_1, \Omega_1] = [T, N]$ and $[\Gamma_n, \Omega_n] = [0, 0]$ for all $n > 1$.

7. POSITIVE EXTENSIONS

We say that $\{R(\pi_n)\}_1^m$ on H has a *positive extension* if there exists a sequence of operators $\{R(\pi_n)\}_{m+1}^\infty$ on H such that $\{R(\pi_n)\}_1^\infty$ is positive definite. In this setting the Caratheodory interpolation problem becomes: Given a sequence of operators $\{R(\pi_n)\}_1^m$ find (if any) all positive extensions of $\{R(\pi_n)\}_1^m$.

Let $\{R(\pi_n)\}_1^\infty$ be positive definite and $\{[\Gamma_n, \Omega_n]\}_1^\infty$ be its choice sequence. One can extend the Levinson algorithm [22,23,27] or the layer peeling algorithm [24,25,26] to recursively compute the choice sequence $\{[\Gamma_n, \Omega_n]\}_1^\infty$ from $\{R(\pi_n)\}_1^\infty$ and visa versa. In this section we extend the ideas in Section 3 and [5,11,20] to recursively compute the choice sequence and solve this Caratheodory interpolation problem.

To this end let $\{U, V\}$ on K in (6.3) (6.4) be the minimal isometric dilation of $\{R(\pi_n)\}_1^\infty$. Let T_m and N_m be the mxm matrices on H_m contained in the upper left hand corner of U and V respectively, i.e.,

$$T_m = P_{H_m}U|H_m = P_{H_m}U_1U_2 \cdots U_m|H_m$$

$$N_m = P_{H_m}V|H_m = P_{H_m}V_1V_2 \cdots V_m|H_m \tag{7.1}$$

$$H_m = H \oplus D_1 \oplus D_2 \oplus \cdots \oplus D_{m-1}\,.$$

Let π'_{m-1} be the operator mapping D_1 into D_{m-1} defined by

$$\pi'_{m-1} = D_{m-1}Q_{m-1}D_{m-2} \cdots Q_2D_1Q_1 \qquad (m \geq 2) \tag{7.2}$$

where $Q_i = E_1[Q_i = E_2]$ if the i-th term on the left of π is $z[\lambda]$ respectively. For example

$$\pi'_1 = D_1E_2 \qquad (\text{if } \pi_1 = \lambda)$$

$$\pi'_7 = D_7E_2D_6E_1D_5E_2D_4E_2D_3E_2D_2E_1D_1E_1 \quad (\text{if } \pi_7 = \lambda z\lambda^3z^2)\,.$$

This and the matrix form of U and V in (6.4) yields

$$R(z\pi_{m-1}) = P_HT_{m-1}\pi_{m-1}(T_{m-1}, N_{m-1})|H + D_{1*} \cdots D_{m-1*}\Gamma_m\pi'_{m-1} \quad (m \geq 2)$$

$$R(\lambda\pi_{m-1}) = P_HN_{m-1}\pi_{m-1}(T_{m-1}, N_{m-1})|H + D_{1*} \cdots D_{m-1*}\Omega_m\pi'_{m-1} \quad (m \geq 2)$$

$$R(z) = \Gamma_1 \text{ and } R(\lambda) = \Omega_1\,. \tag{7.3}$$

The second term consisting of T_{m-1} and N_{m-1} is completely determined by $\{[\Gamma_n, \Omega_n]\}_1^{m-1}$. The operators Γ_m and Ω_m only appear in the last term. Notice that each π'_{m-1} contains only part of the domain of the operators D_i see (7.2). So we need $\{\pi'_{m-1}:\pi_{m-1} \in P_{m-1}\}$ to completely determine Γ_m and Ω_m in (7.3). This and (7.2) (7.3) shows that there is a one to one correspondence between $\{R(\pi_n)\}_1^m$ and the choice sequence $\{[\Gamma_n, \Omega_n]\}_1^m$. In particular $\{R(\pi_n)\}_1^{m-1}$ and

$$\{R(z\pi_{m-1}):\pi_{m-1} \in P_{m-1}\} \quad [\{R(\lambda\pi_{m-1}):\pi_{m-1} \in P_{m-1}\}] \tag{7.4}$$

uniquely determine and are uniquely determined by $\{[\Gamma_n, \Omega_n]\}_1^{m-1}$ and Γ_m $[\Omega_m]$, respectively. This with Theorem 6.1 readily yields the following recursive algorithm.

PROCEDURE 7.1. Equation (7.3) can be used to recursively calculate the choice sequence $\{[\Gamma_n, \Omega_n]\}_1^m$ from $\{R(\pi_n)\}_1^m$. This procedure starts with the initial condition $\Gamma_1 = R(z)$ and $\Omega_1 = R(\lambda)$. Now assume we have computed $\{[\Gamma_n, \Omega_n]\}_1^{m-1}$. The next step is obtained by using $\{[\Gamma_n, \Omega_n]\}_1^{m-1}$ to recursively compute

$\pi_{m-1}(T_{m-1}, N_{m-1})|H$ for all $\pi_{m-1} \epsilon P_{m-1}$. This and (7.3) (7.4) recursively yields $\Gamma_m[\Omega_m]$ respectively by inverting the self adjoint operators D_{i*} and D_i in (7.2) and (7.3). (Note we need (7.4) to completely determine the operators D_i in (7.2) and (7.3).)

REMARK 7.2. If at any step in this procedure $[\Gamma_m, \Omega_m]$ is not a contraction then the procedure stops. In this case $\{R(\pi_n)\}_1^m$ does not admit a positive definite extension, by Theorem 6.1. On the other hand if $\{[\Gamma_n, \Omega_n]\}_1^m$ computed by (7.3) is a choice sequence, then the set of all positive definite extensions $\{R(\pi_n)\}_1^\infty$ of $\{R(\pi_n)\}_1^m$ is the set of all positive definite functions generated by the choice sequence $\{[\Gamma_n, \Omega_n]\}_1^\infty$, where $\{[\Gamma_n, \Omega_n]\}_{m+1}^\infty$ is an arbitrary choice sequence such that $[\Gamma_{m+1}, \Omega_{m+1}]$ is a contraction mapping $D_m \oplus D_m$ into D_{m*}. In particular there is a unique positive definite extension of $\{R(\pi_n)\}_1^m$ if and only if $\{[\Gamma_n, \Omega_n]\}_1^m$ is a choice sequence and $D_m = \{0\}$ or $D_{m*} = \{0\}$.

REMARK 7.3. Assume that $\{R(\pi_n)\}_1^m$ on H admits a positive definite extension and $\{[\Gamma_n, \Omega_n]\}_1^m$ is its choice sequence. Obviously all positive definite extension of $\{R(\pi_n)\}_1^m$ all have the same first m contractions $\{[\Gamma_n, \Omega_n]\}_1^m$ in its choice sequence. So all positive definite extensions determine the same matrices T_m and N_m. Since T_m and N_m are contained in the upper left hand corner of U and V Equation (5.7) gives

$$R(\pi_i) = P_H\pi_i(T_m, N_m)|H \qquad (0 \leq i \leq m \text{ and } \pi_i \epsilon P_i) \tag{7.5}$$

Lemma 5.2 with $C = P_H|X_m$ and $Q = I$ implies that the sequence

$$R^o(\pi_n) = P_H\pi_n(T_m, N_m)|H \qquad (0 \leq n \text{ and } \pi_n \epsilon P_n) \tag{7.6}$$

is positive definite. By following the arguments in Remark 3.3 one can easily show that $\{R^o(\pi_n)\}_1^\infty$ is a *maximal entropy extension* of $\{R(\pi_n)\}_1^m$, that is, the choice sequence for $\{R^o(\pi_n)\}_1^\infty$ is

$$\{[\Gamma_1, \Omega_1], [\Gamma_2, \Omega_2], ..., [\Gamma_m, \Omega_m], 0, 0, 0, \cdots\} . \tag{7.7}$$

As in the linear setting one can show that the maximal entropy extension is the one which maximizes the m step prediction error over the class of all extensions. Finally it is noted that if $\{R(\pi_n)\}_1^m$ admits a unique positive extension ($D_m = \{0\}$ or $D_{m*} = \{0\}$) then this extension is given by the maximal entropy extension in (7.6), where T_m and N_m are commuted recursively by Procedure 7.1.

REMARK 7.4. Assume that $\{R(\pi_n)\}_1^m$ on H admits a positive definite extension and $\{[\Gamma_n, \Omega_n]\}_1^m$ is its choice sequence. Let $\{h(\pi_n)\}_o^m$ be a sequence in H. Let $\hat{R}_m$ be the (positive) self adjoint matrix on $\bigoplus_o^M H$ where $M = 2^{m+1} - 1$ with respect to the basis

$$\{h(\pi_o),\ h(z),\ h(\lambda),\ h(z^2),\ h(z\lambda),\ \ldots,\ h(\lambda^m)\} \tag{7.8}$$

determined by

$$(\hat{R}_m \bigoplus_{o}^{M} h(\pi_n),\ \bigoplus_{o}^{M} h(\pi_n)) = \tag{7.9}$$

$$\sum_{o,o}^{m,m} \sum_{\pi_i \epsilon P_i, \pi_j \epsilon P_j} (R(\pi_i^* \pi_j) h(\pi_j),\ h(\pi_i)) \geq 0\ .$$

Notice that $\hat{R}_m$ is positive because $\{R(\pi_n)\}_1^m$ admits a positive extension, see (5.12). Consider the matrix L_m mapping $\bigoplus_{o}^{M} H$ into H_m defined by

$$L_m = [I|H,\ T_{m+1}|H,\ N_{m+1}|H,\ T_{m+1}^2|H,\ T_{m+1}N_{m+1}|H,\ \ldots,\ N_{m+1}^m|H] \tag{7.10}$$

By extending the arguments in Remark 3.4 to this setting one can show that L_m is an upper triangular factorization of $\hat{R}_m$, that is, $L_m^* L_m = \hat{R}_m$. Therefore Procedure 7.1 provides a recursive algorithm to find a upper triangular factorization of $\hat{R}_m$.

REMARK 7.5. Recall that the stochastic partial bilinear realization problem is: Given a finite sequence of operators $\{R(\pi_n)\}_1^m$ which admits a positive definite extension find a stochastic bilinear system $\{T, N, B, C, Q\}$ such that

$$R(\pi_n) = C\pi_n(T, N)QC^* \qquad (0 \leq n \leq m \text{ and } \pi_n \epsilon P_n)\ . \tag{7.11}$$

Procedure 7.1 gives a recursive solution to this problem by choosing $T = T_m$, $N = N_m$, B equal to the positive square root of $(I - TT^* - NN^*)$ and $C = P_H|H_m$ and $Q = I$. Remark 7.3 shows that this stochastic bilinear system $\{T_m, N_m, B, C, I\}$ generates the maximal entropy extension $\{R^o(\pi_n)\}_1^\infty$ of $\{R(\pi_n)\}_1^m$ (As in the linear case one can show that this B can be identified with D_{m*}.)

We conclude with the following generalization of Theorem 3.5.

THEOREM 7.6 *A sequence* $\{R(\pi_n)\}_1^m$ *on* H *admits a positive extension if and only if the operator* $\hat{R}_m$ *in* (7.9) *is positive* (≥ 0).

SKETCH OF PROOF. The proof is similar to the proof of Theorem 3.5. Assume that (7.9) holds. Consider the Hilbert space F given by sequences of the form $\{h(\pi_n)\}_o^m$ with the inner product defined by (7.9):

$$(\{h(\pi_n)\}_o^m,\ \{h'(\pi_n)\}_o^m) =$$

$$\sum_{o,o}^{m,m} \sum_{\pi_i \epsilon P_i, \pi_j \epsilon P_j} (R(\pi_i^* \pi_j) h(\pi_j),\ h'(\pi_i))\ . \tag{7.12}$$

(As before we identify the zero element f in F by $||f|| = 0$.) Let X be the subspace of F consisting of sequences of the form $\{h(\pi_n)\}_o^{m-1}$. Consider the operators T and N on X defined by

$$T\{h(\pi_n)\}_o^{m-1} = P_X\{h(z\pi_n)\}_o^{m-1} \tag{7.13}$$

$$N\{h(\pi_n)\}_o^{m-1} = P_X\{h(\lambda\pi_n)\}_o^{m-1} .$$

Using (5.9), (5.11) and $R(\phi) = 0$ it is easy to verify that [T, N] is a contraction. Moreover by identifying h in H with $h(\pi_o)$ Equation (7.12) gives (5.16) where T replaces U and N replaces V for $0 \leq n \leq m$ and $\pi_n \epsilon P_n$. Thus

$$R(\pi_n) = P_H \pi_n(T, N)|H \qquad (0 \leq n \leq m \text{ and } \pi_n \epsilon P_n) . \tag{7.14}$$

Therefore {T, N, B, C, I} is a stochastic bilinear realization of $\{R(\pi_n)\}_o^\infty$ where $C = P_H|X$ and B is the positive square root of $(I - TT^* - NN^*)$. Lemma 5.2 implies that $\{R(\pi_n)\}_1^m$ admits a positive extension. The other half of the proof is a simple consequence of the definition of positive definite. This completes the proof.

Finally it is noted that the operator T and N in (7.13) can be identified with T_m and N_m. Moreover the positive definite extension of $\{R(\pi_n)\}_1^m$ generated by this {T, N, B, C, I} is precisely the maximal extropy extension of $\{R(\pi_n)\}_1^m$.

ACKNOWLEDGEMENTS

This research was supported in part by a grant from the National Science Foundation No. ECS 8419354.

REFERENCES

[1] Akaike, H. Stochastic theory of minimal realizations. *IEEE Trans. Auto. Cont.* AC-19 (1974), pp. 667-674.

[2] Akhiezer, N.I. *The Classical Moment Problem.* Edinburgh, Scotland: Olivier and Boyd, 1965.

[3] Anderson, B.D.O. A system theory criterion for positive real matrices. *SIAM. J. Control* Vol. 5, No. 2 (1967) pp. 171-182.

[4] Anderson, B.D.O. and S. Vongpanitlerd, *Network Analysis and Synthesis, A Modern Systems Theory Approach.* Prentice-Hall, Englewood Cliffs, N.J., 1973.

[5] Arsene, Gr., Z. Ceausescu and C. Foias On intertwining dilations VIII. *J. Operator Theory* 4, (1980), pp. 55-91.

[6] Arsene, Gr., Z. Ceausescu and C. Foias On intertwining dilations VII. *Proc. Coll. Complex Analysis,* Joensuw Lecture Notes in Math. 747 (1979) pp. 24-45.

[7] Bruckstein, A.M. and T. Kailath. Inverse scattering for discrete transmission-line models. *Preprint*. Dept. of Electrical Engineering, Stanford University.

[8] Bunce, J.W. Models for N-touples of noncommuting operators, *J. Functional Analysis*, (1984) pp. 21-30.

[9] Ceausescu, Z. and C. Foias. On intertwining dilations V. *Acta Sci. Math.* 40 (1978) pp. 9-32.

[10] Ceausescu, Z. and C. Foias. On intertwining dilations VI *Rev. Roum. Math. Pures Et Appl.* Tome XXIII No. 10 (1978) pp. 1471-1482.

[11] Constantinescu, T. On the structure of the Naimark Dilation. *J. Operator Theory* 12 (1984), pp. 159-175.

[12] Delsarte, PH., Y. Genin and Y. Kamp. Schur parameterization of positive definite block Toeplitz systems. *SIAM J. Appl. Math.* Vol. 36, No. 1 (1979) pp. 34-46.

[13] Desai, U.B. Realization of bilinear stochastic systems. *Proceedings of the 1985 American Control Conference,* Boston pp. 940-945.

[14] Fliess, M. Sur la realization des systemes dynamiques bilineaires *C.R. Academie Science,* Paris Series A, Vol. 277 (1973) pp. 923-926.

[15] Fliess, M. Un outel algebrique: les series formelles noncommutatives in *Mathematical Systems Theory* G. Marchesini, and S. Mitter, eds., Lecture Notes in Economics and Mathematical Systems, Vol. 131 Springer-Verlag, New York pp. 122-148, 1976.

[16] Foias, C. and A.E. Frazho Markov subspaces and minimal unitary dilations. *Acta Sci. Math.* 45 (1983) pp. 165-175.

[17] Frazho, A.E. A shift operator approach to bilinear system theory *SIAM J. Control,* 18 No. 6 (1980) pp. 640-658.

[18] Frazho, A.E. Models for noncommuting operators. *J. Functional Analysis* Vol. 48, No. 1 (1982) pp. 1-11.

[19] Frazho, A.E. Complements to Models for Noncommuting Operators. *J. Functional Analysis,* Vol. 59, No. 3 (1984) pp. 445-461.

[20] Frazho, A.E. Schur contractions and bilinear stochastic systems. *Proceedings of the 1984 Conference on Information Sciences and Systems,* Princeton University, pp. 190-196.

[21] Halmos, P.R. *A Hilbert Space Problem Book.* Princeton: Van Nostrand, 1967.

[22] Kailath, T. A view of three decades of linear filtering theory. *IEEE Trans. Inform. Theory.* Vol. IT-20 (1974) pp. 146-181.

[23] Kailath, T., A. Vieiera and M. Morf Inverses of Toeplitz operators innovations and orthogonal polynomials. *SIAM Review* Vol. 20, No. 1 (1976) pp. 106-119.

[24] Kailath, T. and A.M. Bruckstein. Naimark dilations, state-space generators and transmission lines. *Romanian Conf.* 1984.

[25] Kailath, T., A.M. Bruckstein and D. Morgan. Fast matrix factorizations via discrete transmission lines. *Preprint.* Dept. of Electrical Engineering, Stanford University.

[26] Lev-Ari, H. and T. Kailath. Lattice filter parameterization and modeling of nonstationary processes. *IEEE Trans. on Information Theory,* Vol. IT-30, No. 1, (1984), pp. 2-16.

[27] Levinson, N. The Wiener RMS (root mean square) error criterion in filter design and prediction. *J. Math. Phys.* Vol. 25, No. 4 (1947) pp. 261-278.

[28] Lindquist, A. and G. Picci. State space models for Gaussian stochastic processes. *Proceedings of the NATO-ASI Workshop on Stochastic Systems.* Les Arcs, France, June 1980.

[29] Robinson, E.A. and S. Treitel. *Geophysical Signal Analysis.* Prentice-Hall, Englewood Cliffs, N.J., 1980.

[30] Ruckebusch, G. Theorie Geometrique de la Representation Markovienne, *Ann. Inst. Henri Poincare.* XVI, 3 (1980), pp. 225-297.

[31] Rugh, W.J. *Nonlinear System Theory, The Volterra-Wiener Approach* The Johns Hopkins University Press Baltimore, 1981.

[32] Sarason, D. Generalized interpolation in H^∞. *Trans. Amer. Math. Soc.,* 127 (1967), 179-203.

[33] Schur, I, Uber Potenzreihen, die im Innern des Einheitskreises beschrankt sind, *J. Reine Angew. Math.,* 148 (1918), pp. 122-145.

[34] Sz.-Nagy, B. *Unitary dilations of Hilbert space operators and related topics.* CBMS Regional Conference Series in Math., No. 19, Amer. Math. Soc., (Providence, 1974).

[35] Sz.-Nagy, B. and C. Foias. *Harmonic analysis of operators on Hilbert space.* North Holland. Amsterdam-Budapest, 1970.

10

MARKOV RANDOM FIELDS FOR IMAGE MODELLING & ANALYSIS

FERNAND S. COHEN

Department of Electrical Engineering, University of Rhode Island
Kingston, Rhode Island 02881

ABSTRACT

This chapter deals with the problem of image modelling through the use of 2D Markov Random Field (MRF). The MRF's are parametric models with a noncausal structure where the various dependencies over the plane is described in all directions. We first show how the MRF's are used as texture image models, region geometry models, as well as edge models.Then we show how they have been successfully used for image classification, surface inspection, image restoration, and image segmentation.

INTRODUCTION

Image modelling is a central part of an image understanding system. It holds the key to successful image classification, image segmentation, and scene analysis in general. Images could be described by a hierarchy of models ranging from the description of the basic geometry of the various regions that comprise the image (e.g. shape, size, area, etc.), to the textural content of each of the regions (e.g. image texture, surface shading, surface roughness, image surface contours, etc.), and finally to an inference model that relates the 2D image to 3D scene - an important link in robot vision applications.

For an image model to be successful it should have inherent in it all the attributes describing the observed pattern, yet be as a compact a representation as possible. Within this compact representation an image is characterized by an appropriate parametric model which is generative in nature, i.e., given the model and reasonable estimates of the model parameters, a synthetic image close to the original image may be generated.

In this chapter, image texture, regions and region boundaries are viewed as samples from a parametric probability distribution on the image space. This distribution defines a MRF which describes the statistical dependence of the process at a pixel in an image on the neighboring pixels. As such it is viewed as describing the spatial interaction of the pixels in the image. We begin by defining the structure of the MRF's and their relation to Gibbs distributions. Two specific models - the multi-level Gibbs model and the Gaussian MRF are further introduced and their parameter estimation problem addressed. The image synthesis problem for the above two models as well as the use of these models as texture, region and edge models is discussed next. Finally we show how these models have been successfully used for image classification, surface inspection, image restoration, & image segmentation.

MARKOV RANDOM FIELDS ON A RECTANGULAR LATTICE (IMAGE)

Let r=(i,j) index pixel location (i,j) where i,j specify row and column location on a rectangular MxN lattice L. Let $\{y_r\}$ denote a random field, with y_r the field at pixel r, **Y** a vector specifying the field over the entire lattice, & $\mathbf{Y}_{(r)}$ be the field everywhere but at pixel r. By $\{y_r\}$ a MRF we mean a family of random variables with a joint probability density function on the set of all possible realizations **Y** on the lattice L that satisfies the following

$$p(\mathbf{Y}>0) \text{ for all } \mathbf{Y},$$

$$p(y_r|\mathbf{Y}_{(r)}) = p(y_r|y_v, v\varepsilon\mathbf{D}_p). \tag{1}$$

Here $\mathbf{D}_p$ denotes a neighborhood set about r and $p(y_r|\mathbf{Y}_{(r)})$ denotes the conditional likelihood of y_r given $\mathbf{Y}_{(r)}$. We are only interested here in homogeneous neighborhood systems $D_p = \{v = (m,n): ||r-v||^2 \leq N_p$ and $v \neq r\}$ (see Figure 1). For such a homogeneous field $p(y_r|\mathbf{Y}_{(r)})$ depends only on the configuration of neighbors and is translation invariant.

5	4	3	4	5
4	2	1	2	4
3	1	X	1	3
4	2	1	2	4
5	4	3	4	5

Figure 1

GIBBS DISTRIBUTION AND MARKOV RANDOM FIELD

The MRF $\{y_r\}$ in (1) is characterized by a Gibbs distribution [1,2,16] on the neighborhood set $\mathbf{D}_p$. In order to define the Gibbs distribution it is necessary to introduce the notion of cliques of a neighborhood sysetm. We define a clique here to be set of points that consists either of a single point or has the property that each point in the set is a neighbor (in the sense of (1)) of every other point in the set. For a MRF for which the neighbor set $\mathbf{D}_p$ is the first layer of eight pixels surrounding the center pixel (see Figure 1) the collection of $[\{(i,j)\},\{(i,j)\ (i+1,j)\},\ \{(i,j)\ (i,j+1)\},\ \{(i,j)\ (i+1,j+1)\},\ \{(i,j)\ (i+1,j-1)\}]$, defines the singleton cliques, and doubleton (pair) cliques in the various directions.

The Hammersely-Clifford theorem [2,16] provides the most general form p(**Y**) can take in order to give a valid probability structure that represents a Markov Random Field. This structure is determined by a Gibbs energy function Q ($\mathbf{Y}$) which relates to p($\mathbf{Y}$) as

$$p(\mathbf{Y}) = \exp\{-Q(\mathbf{Y})\} / Z \qquad (2)$$

where Z is a normalizing constant

$$Z = \sum_{\mathbf{Y}} \exp\{-Q(\mathbf{Y})\} \qquad (3)$$

Theorem Q($\mathbf{Y}$) defined as in (4) defines a valid probability distribution for a Markov Random Field

$$Q(\mathbf{Y}) = \sum_{c_i \varepsilon \mathbf{C}} V_i(\mathbf{Y}) = Q(0) + \sum_r G_r y_r + \sum_{r,s} G_{r,s}\, y_r y_s + ... + G_{(1,1),(1,2),...,(N,M)}\, y_{1,1} y_{1,2} \cdots y_{N,M} \qquad (4)$$

provided that the G functions with subscripts (i,j), (k,l),...,(m,n) may only be nonzero iff {(i,j), (k,l),..,(m,n)} form a clique.

Proof: See [2,16]

Q($\mathbf{Y}$) in (4) is decomposed into a sum of contributions from each possible subset of $\mathbf{Y}$. Each such contribution is called an interaction potential. The potentials are grouped into cliques $\mathbf{C} = \{c_i\}_{i=1,,...,MN}$ of size 1,2,...,MN respectively. V_i ($\mathbf{Y}$) corresponds to the contribution of the interaction potentials associated with clique c_i. We limit ourselves to cliques of size 2. Q($\mathbf{Y}$) in (4) becomes then

$$Q(\mathbf{Y}) = Q(\mathbf{0}) + \sum_r G_r\, y_r + \sum_{r,s} G_{r,s}\, y_r y_s. \qquad (5)$$

We are only interested here in models whose parameters $\{G_{r,s}\}$ depend only on the distance and direction from point r to point s as in Figure 1. Because of the Theorem above the local characteristics, i.e., $p(y_r|\mathbf{Y}_{(r)})$ can be determined and is given by

$$p(y_r|\mathbf{Y}_{(r)}) = \exp\{T_r\} / Z' = \exp\{V_r + \sum_{s\varepsilon \mathbf{D}_P} V_{r,s}\} / Z'$$

$$T_r = y_r\, [G_r + \sum_{s\varepsilon \mathbf{D}_P} G_{r,s} y_s]$$

$$Z' = \sum_r \exp\{T_r\} \qquad (6)$$

MULTI-LEVEL GIBBS MODEL

Let y_r (r = (i,j) takes one of C possible values, y_r = 0.1,..,C-1, and let $p(y_r)|\mathbf{Y}_{(r)})$ be as in (6) with

$$V_1(y_r) = a_k \qquad \text{if } y_r = k$$

$$V_2(y_r, y_s) = \begin{cases} b_{r-s} & \text{if } y_r = y_s \\ d_{r-s} & \text{if } y \neq y_s \end{cases} \qquad (7)$$

a_k denotes the fraction of pixels in the image which assumes the value k. $V_2(y_r,y_s)$ depends on the magnitude and direction of the separation between pixel r and s and $\mathbf{D}_p$. $V_2(y_r,y_s)$ is set equal to zero if s is not in the neighborhood of r.

GAUSSIAN MRF

Here $\{y_r\}$ is a Gaussian MRF with $p(y_r|\mathbf{Y}_{(r)}) \sim \eta(\mu_r,\sigma^2)$

$$\mu_r = \mu + \sum_{v \varepsilon D_p} \beta_{r-v}\,(y_v - \mu) \qquad (8)$$

$$p(\mathbf{Y}) = (2\pi\sigma^2)^{-MN/2} |\Psi|^{1/2} \exp\left\{ - [(\mathbf{Y}-\mathbf{U})^t \Psi (\mathbf{Y}-\mathbf{U})] / 2\sigma^2 \right\} \qquad (9a)$$

where **U** is the mean vector and $\sigma^{-2}\Psi$ is the inverse covariance of **Y.**
The argument in the exponent of p(**Y**) could also be put in the form (5)

$$Q(\mathbf{Y}) = \sum y'^2_{ij} + \beta_{10} \sum y'_{ij} y'_{i+1,j} + \beta_{01} \sum y'_{ij} y'_{i,j+1} + \ldots \qquad (9b)$$

where $y'_{ij} = y_{ij} - \mu$. The Gaussian MRF is parametrized by $\alpha = (\mu,\sigma^2, \beta^t)^t$, where $\beta = (\beta_{10}, \beta_{01}, \beta_{11}, \beta_{1,-1}, \ldots)^t$.

Boundary Condition and Missing Data Problem

For a finite image, the pixels at the image boundary have less neighbors than the interior pixels, these pixels constitute missing data. There have been two approximations for the missing data problem. The first, the so-called free boundary condition, sets the interaction potential (6) between any boundary pixel and its missing neighbors to zero. Whereas the second approximation assumes a toroidal lattice that results in a ramdom field which is wrapped around in a torus structure. For the Gaussian MRF the toroidal lattice approximation results in an inverse covariance matrix $\sigma^{-2}\Psi$ that is circulant [15]. Because of the circularity imposed we can easily show that the discrete Fourier Transform Y of **Y** (assumed here to be zero mean) is a white Gaussian field with zero mean and diagonal covariance matrix $[NM\,S_{ij}]$

$$p(\mathrm{Y}) = \prod_{i,j} (1/2\pi\, MNS_{ij})^{1/2} \exp\left\{ - \left(\sum_{i,j} \| \mathrm{Y}_{ij} \|^2 / 2MN\,S_{ij} \right) \right\} \qquad (10)$$

where Y_{ij} is the ij componment of $\mathbf{Y}$ and S_{ij} is the power spectral density $S(\omega_1,\omega_2)$ of the MRF evaluated at $\omega_1 = 2\pi i / M$, and $\omega_2 = 2\pi j / N$.

$$S_{ij} = \sigma^2 / \left[1-2 \sum_{v \varepsilon D_p} \beta_{r-v} \cos\left(2\pi i m / M + 2\pi j n / N\right)\right] \qquad (11)$$

where (m,n) are the components of the vector r-v, and r = (i,j).
For large M and N the choice of the boundary condition becomes unimportant. Since S_{ij} is a variance and hence is a positive quantity, this implies that for a Gaussian MRF the β_{r-v}'s should satisfy the following condition

$$\sum_{v \varepsilon D_p} \beta_{r-v} \cos\left(2\pi i m / M + 2\pi j n / N\right) < 1/2, \quad \text{for } 0 \le i \le M-1, 0 \le j \le N-1 \qquad (12)$$

MRF PARAMETER ESTIMATION

<u>Gibbs Model</u>

For the Gibbs sampler model in (1) - (5) a maximum likelihood estimate of the $\{G_r, G_{r,s}\}$ is complicated by the difficulty associated with computing the normalizing constant Z in (2). For a 64 x 64 image with C equals to 2 levels, there are $2^{(64x64)}$ possible realizations for which exp {-Q(**Y**)} is to be computed. This is obviously a computationally prohibitive task. Besag [2] suggests a more manageable alternative which only uses a subset of the data to estimate $G_{r,s}$'s. This is known as the <u>coding method</u> and it consists of partitioning the image into different sets called codes such that in any one code there are no two points that are in the neighborhood D_p of one another. For a 1st order Gibbs model (see Figure 1) there will only be two such codes as shown in Figure 2.

θ	φ	θ	φ	θ
φ	θ	φ	θ	φ
θ	φ	θ	φ	θ
φ	θ	φ	θ	φ
θ	φ	θ	φ	θ

Figure 2

The MLE estimate is obtained by maximizing

$$\max_{G_r, G_{r,s}} \prod_{r \varepsilon \mathbf{\Theta}} p(y_r | y_s \,\varepsilon \mathbf{\Phi})$$

or

$$\max_{G_r, G_{r,v}} \left[\log Z' + y_r (G_r + \sum_{v \varepsilon D_p} G_{r,v} y_v)\right] = \max_{G_r, G_{r,v}} \left[\log Z' + T_r\right] \tag{13}$$

where **Θ** & **Φ** are the codes that contain the θ & φ pixels respectively.Here the normalizing constant Z' is easily obtainable as the sum is over the C values that y_r can take. The maximization in (13) is nonlinear and an iterative solution is needed. The disadvantage of this method is that only a subset of the data is used. If an MLE is obtained from each different code, it is still unclear as to how to combine them properly. A simple average won't do as the estimators in the different codes are not independent.

For the case where the image is relatively large and when y_r can take only few levels, and for a low order model, a linear and fast estimator [10,11] is easily obtainable. This estimator is based on histograming and approximating $p(y_r = k|\mathbf{Y}_{(r)})$ by its frequency of occurence. Namely for the defined neighborhood D_p, there are a finite set of realizations that the y's in D_p can take. Each such realization is called a neighborhood configuration C_k. For each neighborhood configuration (say configuration k) the pixels with this configuration are found and the number N_k counted. For these pixels, we count how many assumed the values 0,1,..,C-1 respectively. Let N_{ki}, i=0,1,..,C-1 be those numbers respectively.Then $p(y_r{=}i|C_k)$ is approximated by N_{ki}/N_k. The linearization is obtained by eliminating the constant Z' in (13) via taking the ratio

$$\log \left[p(y_r = i|C_k) \,/\, p(y_r = j|C_k)\right] = T_i(C_k) - T_j(C_k) \tag{14}$$

The LHS of (13) is N_{ki}/N_{kj} and hence (14) becomes a linear equation in the parameters.

By taking different configuration C_k, and different i and j, we obtain a set of linear equations and a least square (or weighted least square) solution that yields and estimate of the $\{G_r, G_{r,s}\}$ can be obtained. This method heavily depends on how well the $p(y_r|C_k)$ are estimated by histogramming.

The Gaussian MRF

For the Gaussian MRF the normalized constant is easily obtainable (9), and hence an MLE is obtained by maximizing either (9) or (10). The estimator is nonlinear and again an iterative solution is needed. The estimates obtained for the β's are guarantee to satisfy (12). Another estimator which is computationally attractive, and which is asymptotically consistent is the pseudo-likelihood estimator [4,15,17]. This estimator is shown to have an asymptotic variance [15] which is tighter than that obtained with the coding method. The pseudo-likelihood estimator is obtained by maximizing the pseudo-likelihood function

$$\prod_{r \varepsilon L} p(y_r | \mathbf{Y}_{(r)}) \tag{15}$$

where the product is over the entire lattice L.The pseudo-likelihood function was shown [4] to be an effective performance functional for MRF classification. The pseudo-likelihood estimator for β is a least square type estimator and takes the form

$$\beta = \left[\sum_{r \varepsilon L} \mathbf{X}_r \mathbf{X}_r^t\right]^{-1} \sum_{r \varepsilon L} \mathbf{X}_r y'_r \quad , \qquad r = (i,j) \tag{16}$$

where $\mathbf{X}_r = (y'_{i-1,j} + y'_{i+1,j}, y'_{i,j-1} + y'_{i,j+1}, \ldots)$ and $y'_{ij} = y_{ij} - \mu$.
The only drawback associated with β is that there is no guarantee that it will satisfy (11). In this case the random field $\{y_r\}$ is no longer Gaussian, though it might still define a MRF!

ASYMPTOTIC NORMALITY OF THE MLE

In this section we state some classical results concerning the asymptotic form of the MLE, then point out how to make use of these results for inspection of a textured surface, & unsupervised segmentation. These two applications are further discussed in upcoming sections. Because of the asymptotic normality of the MLE, defect detection is cast as a likelihood ratio test that reduces to a χ^2 test [9] (see section on MRF for surface inspection). Moreover, as we can express the joint likelihood of the different windows (say L windows) that comprise a textured region in terms of the joint likelihood of the windows MLE's (α_1^*, α_2^* ,..., α_L^*), this allows the segmentation of a textured image into its constituent cluster textured regions to be cast as the problem of the clustering of unlabelled samples from a mixture of Gaussian densities [8] (see section on unsupervised segmentation).

Let $\mathbf{Y}$ be the whitened field obtained from the MRF data vector $\mathbf{Y}$ associated with a textured region. If $\mathbf{Y}$ is a GMRF, then $\mathbf{Y}$ is the discrete Fourier transform of the data vector $\mathbf{Y}$. Let assume that the textured region is partitioned into L disjoint windows. Let $\mathbf{Y}_k$ be the Fourier data vector in the kth window. Because of the independence of the $\mathbf{Y}_k$'s ($\mathbf{Y}$ being a white field (10)), $p(\mathbf{Y} | \alpha)$ can be written as

$$p(\mathbf{Y} | \alpha) = \prod_k p(\mathbf{Y}_k | \alpha) \tag{17}$$

For large window size $p(\mathbf{Y}_k | \alpha)$ is approximately Gaussian in α [18]. That is

$$p(\mathbf{Y}_k | \alpha) \approx p(\mathbf{Y}_k | \alpha^*) \exp\{-(\alpha_k^* - \alpha)^t \, \Psi(\alpha_k^*) (\alpha_k^* - \alpha) / 2\} \tag{18}$$

where α^* is the MLE of α based on the data in window k, & $E[\Psi(\alpha_k^*)]$ is the Fisher information matrix.

$$\Psi(\alpha_k^*) = [\Psi_{ij}(\alpha_k^*)] \tag{19}$$

where

$$\Psi_{ij}(\alpha_k^*) = -\partial^2 \log p(\mathbf{Y}_k \mid \alpha) / \partial\alpha_i \partial\alpha_j \mid \alpha_k^* \tag{20}$$

Using the approximation in (18), $p(\mathbf{Y} \mid \alpha)$ can then be written as

$$p(\mathbf{Y} \mid \alpha) \approx \Big[\prod_k p(\mathbf{Y}_k \mid \alpha_k^*)\Big] \exp\Big\{-\sum_k (\alpha_k^* - \alpha)^t \Psi(\alpha_k^*)(\alpha_k^* - \alpha) / 2\Big\} \tag{21}$$

The approximation in (21) enabled the likelihood function of the data in a region to be expressed in terms of the joint likelihood of the reduced data set ($\alpha_1^*, \alpha_2^*, \ldots, \alpha_L^*$).

SUFFICIENT STATISTICS FOR THE MRF

The class of MRF introduced earlier belongs to the exponential type family, & therefore their likelihood function is factorizeable in the Neyman-Fisher sense [18], i.e.,

$$p(\mathbf{Y}|\alpha) = \exp\Big\{ \mathbf{T}^t(\mathbf{Y})\,\alpha + c(\alpha) \Big\} \tag{22}$$

That means that the sufficient statistics $\mathbf{T}(\mathbf{Y})$ exists, & the likelihood $p(\mathbf{T}(\mathbf{Y})|\alpha)$ is equivalent to $p(\mathbf{Y}|\alpha)$. The advantage of working with $\mathbf{T}(\mathbf{Y})$ becomes evident when a textured surface is to be inspected in "real- time" (see section on inspection), or when a textured image is to be segmented using unsupervised clustering (see section on unsupervised clustering & segmentation). As we have seen, MLE of the MRF parameters involved nonlinear iterative maximization. In contrast the components of $\mathbf{T}(\mathbf{Y})$ are easily computable. For the multi-level MRF in section 2.2, the computation of $\mathbf{T}(\mathbf{Y})$ involves few additions & comparisons, while for the GMRF it involves few additions & multiplications.

For the GMRF, the sufficient statistics $\mathbf{T}(\mathbf{Y})$ for the parameter set $(\mu, \sigma^2, \beta_{10}, \beta_{01}, \beta_{11}, \beta_{1,-1}, \ldots)$ is given by $(\mu^*, R^*_{00}, R^*_{10}, R^*_{01}, R^*_{11}, R^*_{1-1}, \ldots)$, where μ^*, R^*_{mn} are the sample mean, & sample autocovariance function in the (m,n) direction respectively, with R^*_{00} being the sample variance. Assuming a square NxN lattice L, R^*_{mn} is given in (23) as

$$R^*_{mn} = \left[\sum (y_{ij} - \mu^*)(y_{i+m,j+n} - \mu^*)\right] / N^2 \tag{23}$$

The expected value of the R^*_{mn} is the theoritical autocovariance function, & satisfies the difference equation in (24) [17],

$$R_r = \sum_{r \,\varepsilon\, Dp} \beta_{r-v} R_v + \sigma^2 \delta_r \tag{24}$$

The R^*_{mn}'s & μ^* are statistically independent. The variance of μ^* is equal to S_{00} / N^2, with S_{00} given in (11). The covariance function associated with $\mathbf{R}^* = (R^*_{00}, R^*_{10}, R^*_{01}, R^*_{11}, R^*_{1,-1}, \ldots)$ is given by [7,8],

$$\mathrm{cov}(R^*_s, R^*_v) \approx \sum_r (R_r R_{r+v-s} + R_{r+v} R_{r-s}) / N^4 \tag{25}$$

where R_s satisfies (24). Asymptotically, **T(Y)** is shown to be Gaussian [7,8].

For the multi-level MRF given in (9), the sufficient statistics $\mathbf{T(Y)} = (a^*_1, a^*_2, a^*_3, a^*_4, b^*_{01}, b^*_{10}, b^*_{11}, b^*_{1,-1}, \ldots)$, where a^*_k is the fraction of pixels that assume the value k, & b^*_{kl} is the total number of pixels that assume the same intensity value & are neighbor of one another in the (k,l) direction . It can also be shown that **T(Y)** is asymptotically Gaussian [7,8,9].

MRF SYNTHESIS

For the general MRF the synthesis problem is addressed as a probabilistic relaxation problem where p(**Y**) is cast as the equilibrium distribution of an irreducible, aperiodic Markov chain {**Y**(t), t=1,2,..}. A realization **Y** of the MRF is obtained by visiting each state of the chain infinitely often. Let **Y**(t-1) be the state of the chain at time (t-1) and and **Y**(t) be the state at time t. **Y**(t) is exactly the same as **Y**(t-1) except that at pixel r, i.e., $y_r(t-1)$ has been changed into $y_r(t)$. The evolution of the chain from state **Y**(t-1) to **Y**(t) is given by

$$p(\mathbf{Y}(t)) = p(y_r(t) \mid \mathbf{Y}_{(r)}(t-1))\, p(\mathbf{Y}(t-1)) / p(y_r(t-1) \mid \mathbf{Y}_{(r)}(t-1)) \tag{26}$$

From (6)

$$p(\mathbf{Y}(t)) / p(\mathbf{Y}(t\text{-}1) = \exp\left\{ [y_r(t)\text{-}y_r(t\text{-}1)]\, [G_r + \sum_{s \varepsilon D_p} G_{r,s} y_s(t\text{-}1)] \right\} \qquad (27)$$

and

$$p(\mathbf{Y}) = \lim_{t \to \infty} \; p(\mathbf{Y}(t) = \mathbf{Y} | \mathbf{Y}(0) = \mathbf{Y}_0) \quad \text{for an arbitrary } \mathbf{Y}_0 \qquad (28)$$

We refer the reader to [10,14] for a more detailed formulation about p(**Y**) being an equilibrium distribution of a Markov chain.

The Metropolis Algorithm

Briefly the algorithm consists of the following steps: given the state of the system **Y**(t) at time t, another configuration **X** is chosen at random and the ratio $r = p(\mathbf{Y}(t)) / p(\mathbf{X})$ is computed. If $r > 1$, then $\mathbf{Y}(t+1) = \mathbf{X}$, whereas if $r \leq 1$, the transition is made with probability r. That means that a variable ζ selected from a uniform U[0,1] distribution is compared to r and if $\zeta \leq r$ we set $\mathbf{Y}(t+1) = \mathbf{X}$ otherwise $\mathbf{Y}(t+1) = \mathbf{Y}(t)$.

In the so-called "Exchange Algorithm" [10,11], **X** is obtained from **Y**(t) by randomly picking two pixels and exchanging their values if they are different. The drawback of this algorithm is that the intensity histogram is constant throughout the iterations which makes the system dependent on the initial configuration. A better algorithm is the so-called "single-Flip" where **X** is exactly **Y**(t) except that at a pixel r, $y_r(t)$ is changed into another permissible value.

Gaussian MRF Generation

For the Gaussian MRF one can either use the Metropolis algorithm or assume a Toroidal lattice approximation as in (10). The components of the white Gaussian Field Y is generated by randomly selecting Y_{ij} from a Gaussian random number generator with zero mean and variance MNS_{ij}. The random field **Y** is then generated by inverse Fourier transforming (**Y**+ **U'**), where **U'** = DFT(**U**).

MARKOV RANDOM FIELD AS TEXTURE MODEL

The Gaussian MRF is used here to synthesize and model many textures of interest from the Brodatz album as well as naturally occurring scenes. In Figure 3 the original textures are shown in the left windows, while the right windows show the synthesized images for a class of textures from the Broadatz album. Figure 4 [6] shows a natural scene where the image is largely textured in structure. Six texture types were chosen, the left tree foliage with roughly isotropic spatial variation, the foliage of the large tree on the right with greater spatial correlation in the vertical direction, the house roof shingles, the road and the grass. A set of Gaussian MRF model parameters were estimated for each of these texture-types from a small window in each case. Then these models were used to generate slightly larger windows of artificial images. The generated images were inserted in the picture class at the locations from which the data used for estimating the model parameters were taken. What is surprising is how the models appear to capture the different texture structures. We leave it to the reader to find where the artificial textures are in Figure 4. Figures 5 & 6 show textile fabrics which have been modelled by a 4-level MRF. The original textured fabrics are shown in the left windows, while the right windows show the synthesized images using the models.

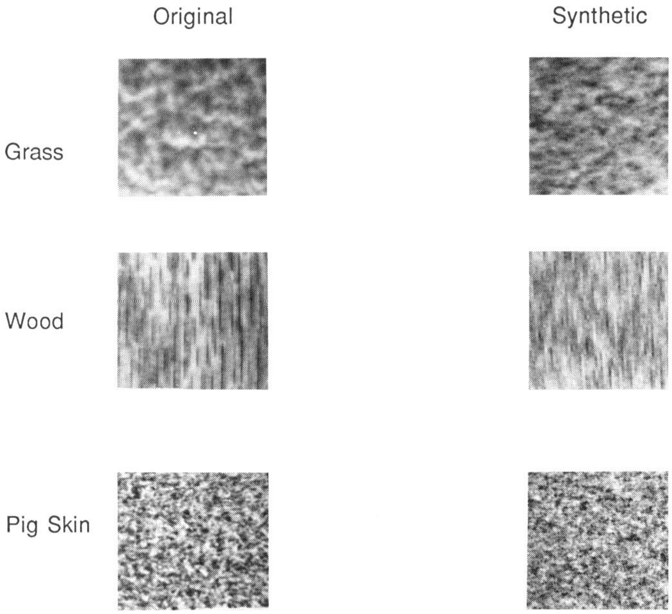

Figure 3

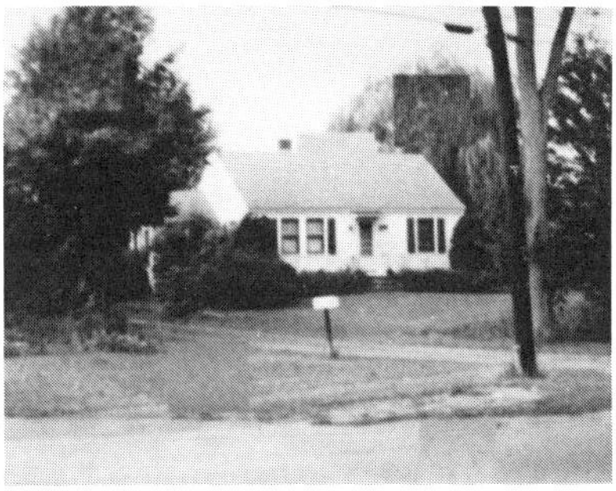

Figure 4

Original

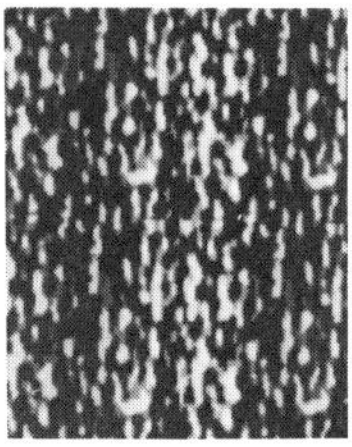

Synthetic

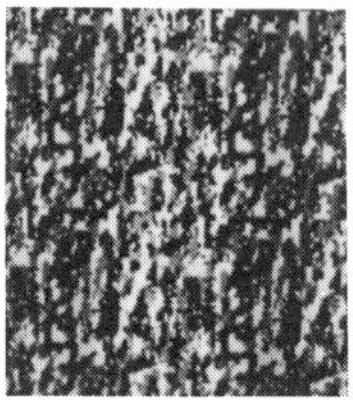

Denim

Figure 5

Original

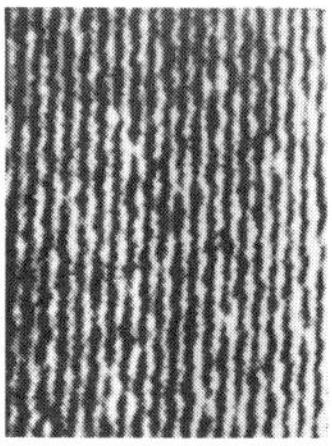

Synthetic

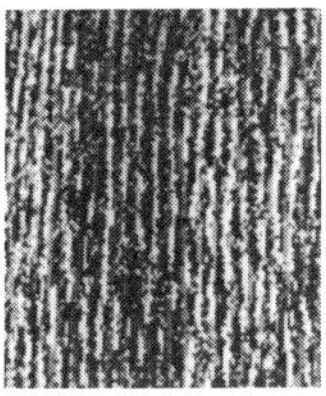

Sateen

Figure 6

MRF AS REGION MODELS

For an image which consists of a finite number of regions in each of which the image data comes from one of C textures classes, one can use the MRF (Gibbs model) to model the global image-geometry, i.e., the geometry of the different regions in the image. These regions may be described by their cluster size which is controlled by how strongly points in the same texture category attract or repel one another. Directionality of the clusters is another feature that describe the different regions in the image. Other region features such as blob-likeness, regularity and line-likeness can also be incorporated in the MRF region model. Figure 7 shows different regions features for a binary image.

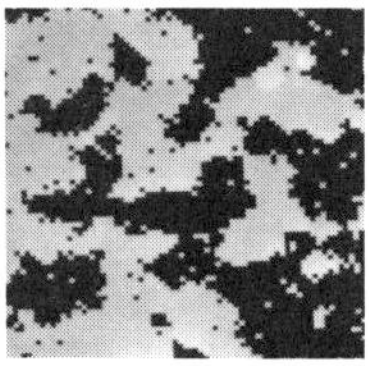

Small Clusters

(a)

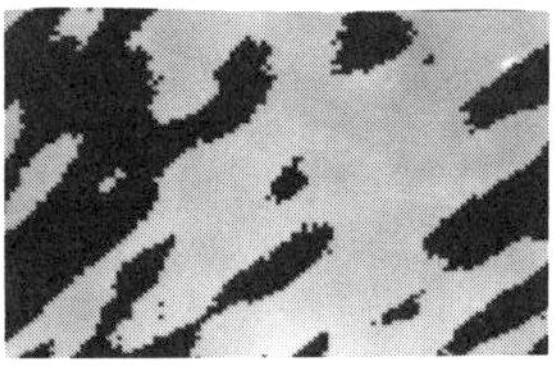

Big Clusters with Directionality

(b)

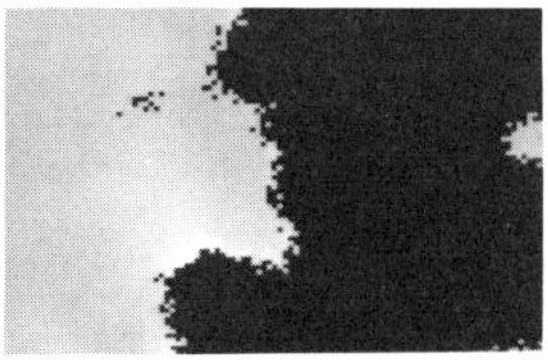

Isotropic Huge Clusters

(c)

Figure 7

MRF AS EDGE MODELS

The pixels in the lattice L are viewed as sites and edges between pixels are branches between the sites. The simplest edge process **E** is described in terms of the presence of absence of an edge between two sites, along with a finite neighborhood that describes the stochastic dependence of the absence or presence of an edge at a site r, on the presence or absence of edges at sites surrounding r. Figure 8 shows the pixels and edge sites in a lattice L. The pixel sites are denoted by a "O", whereas "X" are the edge sites. The edge neighborhood about an edge site X is the same as in Figure 1.

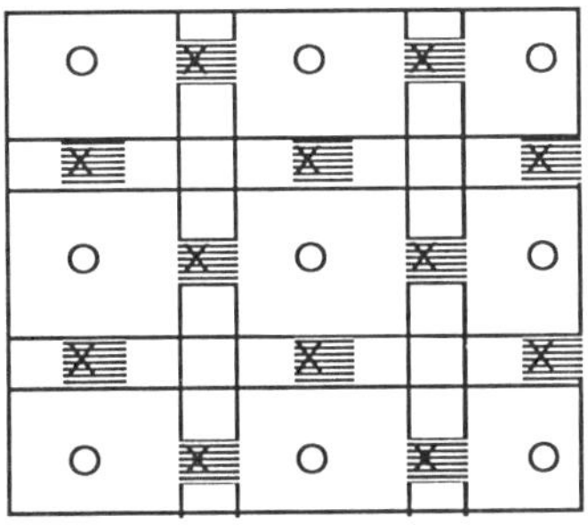

Figure 8

If straight boundaries between regions are favored, then one should choose the various potentials in $p(e_r \mid \mathbf{E}_{(r)})$ so that the $p(e_r = 1 \mid \mathbf{E}_{(r)})$ is high for the case shown in Figure 9a, and low for the case shown in Figure 9b. The pixel sites are denoted by "O", and a branch between two pixel sites indicates the presence of an edge.

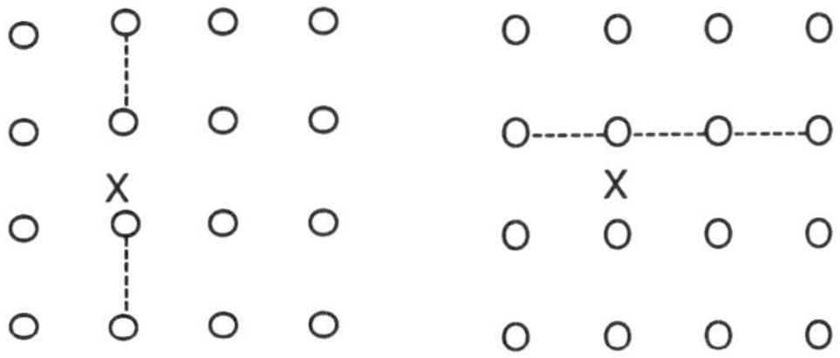

(a) (b)

Figure 9

For curvy boundaries one should consider an edge process with more than just levels. Here it is necessary to consider edge elements with varying orientations. We refer the reader to [14] for more details on the use of the MRF for modelling edges.

MRF FOR IMAGE CLASSIFICATION

MRF has been successfully used for recognizing images in which the data comes from one of C MRF [15]. For the Gaussian MRF three different functionals were used to classify an image. The first is the ML function in (9) or (10). The choice of an appropriate model from a class of competing models is achieved according to the decision rule max $p(\mathbf{Y}|\,\alpha_C)$ where α_C is the parameter vector associated with class C. In [6] a pseudo-likelihood function (15) was used as an alternative to the ML function. This functional was proved to be an effective texture recognizer [6]. For the natural scene in Figure 4, the pseudo-likelihood functional for classifying data in small windows as one of six texture-types, the left tree foliage, the foliage of the tree on the right, the house roof shingles, the sky, the road, and the grass resulted in the classification shown in Figure 10.

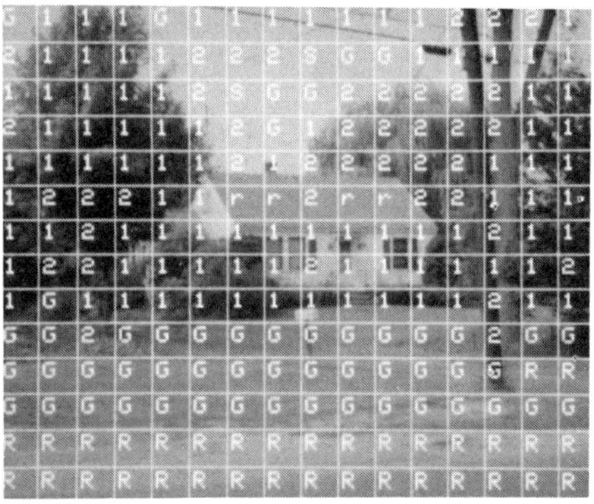

Figure 10

The third classifier is based on using the sufficient statistics vector $\mathbf{T}(\mathbf{Y})$ given in a previous section. The use of this classifier is illustrated in the next section for the purpose of surface inspection.

MRF FOR SURFACE INSPECTION

The MRF's have been used by Cohen, et. al. [9], for inspecting textile & lumber surfaces. For textile they have used the multi-level MRF, whereas for the wood inspection problem, the GMRF was used. The goodness of the models, made possible by the synthesis nature of the MRF, could be seen in Figures 5 & 6 for two textile fabrics, & in Figure 3 for the lumber surface.

The inspection station consisted mainly of two units. The first is the feature extractor or modeller that learns about the surface to be inspected & the pertinent features to be used in the defect detection process. This learning process is usually done off-line, & has no time limitation constraint.The second unit is the classifier that uses the appropriate models found previously to design a classifier that inspects in "real-time" a sample taken from this surface.

In the learning phase, a MRF is fitted to a nondefective surface, & the MRF parameters are estimated.Surface inspection is cast as a statistical hypothesis test. The image of the surface to be inspected is partitioned disjoint square windows, & a MLE of the MRF parameters, or the sufficient statistics are computed for each window. Exploiting the asymptotic normality of the MLE &/or the sufficient statistics, a likelihood ratio test [9] is carried out to check whether or not the data in a window confines to the model. It is shown in [9] that because of the asymptotic normality of the MLE &/or sufficient statistics, the likelihood ratio test reduces to a χ^2 test on the variable

$$d^2 = -(\boldsymbol{\alpha}^* - \boldsymbol{\alpha})^t \boldsymbol{\Psi} (\boldsymbol{\alpha}^* - \boldsymbol{\alpha}) \qquad (29)$$

$\boldsymbol{\alpha}$ & $\boldsymbol{\Psi}$ are known function of the MRF parameters, & are computed beforehand in the learning phase. Hence, to compute d^2, one has to compute $\boldsymbol{\alpha}^*$ first. When $\boldsymbol{\alpha}^*$ is the MLE, a nonlinear maximization is to be performed, & is somewhat time consuming. On the other hand, the sufficient statistics are easily computable as discussed before. To assure "real-time" inspection, we have used the sufficient statistics as our feature vector.

A set of experiments have been run on samples of the textile fabrics shown in Figures 5 & 6. The image of the fabrics had their gray level histograms equalized to four levels. Figure 11a shows the result of the model based classifier on a defective piece of denim fabric. These defective parts have been successfully depicted. The experiment has also been run on the sateen fabric, & the result is shown in Figure 11b. The windows which were found to be defective are labelled "D", & the nondefective ones are labelled "R" for regular. We refer the readers to [9] for an extensive treatment of the subject.

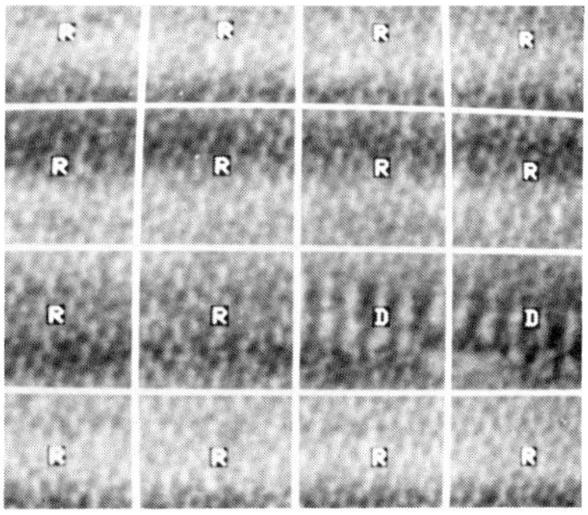

Denim

(a)

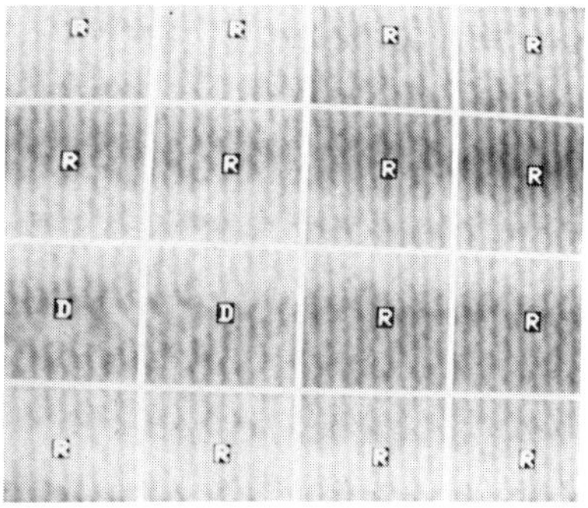

Sateen

(b)

Figure 11

MRF FOR IMAGE RESTORATION

Image restoration deals with reconstructing images that has been degraded by one or more sources of degradations. One source of degradation can be a blurring function which corresponds to a shift-invariant point spread function **H.** The blurred image **H**(**Y**) may undergo a nonlinear transformation when recorded by a sensor. Let **N** be this nonlinear transformaiton. Finally the sensor may produce a random sensor noise **V**. **V** is assumed a white Gaussian process $\mathbf{V} \sim \eta(\mathbf{U},\sigma^2 I)$. The degraded image **G** can be then written as

$$\mathbf{G} = \mathbf{N}(\mathbf{H}(\mathbf{Y})) + \mathbf{V} \tag{30}$$

The ideal image process is a MRF which incorporates both an unobservable hidden region (or edge) model **S**, and an observable model **Y** that describes the textured image data in each region. Under this scenario the ideal image has been generated by first generating the texture-type regions, and then fills in each region with an image having the appropriate texture. Let **X** denote the ideal image process **X**=(**Y**,**S**). The noise process **V** is assumed independent of **Y** and **S**. For preserving the local neighborhood associated with **X** the blur **H** is a simple convolution over a small window. As stated the restoration problem is that of finding that **X** maximizes

$$p(\mathbf{G} = \mathbf{g} \mid \mathbf{X} = \mathbf{x})\, p(\mathbf{X} = \mathbf{x})\, p(\mathbf{G} = \mathbf{g} \mid \mathbf{Y} = \mathbf{y}, \mathbf{S} = \mathbf{s})\, p(\mathbf{Y} = \mathbf{y}, \mathbf{S} = \mathbf{s}) \tag{31}$$

The term $p(\mathbf{G} = \mathbf{g} \mid \mathbf{Y} = \mathbf{y}, \mathbf{S} = \mathbf{s})$ is nothing but $p(\mathbf{V} = \mathbf{v})$ where **v** is a known function of **g** and **N**(**H**(**Y**)). The term $p(\mathbf{Y} = \mathbf{y}, \mathbf{S} = \mathbf{s})$ is a MRF with a known Gibbs distribution. It follows then that $p(\mathbf{X} = \mathbf{x} \mid \mathbf{G} = \mathbf{g})$ is a gibbs distribution [14] with energy function $Q(\mathbf{X} \mid \mathbf{G})$ given by

$$Q(\mathbf{X} \mid \mathbf{G}) = ||\mathbf{U}\text{-}\mathbf{v}||^2 / 2\sigma^2 + Q(\mathbf{X}) \tag{32}$$

The restoration problem is then that of minimizing $Q(\mathbf{X} \mid \mathbf{G})$ over all **X**. This is a horrendeous computational optimization problem. In the next sections we discuss various methods for obtaining **X** that minimizes (32).

Iterative Relaxation

The relaxation method introduced in this section has been used by Cohen and Cooper [4-6] and is based on iteratively improving the posterior distribution $p(\mathbf{X} \mid \mathbf{G})$ through local changes. If **X** is the estimated image at the nth stage, then any pixel r is chosen, and x_r is kept as is or change into a new permissible value depending on

$$\max_{x_r} p(\mathbf{G} \mid \mathbf{X})\, p(\mathbf{X}) \tag{33}$$

This provides a new estimate. The next iteration is carried out by choosing another pixel s and repeat the same process until there is no further changes at all pixels. The computation involved with examining each pixel is simple. (33) can be written as

$$\max_{x_r} p(\mathbf{G}_{(r)}, g_r \mid x_r, \mathbf{X}_{(r)})\, p(x_r, \mathbf{X}_{(r)}) \tag{34a}$$

But this equivalent to maximizing

$$\max_{x_r} p(g_r \mid \mathbf{G}_{(r)}, \mathbf{X})\, p(x_r \mid \mathbf{X}_{(r)}) \tag{34b}$$

(34b) involves only a local neighborhood about r. The parallel version of the above algorithm is based on dividing the image into codes where any two elements in a code are not in the neighborhood of one another. This implies that one can process all the pixels in one code simultaneously with parallel hardware. This relaxation algorithm is guaranteed to converge to a local maximum of $p(\mathbf{G},\mathbf{X})$. Convergence to a global maximum will depend on how good the initial guess for **X** is.

Examples In this section we show few examples where the relaxation algorithm has been used [4-6]. In this work no blur or nonlinear transformation on **Y** is assumed. The observed data **G** obeys the model

$$\mathbf{G} = \mathbf{Y} + \mathbf{V} \tag{35}$$

where **Y** is a multi-level MRF, **V** is a white Gaussian field. The problem here is to restore **Y**.

Figure 12a shows an original image which is a sample for a MRF with $p(y_r \mid \mathbf{Y}_{(r)})$ given in (7) for a fifth order isotropic model having parameters $V_1(y_r) = 0.25$, $y_r = 0,1,2,3$; $V_2(y_r,y_s) = 1.5$ for $y_r = y_s$, and -1.5 for $y_r \neq y_s$. $V_2(y_r,y_s) = 0$ if s is not in the neighborhood Dp of r. The degraded image which is simply Figure 12a plus additive white Gaussian noise with $\sigma=1$ relative to gray levels $0 \leq y \leq 3$ is shown in Figure 12b. Figure 12c,d shows the initial **Y** that the relaxation algorithm used and the restored image after 10 passes through the image. Figure 13a shows a hand-drawn image of three overlapping squares on a background. The squares and the background are of different gray levels. Figure 13b is the degraded image with additive noise with $\sigma=1$ relative to gray levels $0 \leq y \leq 3$.

Figures 13c,d show the initial and final (restored) image **Y**. **Y** was approximated here by a fifth order MRF as in (7) with $d_{r-s} = -b_{r-s} = 2.5$ for $s \in Dp$. The MRF model approximation was designed to stress region connectivity and encourage big blob formation. A realization of the above MRF is shown in Figure 14.

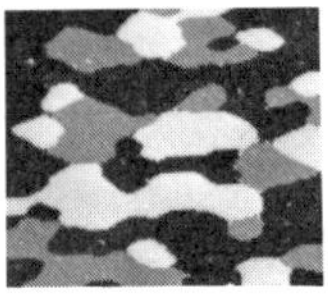

(a)

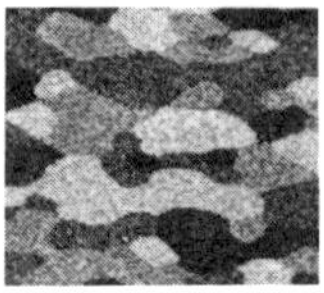

(b)

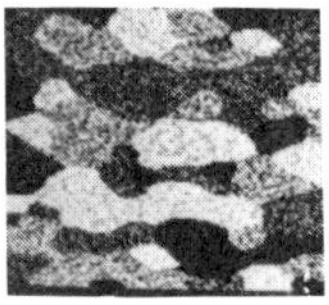

(c)

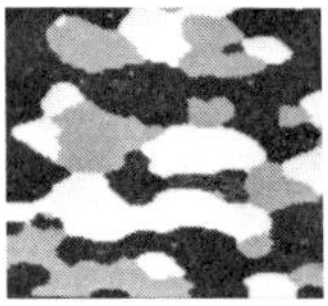

(d)

Figure 12

(a)

(b)

(c)

(d)

Figure 13

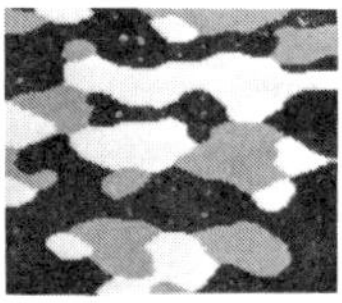

Figure 14

<u>Stochastic Relaxation and Annealing</u>

The stochastic relaxation method introduced in this section has been used by Geman and Geman [14]. It is similar to the "Iterative Relaxation" algorithm used by Cohen and Cooper [4-6] except that the change at a pixel is random, and is generated by sampling from a local conditional probability distribution. As such it is very similar to the Metropolis algorithm introduced earlier. To avoid convergence to local maxima of $p(\mathbf{G},\mathbf{X})$, the $p(\mathbf{X})$ and hence $p(x_r \mid \mathbf{X}_{(r)})$ were made dependent on a global control parameter "T" called temperatue such that $p(\mathbf{X})$ and $p(x_r \mid \mathbf{X}_{(r)})$ are

$$p(\mathbf{X}) = \exp\{-Q(\mathbf{X}) / T\} / Z \tag{36}$$

$$p(x_r \mid \mathbf{X}_{(r)}) = \exp\{-T_r / T\} / Z \tag{37}$$

where T_r is given in (6) and

$$Z = \sum_{x_r} \exp\{-T_r / T\} \tag{38}$$

At the beginning of the stochastic relaxation process there is a loose coupling between neighboring pixels and a chaotic (high noise level) appearance to the image. That means that the local distribution associated with a stochastic change at a pixel should be uniform. Having T high in (36) enforces that, and allows for changes that might result in a decrease in $p(\mathbf{G},\mathbf{X})$. As the relaxation proceeds the coupling between neighboring pixels becomes more regular and the stochastic relaxation process starts looking for local changes that increase $p(\mathbf{G},\mathbf{X})$. T in (36) is lowered and the local distribution becomes more picked. This gradual reduction in temperature simulates "annealing" a procedure by which certain chemical systems can be driven to their low energy, highly regular states. The process is very delicate and if T is lowered too rapidly the system might be locked in a nonequilibrium state. In our context the system may converge to a local maximum of $p(\mathbf{G},\mathbf{X})$. One of the main contributions of Geman and Geman is to prove the existence of annealing schedules which guarantee convergence to global maximum of $p(\mathbf{G},\mathbf{X})$ and identify the rate of decrease relative to the number of sweeps. We refer the readers to [14] for further details.

<u>Examples</u> Figure 15a shows a sample of 5 level MRF with a second-order model with $V_2(y_r,y_S) = 1/3$ if $y_r = y_s$, and $-1/3$ if $y_r = y_S$. $V_2(y_r,y_S) = 0$ if $s \in Dp$. Figure 15b is simply Figure 15a plus additive Gaussian noise with $\sigma=1.5$ relative to $0 \leq y \leq 4$. Figure 15c is the restoration after 25 iterations in the annealing process, while Figure 15d is after 300 iterations.

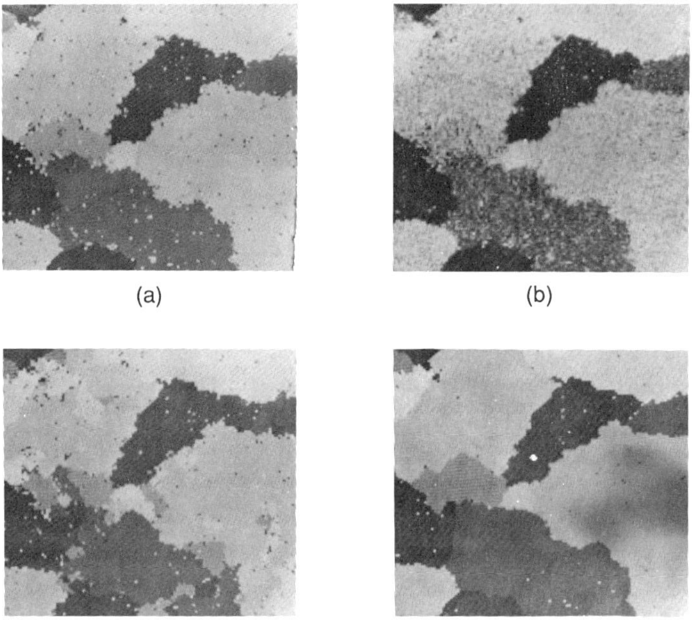

(a) (b)

(c) (d)

Figure 15

Another example where the degraded image uses the model

$$\mathbf{G} = [\mathbf{H}(\mathbf{Y})]^{1/2} \,.\, \mathbf{N} \qquad (39)$$

is shown in Figures 16a-d. The mean and variance of the noise were 1 and 0.1 respectively and **H** was given by

$$H(m,n) = \begin{cases} 1/2 \quad , \quad m = 0, n = 0 \\ 1/16 \quad , \quad |m|, |n| \leq 1, (m,n) \neq (0,0) \end{cases} \qquad (40)$$

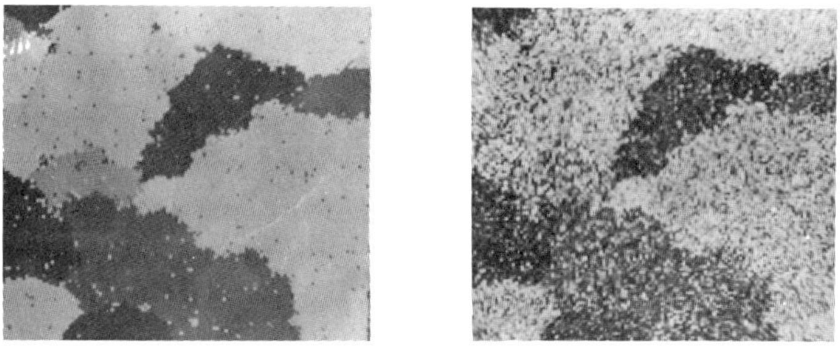

(a) (b)

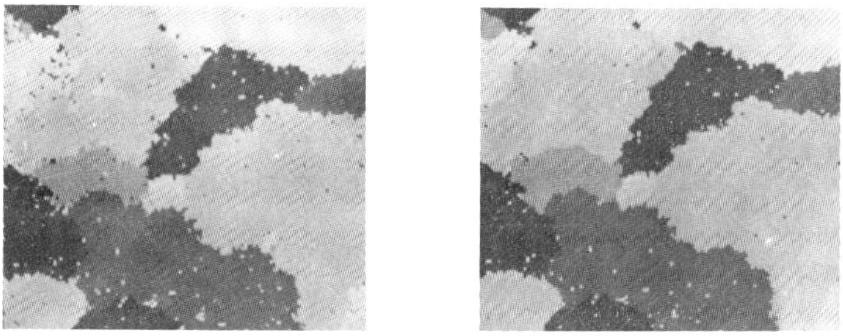

(c) (d)

Figure 16

MRF FOR IMAGE SEGMENTATION

This section is concerned with the problem of segmenting an image into regions in each of which the data is modelled as one of known C MRF's. The image data in texture-type regions i and j (i,j = 0,...C-1) are assumed statistically independent. Let y_{ij} be a variable taking values 0,1,..,C-1, and let **Y** denote the MN dimensional vector having components y_{ij}. Then **Y** specifies the partition of the MxN image into the texture type 0,1,..C-1 regions. Let g_{ij} denote the picture function at pixel (i,j), and **G** be the MN vector having components g_{ij}. If **Y** is viewed as a constant but unknown vector, then the segmentation problem is to find **Y** that maximizes p(**G** |**Y**). Here p(**G** | **Y**) is given by

$$p(\mathbf{G}\,|\mathbf{Y}) = \prod_{k=0}^{C-1} p(\mathbf{G}_k\,|\,k) \tag{41}$$

where $\mathbf{G}_k$ denotes the picture function vector for texture-type k region, and $p(\mathbf{G}_k|k)$ denotes the likelihood of $\mathbf{G}_k$ given that $\mathbf{G}_k$ comes from the kth MRF's. If **Y** is viewed as a random vector multi-level MRF, then the segmentation problem is to determine **Y** for which the posterior likelihood p(**Y** | **G**) is a maximum. **Y** here would reflect our prior knowledge about the geometry of the different regions that comprise the scene. As in most scenes of interest, different regions try to cluster, **Y** is chosen to model the spatial continuity within regions. That was the rational of the choice of **Y** in Figure 13. Both the iterative and the annealing algorithms described earlier can be used for image segmentation with **Y** chosen to be a MRF that reflect our prior knowledge of the geometry of the regions comprising the image. Both algorithms require more than one sweep through the image data. The two algorithms introduced in the next secitons require a single sweep through the data. The first one views **Y** as a MRF, assumes prior knowledge of p(**Y**), and uses dynamic programming [13] based segmentation to maximize p(**Y** | **G**). The second algorithm views **Y** as an unknown constant, find **Y** that maximizes p(**G** | **Y**) using hierarchical based segmentation [4-7].

Dynamic Programming Segmentation

The model that Elliot , et. al. [13] used is

$$g_{ij} = f_{ij}(y_{ij}) + v_{ij},$$

$$\mathbf{G} = \mathbf{F}[\mathbf{Y}] + \mathbf{V} \tag{42}$$

where $\mathbf{Y}$ is a C-level MRF, and the mapping f_{ij} is used to characterize the texture model. A region here is characterized by a constant intensity so that

$$f_{ij}(y_{ij}) = \mu_k, \quad \text{if } y_{ij} = k,\ k=0,1,..,C-1 \tag{43}$$

v_{ij} as before is a white gaussian field with zero mean and variance σ^2.

The segmentation problem is first posed for processing DxN_2 subimage that consists of D rows and N_2 columns. It is then applied to overlapping stripes of the large N_1xN_2 image. For the DxN_2 subimage log p(G,Y) or simply l(G,Y) is given by

$$\log p(\mathbf{G},\mathbf{Y}) = -(DN_2/2)\log 2\pi\sigma^2 - \sum_{m=0}^{C-1}\ \sum_{(i,j)\varepsilon S_m} (y_{ij}-\mu_m)^2/2\sigma^2 - \log Z + \sum_{c\varepsilon C} V_c(\mathbf{Y}) \tag{44}$$

where $S_m = \{(i,j): y_{ij} = m\}$

(44) can be recursively computed as

$$l_o = (DN_2/2)\log 2\sigma^2 - \log Z$$

and

$$l_k = l_{k-1} + \sum_{c\varepsilon C_{k,k-1}} V_c(Y) - \sum_{m=0}^{C-1}\ \sum_{(i,j)\varepsilon S_{km}} (Y_{ij}\ \mu_m)^2/2\sigma^2 \tag{45}$$

where $C_{k,k-1}$ is a clique collection that contains only pixels in column k and k-1 and $S_{km} = \{(i,j): y_{ij} = m, j = k\}$, for $1 \le i \le D$. The recursion allows a forward dynamic programming to be implemented for finding $\mathbf{Y}$. The state space has dimension M^D and there are M^D possible segmentations of each column of the DxN_2 subimage. There are N_2 iterations associated with this algorithm. The algorithm is only tractable for small values of M and D. Moreover, it restricts the MRF $\mathbf{Y}$ to have negligible correlation beyond the strip width D, i.e. y_{ij} and $y_{i+D,j}$ for all (i,j) are assumed to have negligible correlation or covariance. This is essential for preserving the finite dimensionality of the state space in the dynamic programming. The segmentation of a row i in the N_1xN_2 image via using the DxN_2 stripes yields close to optimal solution only when the segmentation of the ith row is negligibly impacted by the segmentation of row i-D. Two examples are shown in Figures 17 and 18. Figure 17 shows the clean and noisy images as well as the segmented image (lower left) and a median filtered version of the segmented image of a cube. Figure 18 shows the results of segmenting a section of a SAR image into two regions for three different sets of clique parameters.

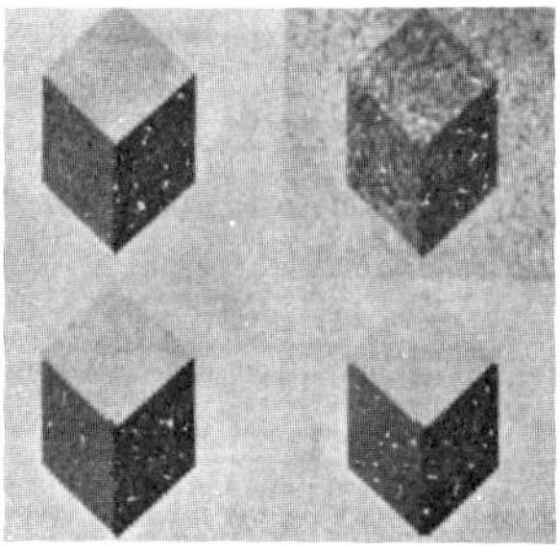

Figure 17

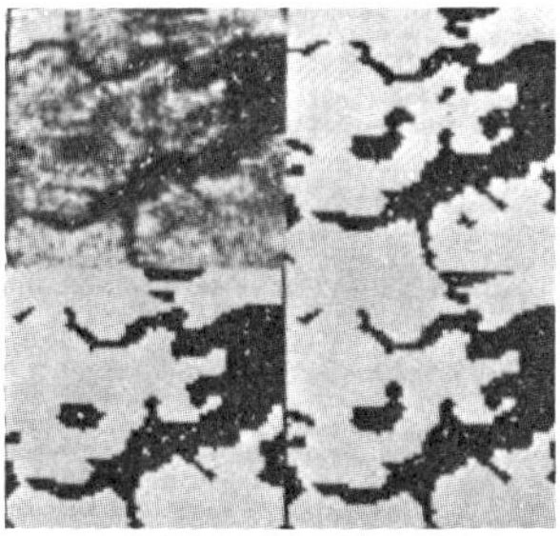

Figure 18

Hypothesis-Testing and Hierarchical Segmentation

Y here is viewed as an unknown to be estimated, and **G** is assumed to come from C known Gaussian MRF's. The segmentation problem consists of two stages. First the image is partitioned into relatively small disjoint windows. Because of the relative small size of the windows it is assumed that each window contains atmost two different texture-types in it.

Window Classification The first stage of the image segmentation process is to classify a window as being homogeneous or mixed, and decide on the texture types present in the window. This is done via a multiple-hypothesis testing process [7] which considers a pair of texture types at a time. The proportion γ^*_{mn}, of say texture m is estimated and the likelihood of the statistics $\mathbf{R}^* = (\mu^*, R^*_s$ for all possible s=(i,j)) under the assumption of γ^*_{mn}, i.e., $p(\mathbf{R}^*|\gamma^*_{mn})$ is computed. μ^*, R^*_s are the sample mean, & sample autocovariance function. A generalized likelihood ratio test is then used to choose the most likely hypothesis. A window is classified as mixed if γ^*_{mn} is close to 0 or 1. The reason for using $p(\mathbf{R}^*|\gamma^*_{mn})$ is that $p(\mathbf{G}|\gamma^*_{mn})$ is not known a priori and therefore can't be used as a decision function. Moreover, $\mathbf{R}^*$ directly depends on γ^*_{mn}. $E[\mu^*]$ & $E[R^*_s]$ are shown [7] to be given as,

$$E[\mu^*] = (1 - \gamma_{mn})\mu_m + \gamma_{mn}\mu_n = \mu_m + \gamma_{mn}(\mu_n - \mu_m)$$

$$E[R^*_s] \approx \mu^2_m + R^m_s + \gamma_{mn}(R^n_s - R^m_s + \mu^2_m - \mu^2_n) \qquad (46)$$

where R^n_s & R^m_s are the theoritical autocovariance functions of the two GMRF's computed at lag s. It is shown in [7] that $p(\mathbf{R}^*|\gamma^*_{mn})$ is asymptotically Gaussian, & that the classifier using

$p(\mathbf{R}^*|\gamma^*_{mn})$ incurs a probability of misclassification that goes to zero as the window size goes to infinity. γ^*_{mn} is estimated using generalized least square (GLSE) method derived from the set of equations in (46). The GLSE is shown to be asymptotically efficient & consistent. All mixed windows are further segmented using the Maximum Likelihood Hierarchical Segmentation algorithm introduced by Cohen and Cooper [4-6].

Window Segmentation The segmentation of a window into two known texture types (say texture 0 and 1) is hierarchical and seeks to maximize $p(\mathbf{G}|\mathbf{Y}) = p(\mathbf{G}_0|\mathbf{Y})\, p(\mathbf{G}_1|\mathbf{Y})$. The window is partitioned into four quadrants and the best segmentation is obtained. Then each of the four blocks is itself partitioned into four and the process is repeated until the block size is such that each block consists of one pixel. Central to the algorithm is the computation of the joint likelihood of a number of equal-size square blocks of data, as well as conditional likelihood of one such block given the others. These are normally computationally prohibitive tasks. However by exploiting the inherent structure of the MRF's we can show that these can be decomposed into the contribution of the data within each block and the contribution of the interaction of the data in each pair of blocks. These will be computed and stored before starting the segmentation process. The joint conditional likelihood of a group of equal size blocks of data can be determined by table lookup. If a block $\mathbf{G}$ with n data points is divided into four quadrants $\mathbf{G}_1, \mathbf{G}_2, \mathbf{G}_3, \mathbf{G}_4$, then it is shown [4-6] that $p(\mathbf{G})$ can be expressed in terms of the sum of contribution of the data in each quadrants and the contribution of the interaction of the data in each pair of blocks

$$p(\mathbf{G}) = (2\pi\sigma^2)^{n/2}\ |\Psi|^{1/2} \exp\left\{ - \left[\Sigma I_W(\mathbf{G}_j) + 2 \Sigma\ I_B(\mathbf{G}_i,\mathbf{G}_j) \right] / 2\sigma^2 \right\} \tag{47}$$

where $I_W(\mathbf{G}_j) = \mathbf{G}_j \Psi_j \mathbf{G}_j$, and $I_B(\mathbf{G}_i,\mathbf{G}_j) = \mathbf{G}_i \Psi_{ij} \mathbf{G}_j$. The $I_W(\mathbf{G}_j)$ and $I_B(\mathbf{G}_i,\mathbf{G}_j)$ are recursively computed at the various resolutions and are stored. The conditional likelihood of nxn data block $\mathbf{G}_X$ given surrounding blocks $\mathbf{G}_Y$ is

$$p(\mathbf{G}_X|\ \mathbf{G}_Y)= (2\pi\sigma^2)^{n/2}|\Psi_X|^{1/2} \exp\{-[\ I_W(\ \mathbf{G}_X) + 2\ I_B\ (\mathbf{G}_X, \mathbf{G}_Y\ ) + \mathbf{G}_Y \Psi_{XY} \Psi^{-1}{}_X \Psi_{XY}\ \mathbf{G}_Y]/2\sigma^2\} \tag{48}$$

In general $\mathbf{G}_Y \Psi_{XY} \Psi^{-1}{}_X \Psi_{XY}\ \mathbf{G}_Y$ is smaller than $2\ I_B\ (\mathbf{G}_X, \mathbf{G}_Y\ )$ and one might use the cruder approximation

$$p(\mathbf{G}_X|\ \mathbf{G}_Y)= (2\pi\sigma^2)^{n/2}|\Psi_X|^{1/2} \exp\{-[\ I_W(\ \mathbf{G}_X) + 2\ I_B\ (\mathbf{G}_X, \mathbf{G}_Y\ )\ ]/2\sigma^2\} \tag{49}$$

Again the exponent in (49) is determined by lookup table.

Consider Figure 19. At the end of the 2nd stage of the hierarchical segmentation algorithm, the window has been partitioned into 16 blocks (i,j;2) with i=1,2,3,4 and j=1,2,3,4. These blocks have been segmented into two texture-type classes, denoted here as types 0

& 1 respectively. For the 3rd stage, each of these blocks is to be partitioned into four subblocks, and the four subblocks are to be allocated to one of the two classes. One possible segmentation is shown in Figure 19 for the subblocks associated with the 2nd stage block (3,2;2). The optimal 3rd stage segmentation of block (3,2;2) is that for which segmented data in the block has the highest conditional likelihood. Consider the segmentation shown in Figure 19.

0	0	0	0
0	0	1	1
0	0 1 0 1	0	1
0	1	1	1

Figure 19

Using the approximation described in (49) to the conditional likelihood of this segmentation, the quadratic in the exponent of the likelihood is

$$-(1/2\,\sigma^2(0))\,\{ I^0_W(5,3;3) + I^0_W(6,3;3) + 2\,[I^0_B(5,3,6,3{:}3) + I^0_B(5,3,4,2;3) + I^0_B(5,3,4,3;3) + I^0_B(5,3,4,4;3) + I^0_B(5,3,5,2;3) + I^0_B(5,3,6,2{:}3) + I^0_B(6,3,5,2;3) + I^0_B(6,3,6,2;3) + I^0_B(6,3,7,2;3)\} - (1/2\,\sigma^2(1))\,\{ I^1_W(5,4,;3) + I^1_W(6,4;3) + 2[I^1_B(5,4,6,4{:}3) + I^1_B(5,4,4,5;3) + I^1_B(6,4,7,3;3) + I^1_B(6,4,7,4;3) + I^1_B(6,4,7,5;3)]\} \quad (50)$$

Little computation is required for (50). The $\{ I_W(.,.;3), I_B(.,.,.,.;3)\}$ are acquired by table lookup. Hence the conditional likelihood of this segmentation is computed with a small number of multiplications and additions. We refer the reader to [4-6] for a more detailed discussion.

Examples Figure 20 is a 128x128 image from an outdoor scene of a textured image which consists of 4 distinct texture classes -- earth region, grass region, bush region, and dark bush region. We assumed here that the textures are modelled by a second order noncausal Gaussian Markov Random Field (GMRF) of a grass region abutting an earth region. The parameters of the different regions were estimated from the coarse segmentation shown in Figure 21. This was obtained via unsupervised clustering as explained in the next section. These parameters were then used in the hierarchical segmenter. In each window classified as mixed, the hierarchical segmenter was run. The ML segmentation is shown in Figure 22.

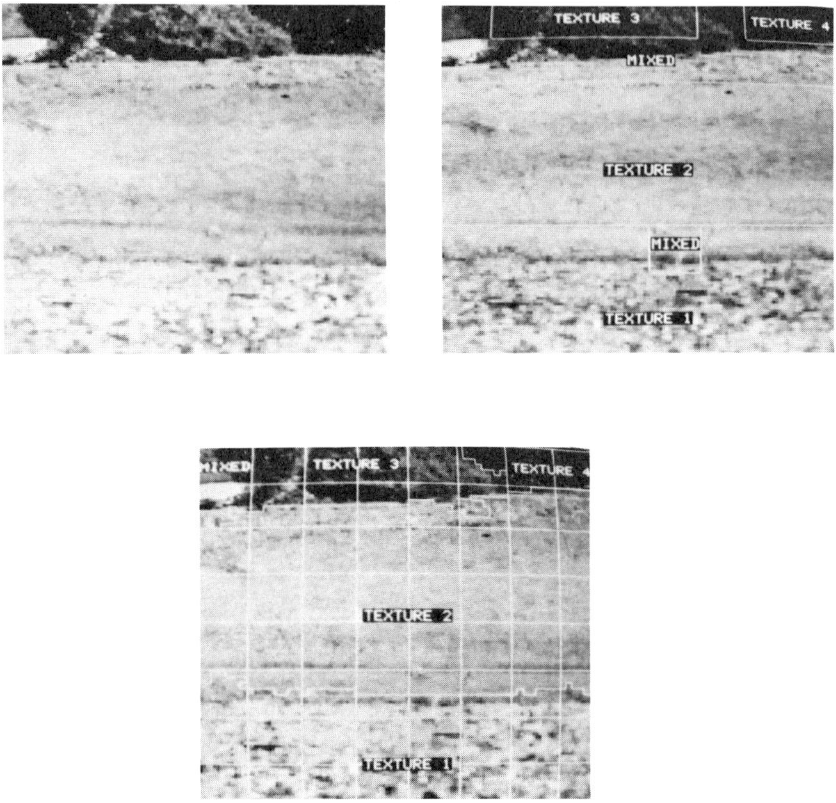

Figure 22

Figure 23a is a 256x256 image having four different regions in each of which the data from one of four different Gaussian MRF's. The boundary between the regions is displayed in white and the regions are labelled 1,2,3 and 4. Again the parameters of the 4 textured classes were obtained from the coarse segmentation shown in Figure 23b (see next section). Figure 23c shows the classification of 32x32 windows as homogeneous (H) or mixed (M) using the hypothesis test described earlier. Figure 23d shows the fraction of the first of the two textures types that comprise each window classified as mixed, and the texture type for the homogeneous windows. Figure 23e shows the hierarchical segmentation obtained in mixed windows using the texture type parameters to which each mixed window has been allocated.

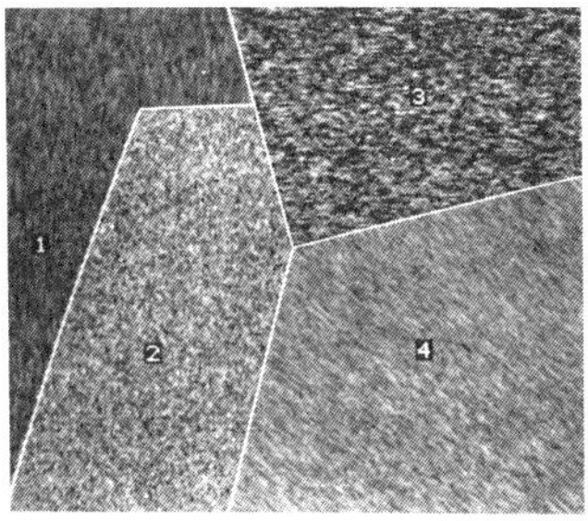

(a)

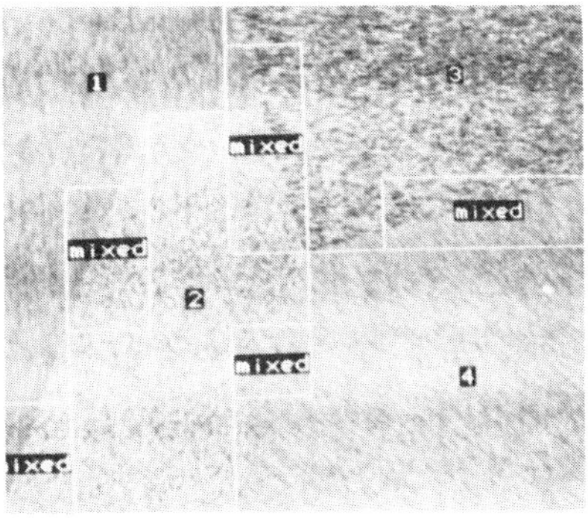

(b)

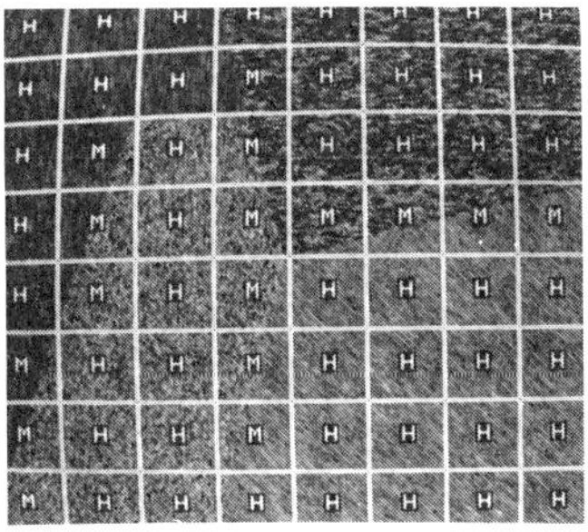

(c)

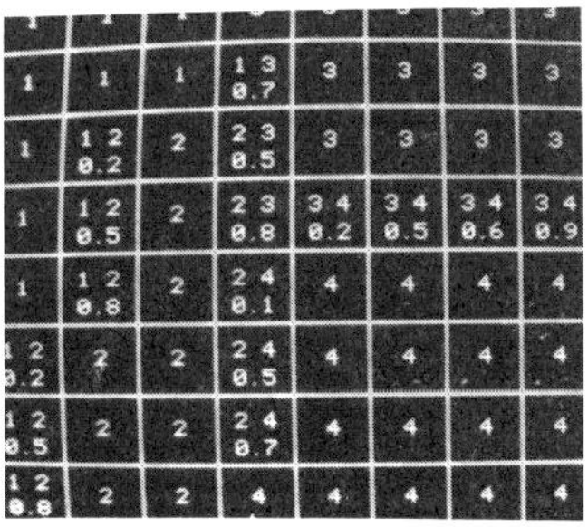

(d)

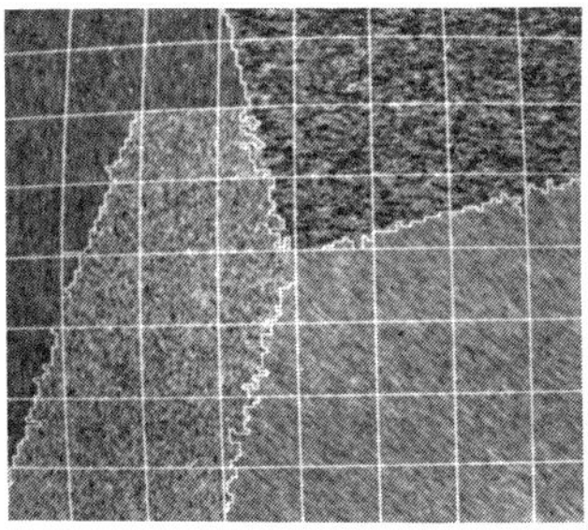

(e)

Figure 23

Unsupervised Segmentation

All of the above segmentation algorithms assumed partial or total prior knowledge of the MRF parameters. In this section we show we could relaxe this restriction. As such, the problem here is to segment a textured image with the only knowledge that any textured region in the image is modelled by a MRF of fixed order, & known structure. No prior knowledge (partial or total) of the parameter set , or the number of available texture regions is assumed. This problem is addressed by Cohen & Fan [8], who used clustering analysis as a solution.The image is divided into relatively small disjoint square windows and a maximum likelihood estimate (MLE), or a sufficient statistics α^* of α (for a fixed order model) is obtained in each window. Because of the relatively small size of the windows, most of the windows are homogeneous. The components of α^* are viewed as features, and α^* as a feature vector. The windows are grouped into different texture regions based on clustering analysis. To simplify the clustering process, the dimensionality of the feature vector is reduced via a Karhunen-Loeve decomposition of the scatter matrix of the α^* 's. Each α^* is projected onto the dominant mode (eigenvector) of the scatter matrix. The projected data is used in the clustering process. When the window size is somewhat large, the MLE (or sufficient statistics) is approximately Gaussian (see previous sections) . We exploit this asymptotic property, and regard the MLE (or sufficient statistics) set as the unlabelled samples. In that case, the clustering problem becomes that of the identification of the mixture of Gaussian densities.We use the generalized mixture likelihood function as the clustering metric. This was found to perform very close to the Bayes segmenter [8], while avoiding the computational burdens associated with the mixture likelihood maximization. When the window size is relatively small, or when the MRF parameters for a texture region are slowly varying, or when some of the clusters are sparse, the generalized mixture likelihood function tend to favor clustering the data into higher number of groups than it really is. To overcome this problem, we use a within-group variance metric (a Mahalanobis distance) weighted by a weight that directly depends the number of groups.This turns out to be equivalent to a weighted MLE clustering metric for large window size.The clustering is achieved by minimizing this weighted within-group variance metric. The metric is computationally simpler than the generalized mixture likelihood function metric, & less sensitive to deviation from the normality assumption. Its performance, however, is inferior to the generalized mixture likelihood function metric, when the unlabelled samples are Gaussian.

Examples For the textured image shown in Figure 20, the clustering obtained using the clustering procedure outlined above led to the coarse segmentation shown in Figure 21. Figure 22 shows the high resolution segmentation obtained by segmenting the mixed windows using the classification & parameters obtained from the coarse segmentation, & the hierarchical algorithm in [4-6]. The ability of the clustering algorithm to cluster the data without any supervision is impressive.

The same set of experiment has been also tried on the textured image (consisting of 4 synthesized GMRF's) shown in Figure 23a. Figure 23b shows the clustering obtained based on 32x32 sized windows using the generalized mixture likelihood metric based on MLE & sufficient statistics feature vector respectively. Figure 23e shows the high resolution segmentation obtained by segmenting the mixed windows using the classification & parameters obtained from the coarse segmentation.

Finally,we ran experiments with a textured image which consists of 4 four-level synthesized MRF's. Figure 24a shows the textured image, while the coarse segmentation is shown in Figure 24b.

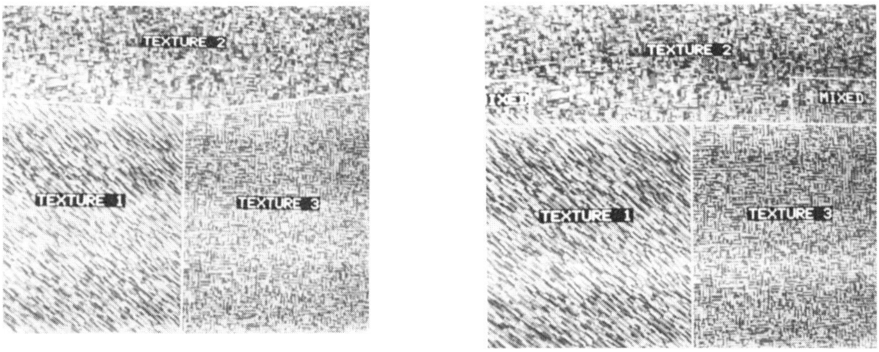

(a) (b)

Figure 24

CONCLUSION

MRF's appear to be very powerful for modelling images. Because their global structure is fully determined by their local structure, that makes them very attractive to use for image classification, restoration, and segmentation as this local structure allows for huge computational savings.

ACKNOWLEDGEMENTS

This work was supported by the National Science Foundation under grant Number ECS-840-477, and the Robotics Research Center at the University of Rhode Island.

REFERENCE

1. M. S. Bartlett, London: Chapman and Hall, 1976.
2. J. Besag, J. Royal Statis. Soc., Series B, Vol. 36, pp 192-326, 1974.
3. R. Chellappa and R. L. Kayshap, IEEE Trans. ASSP, Vol. ASSP-30, pp. 461-472, 1982.
4. F. S. Cohen and D. B. Cooper, Proc. of the 3rd Int. Conf. on Robot Vision & Sensory Control, Cambridge, MA, SPIE 449, pp. 17-28, Nov. 1983.

5. F. S. Cohen, D. B. Cooper, J. F. Silverman, & E. B. Hinkle, Proc. of the 7th Int. Conf. on Pattern Recognition, Montreal, Canada, pp. 1104-1107, 1984.

6. F. S. Cohen and D. B. Cooper, Accepted for Publication in IEEE Trans. PAMI. Also appears as Tech. Rep., Division of Eng., Brown Univ., Prov., RI 02912, July 1984.

7. Z. Fan and F. S. Cohen , Tech-Rep., Electr. Eng. Dept., Univ. of Rhode Island, Kingston, RI 02881, Jan. 1986.

8. F. S. Cohen and Z. Fan, Tech-Rep., Electr. Eng. Dept., Univ. of Rhode Island, Kingston, RI 02881, March 1986.

9. F. S. Cohen, S. F. Attali, and Z. Fan, Tech-Rep., Electr. Eng. Dept., Univ. of Rhode Island, Kingston, RI 02881, May 1986.

10. G. C. Cross, Tech. Rep. 80-02, Dept. of Comp. Sc., College of Eng., Michigan State Univ., 1980.

11. G. C. Cross and A. K. Jain, IEEE Trans. Pami, Vol. PAMI-5, pp. 25-39, 1983.

12. H. Elliot, H. Derin, R. Cristi and D. Geman, Proc. IEEE 1985 ICASSP, March 1985.

13. H. Elliot and H. Derin, Tech. Rep., Elect. Comp. Eng. Dept., Univ. of Mass., Amherst, MA 01003, Sept. 1984.

14. S. Geman and D. Geman, IEEE Trans. PAMI, Vol. 6, No. 6, pp. 721-741, Nov. 1984.

15. R. L. Kashyap and R. Chellappa, IEEE Trans. Inf. Theory, Vol. IT-29, pp 60-72, 1983.

16. R. Kinderman and J. L. Snell, Amer. Math. Soc., 1980, Providence, RI.

17. J. Woods, IEEE Trans. Inf. Theory, Vol. 18, pp. 232-240, March 1972.

18. S. Zacks, Pergamon 1981.

11

SMOOTHING WITH BLACKOUTS†

Howard L. Weinert* and Edward S. Chornoboy**

ABSTRACT

Motivated by recent work of Pavon [1], we develop a new, efficient algorithm for a smoothing problem with missing data. Our approach is based on the method of complementary models introduced by Weinert and Desai [2]. We also discuss parallel implementation of our algorithm.

INTRODUCTION

In a recent paper [1], Pavon considered the following smoothing, or interpolation, problem. A state model is defined on the interval $[0, T]$ by

$$\dot{x}(t) = Ax(t) + Bu(t), \ x(0) = x_0 \tag{1}$$

where A and B are matrices of dimensions nxn and nxm, respectively. The initial state x_0 and process noise u are zero-mean, uncorrelated with each other, and have the following second moments:

$$E\, x_0 x_0' = \Pi_0$$
$$E\, u(t) u'(s) = I_m \delta(t-s) .$$

Due to hardware failure or some other cause, we have measurements of the state only on the intervals $[0, T_1]$ and $[T_2, T]$ where $0 < T_1 < T_2 < T$. We thus wish to find the linear least-squares estimate of $x(t)$, $t \epsilon [0, T]$, based on the measurements

$$y(t) = C\,x(t) + v(t), \ t \epsilon [0, T_1] \bigcup [T_2, T]. \tag{2}$$

Here, C is a pxn matrix and the measurement noise v is zero-mean, uncorrelated with x_0 and u, and is white with unit intensity. Using results from stochastic realization theory, Pavon derived a two-filter type algorithm for the estimate $\hat{x}$ which entails the solution of two Riccati equations.

In this paper, we use the method of complementary models [2]-[7] to derive a new algorithm for $\hat{x}$, requiring the solution of a single Riccati equation. Our derivation is considerably simpler than Pavon's, and generalizes easily to the case of multiple blackout intervals. Furthermore, our estimate can be easily updated without reprocessing the data if the initial state variance Π_0 is changed.

---

† The work of the first author was supported by the Office of Naval Research under Contract N00014-85-K-0255.

* Department of Electrical Engineering and Computer Science, The Johns Hopkins University, Baltimore, MD 21218.

** Department of Biomedical Engineering, The Johns Hopkins University, Baltimore, Maryland 21205.

THE COMPLEMENTARY MODEL AND HAMILTONIAN SYSTEM

Suppose Π_0 has rank q and consider a full rank factorization

$$\Pi_0 = DD'.$$

Define a q-vector ξ of orthonormal random variables by

$$x_0 = D\xi. \tag{3}$$

Let H be the Hilbert space spanned by the components of ξ, $u(t)$, $t\epsilon[0,T]$, and $v(t)$, $t\epsilon[0,T_1]\cup[T_2,T]$. Let M be the closed subspace of H spanned by the components of $y(t)$, $t\epsilon[0,T_1]\cup[T_2,T]$. The complementary model of (1)-(2) generates variables which span the orthogonal complement $\overline{M}$ of M. The derivation of the complementary model in the present context follows the same lines as previous derivations - see particularly [4]-[7]. The details will be omitted. The complementary model is

$$\begin{aligned}\dot{x}_c(t) &= \begin{array}{ll}-A'x_c(t) - C'v(t) & , t\epsilon[0,T_1]\cup[T_2,T] \\ -A'x_c(t) & , t\epsilon(T_1,T_2)\end{array} \\ x_c(T) &= 0 \ , \ x_c(T_i-) = x_c(T_i+) \ , \ i=1,2 \\ y_c(t) &= u(t) - B'x_c(t) \ , \ t\epsilon[0,T] \\ \theta &= \xi - D'x_c(0).\end{aligned} \tag{4}$$

The components of θ and $y_c(t)$, $t \epsilon [0,T]$, span $\overline{M}$.

Solving for u and ξ in (4) and substituting into (1) and (3) yields

$$\begin{aligned}\dot{x}(t) &= Ax(t) + BB'x_c(t) + By_c(t) \ , \ t\epsilon[0,T] \\ x_0 &= D\theta + \Pi_0 x_c(0).\end{aligned} \tag{5}$$

Then solving for v in (2) and substituting into (4) gives

$$\dot{x}_c(t) = \begin{array}{ll}-A'x_c(t) + C'Cx(t) - C'y(t) & , t\epsilon[0,T_1]\cup[T_2,T] \\ -A'x_c(t) & , t\epsilon(T_1,T_2).\end{array} \tag{6}$$

Projecting all random quantities onto the subspace M gives the estimate Hamiltonian

$$\begin{bmatrix}\dot{\hat{x}}(t) \\ \dot{\hat{x}}_c(t)\end{bmatrix} = \begin{bmatrix}A & BB' \\ C'C & -A'\end{bmatrix}\begin{bmatrix}\hat{x}(t) \\ \hat{x}_c(t)\end{bmatrix} + \begin{bmatrix}0 \\ -C'y(t)\end{bmatrix}, \ t\epsilon[0,T_1]\cup[T_2,T]$$

$$\begin{bmatrix}\dot{\hat{x}}(t) \\ \dot{\hat{x}}_c(t)\end{bmatrix} = \begin{bmatrix}A & BB' \\ 0 & -A'\end{bmatrix}\begin{bmatrix}\hat{x}(t) \\ \hat{x}_c(t)\end{bmatrix} \quad , t\epsilon(T_1,T_2) \tag{7}$$

$$\begin{aligned}\hat{x}_o &= \Pi_0\hat{x}_c(0) \\ \hat{x}_c(T) &= 0, \ \hat{x}_c(T_i-) = \hat{x}_c(T_i+) \ , \ i=1,2 \ .\end{aligned}$$

Here $\hat{x}$, $\hat{x}_c$ denote the linear least-squares (smoothed) estimates of x, x_c, respectively.

THE SMOOTHING ALGORITHM

To solve the estimate Hamiltonian (7) for $\hat{x}$, we perform the following change of variables

$$\rho(t) = N(t)\hat{x}(t) + \hat{x}_c(t) \quad , \quad t\epsilon[0,T] \tag{8}$$

where

$$\dot{N}(t) = \begin{cases} -A'N(t) - N(t)A + N(t)BB'N(t) - C'C \quad , \quad t\epsilon[0,T_1]\bigcup[T_2,T] \\ -A'N(t) - N(t)A + N(t)BB'N(t) \quad , \quad t\epsilon(T_1,T_2) \end{cases} \tag{9}$$
$$N(T) = 0 \; , \; N(T_i-) = N(T_i+) \; , \; i=1,2 \; .$$

Differentiating (8), and using (7) and (9), we get

$$\dot{\rho}(t) = \begin{cases} -(A - BB'N)'\rho(t) - C'y(t) \quad , \quad t\epsilon[0,T_1]\bigcup[T_2,T] \\ -(A - BB'N)'\rho(t) \quad , \quad t\epsilon(T_1,T_2) \end{cases} \tag{10}$$
$$\rho(T) = 0 \; , \; \rho(T_i-) = \rho(T_i+) \; , \; i=1,2 \; .$$

In terms of ρ , the smoothed estimate is given by

$$\dot{\hat{x}}(t) = (A - BB'N)\hat{x}(t) + BB'\rho(t) \; , \; t\epsilon[0,T]$$
$$\hat{x}_0 = \Pi_0(I + N(0)\Pi_0)^{-1}\rho(0) \; . \tag{11}$$

Our smoothing algorithm thus has the following steps: solve for N using (9); solve for ρ using (10); solve for $\hat{x}$ using (11). Note that Π_0 enters the algorithm only through (11). Therefore, once the data has been used to compute $\hat{x}$ for a given Π_0 , it does not need to be re-used to update the estimate if Π_0 is changed. All that is needed is the solution of the homogeneous version of the differential equation in (11), with initial condition equal to the difference of the two values of the estimated initial state.

To find the smoothing error covariance, let $\tilde{x} = x - \hat{x}$ and $\tilde{x}_c = x_c - \hat{x}_c$, project Eqs. (5)-(6) onto $\overline{M}$, and change variables via

$$\psi(t) = N(t)\tilde{x}(t) + \tilde{x}_c(t) \; , \; t\epsilon[0,T] \; .$$

Then

$$\dot{\psi}(t) = -A'\psi(t) + NB(y_c(t) + B'\psi(t)) \; , \; t\epsilon[0,T]$$
$$\psi(T) = 0$$

and

$$\dot{\tilde{x}}(t) = (A - BB'N)\tilde{x}(t) + B(y_c(t) + B'\psi(t)) \; , \; t\epsilon[0,T]$$
$$\tilde{x}_0 = (I + \Pi_0 N(0))^{-1}(D\theta + \Pi_0\psi(0)) \; .$$

By analogy with the usual Kalman filter, one can see that $\psi(t)$ is the linear least-squares estimate of $x_c(t)$ based on $y_c(s)$, $t \leq s \leq T$. Therefore, $(y_c + B'\psi)$ is the innovation process of y_c and is thus white. Since $(y_c + B'\psi)$ is uncorrelated with both x_0 and $\hat{x}_0$, the smoothing error covariance $P(t) = E[\tilde{x}(t)\tilde{x}'(t)]$ satisfies the linear differential equation

$$\dot{P}(t) = (A - BB'N)P(t) + P(t)(A - BB'N)' + BB' \ , \ t \epsilon [0,T] \ . \qquad (12)$$

The initial condition is

$$\begin{aligned} P(0) = E[\tilde{x}_0\tilde{x}_0'] = E[\tilde{x}_0\xi']D' &= (I + \Pi_0 N(0))^{-1} D(E[\theta\xi'])D' \\ &= (I + \Pi_0 N(0))^{-1}\Pi_0 \\ &= \Pi_0(I + N(0)\Pi_0)^{-1} \ . \end{aligned} \qquad (13)$$

Note that Eqs. (12)-(13) have exactly the same form as in the case of no blackout regions [2]. Also, P depends on Π_0 only via (13), so it can be easily updated if Π_0 is changed.

N BLACKOUT REGIONS

The generalization of our approach to N blackout regions presents no conceptual or analytical difficulties. Let

$$0 < T_1 < T_2 < \cdots < T_{2N} < T$$

and define

$$\begin{aligned} J &= [0, T_1] \cup [T_2, T_3] \cup \cdots \cup [T_{2N}, T] \\ \bar{J} &= (T_1, T_2) \cup (T_3, T_4) \cup \cdots \cup (T_{2N-1}, T_{2N}) \ . \end{aligned}$$

Note that $J \cup \bar{J} = [0, T]$. We have measurements of the state on J , and $\bar{J}$ is the union of the N blackout regions. The smoothing algorithm is then

$$\dot{N}(t) = \begin{matrix} -A'N(t) - N(t)A + N(t)BB'N(t) - C'C \ , & t \epsilon J \\ -A'N(t) - N(t)A + N(t)BB'N(t) \ , & t \epsilon \bar{J} \end{matrix} \qquad (14)$$

$$N(T) = 0 \ , \ N(T_i -) = N(T_i +) \ , \ i = 1,2,\ldots,2N \ .$$

$$\dot{\rho}(t) = \begin{matrix} -(A - BB'N)'\rho(t) - C'y(t) \ , & t \epsilon J \\ -(A - BB'N)'\rho(t) \ , & t \epsilon \bar{J} \end{matrix} \qquad (15)$$

$$\rho(T) = 0 \ , \ \rho(T_i -) = \rho(T_i +) \ , \ i = 1,2,\ldots,2N \ .$$

$$\begin{aligned} \dot{\hat{x}}(t) &= (A - BB'N)\hat{x}(t) + BB'\rho(t) \ , \ t \epsilon [0,T] \\ \hat{x}_0 &= \Pi_0(I + N(0)\Pi_0)^{-1}\rho(0) \ . \end{aligned} \qquad (16)$$

The smoothing error covariance is again given by Eqs. (12)-(13).

PARALLEL IMPLEMENTATION

It is clear from an examination of Eq. (15) that the observations in each subinterval can be processed in parallel. Defining $T_0 = 0$, $T_{2N+1} = T$, we can write

$$\rho(t) = \sum_{j=1}^{N+1} \rho_j(t)$$

where ρ_j depends only on $y(t)$, $t \epsilon [T_{2j-2}, T_{2j-1}]$. Each ρ_j is computed in a separate processor via

$$\dot{\rho}_j(t) = \begin{cases} 0 & , T_{2j-1} < t \leq T \ , \rho_j(T) = 0 \\ -(A - BB'N)'\rho_j(t) - C'y(t) & , t \epsilon [T_{2j-2}, T_{2j-1}] \ , \rho_j(T_{2j-1}) = 0 \\ -(A - BB'N)'\rho_j(t) & , 0 \leq t < T_{2j-2} \ , \rho_j(T_{2j-2}-) = \rho_j(T_{2j-2}+) \ . \end{cases} \tag{17}$$

Once again using the concept of superposition, one can see from Eq. (16) that $\hat{x}$ can be decomposed as

$$\hat{x}(t) = \sum_{j=1}^{N+1} x_j(t) \tag{18}$$

where

$$\begin{aligned} \dot{x}_j(t) &= (A - BB'N)x_j(t) + BB'\rho_j(t) \ , \ t \epsilon [0, T] \\ x_j(0) &= \Pi_0(I + N(0)\Pi_0)^{-1}\rho_j(0) \ . \end{aligned} \tag{19}$$

Each processor can carry out its task of computing x_j via (14), (17), (19) independently of the other processors, so no communication between processors is needed. The final step in the parallel implementation involves sending the x_j to another processor which adds them to produce $\hat{x}$ (see Eq. (18)).

The quantity x_j is not the linear least-squares estimate of x based on $y(t)$, $t \epsilon [T_{2j-2}, T_{2j-1}]$. However, one can still provide a geometric interpretation for x_j as follows. Let M and M_j be the closed subspaces of H spanned by the components of $y(t)$, $t \epsilon J$, and $y(t)$, $t \epsilon [T_{2j-2}, T_{2j-1}]$, respectively. Suppose that the M_j are pairwise disjoint. Then M has the direct sum decomposition

$$M = M_1 + \cdots + M_{N+1}$$

and x_j is the oblique projection [8] of x onto M_j along $M_1 + \cdots + M_{j-1} + M_{j+1} + \ldots + M_{N+1} + \overline{M}$.

CONCLUSIONS

We have derived a new algorithm for smoothing the state of a linear system when part of the observation record is missing. If the cross-covariance between the process and measurement noises is a delta function (the case treated by Pavon), our approach generalizes as shown in [3]. When the width of the longest observation subinterval goes to zero, our algorithm reduces to the one given in [4]. When there are no blackout intervals, we get the algorithm in

[2].

REFERENCES

[1] M. Pavon, "Optimal interpolation for linear stochastic systems," *SIAM J. Contr. Optimiz.*, vol. 22, 618-629, 1984.

[2] H. L. Weinert and U. B. Desai, "On complementary models and fixed-interval smoothing," *IEEE Trans. Auto. Contr.* , vol. AC-26, 863-867, 1981. Auto. Contr. , vol. AC-26, 863-867, 1981.

[3] U. B. Desai, H. L. Weinert, and G. J. Yusypchuk, "Discrete-time complementary models and smoothing algorithms: the correlated noise case," *IEEE Trans. Auto. Contr.* , vol. AC-28, 536-539, 1983.

[4] H. L. Weinert, "The complementary model in continuous/discrete smoothing," pp. 353-363 in *Time Series Analysis of Irregularly Observed Data* , E. Parzen, Ed., Springer, 1984.

[5] H. L. Weinert, "On the inversion of linear systems," *IEEE Trans. Auto. Contr.* , vol. AC-29, 956-958, 1984.

[6] M. B. Adams, A. S. Willsky, and B. C. Levy, "Linear estimation of boundary value stochastic processes - Parts I and II," *IEEE Trans. Auto. Contr.* , vol. AC-29, 803-821, 1984.

[7] T. Kailath and M. Wax, "A note on the complementary model of Weinert and Desai," *IEEE Trans. Auto. Contr.* , vol. AC-29, 551-552, 1984.

[8] P. R. Halmos, *Finite-Dimensional Vector Spaces* , Springer, 1974.

12

STOCHASTIC BILINEAR MODELS AND ESTIMATORS WITH NONLINEAR OBSERVATION FEEDBACK*

R.R. MOHLER and W.J. KOLODZIEJ

Department of Electrical & Computer Engineering
Oregon State University, Corvallis, OR 97331

ABSTRACT

An overview is presented on modelling and state estimation of a significant class of nonlinear stochastic processes. This includes certain bilinear systems with observation feedback. Appropriate applications include immunology, nuclear power generation, heat transfer and track signal processing.

INTRODUCTION

The purpose of this paper is to present an overview of recent developments on the modelling and state estimation of bilinear stochastic systems, BLSS, which have observation feedback. It is shown that such models are useful in immunology, energy, and tracking. The observation feedback, with some restriction, may be nonlinear. The resulting closed-loop system, which is linear in the unobserved (unmeasured) states, is termed a conditionally-linear system, CLS. i.e., the system is linear in the unmeasured states, or linear on the condition that certain other variables are measured (observed).

Here, such processes are defined by the following stochastic differential equations:

$$\begin{aligned} dx_t &= [f(t,y) + F(t,y)x_t]dt + G(t,y)dw_t, \\ dy_t &= [h(t,y) + H(t,y)x_t]dt + R(t,y)dv_t, \end{aligned} \qquad (1)$$

*Research funded by ONR Contract No. N00014-81-K-0814 and NSF Grant 8215724.

where x_t, y_t, are the n and m dimensional vectors respectively; w_t, v_t are mutually independent vector Wiener processes of appropriate dimension which are also independent of x_0, y_0; and f(.), F(.), G(.), h(.), H(.), R(.) are nonanticipative matrix functionals of appropriate dimensions with certain time-growth limitations to ensure existence of solutions. See references (1) and (2).

It is readily seen that, in a formal manner, equation (1) represents a stochastic integral formulation of the traditional Gaussian white-noise-driven differential equation with the appropriate correction term required in the general case of (1) for the Ito representation. Also, it is seen that equation (1) represents a BLSS whereby the feedback control policy is synthesized from its measured variables. This may take the form of:

$$dx_t = (A(t)x_t + \sum_{i=1}^{k} B_i(t)u_i(t,y)x_t + C(t)u(t,y))dt + G(t,y)dw_t,$$

$$(2)$$

$$dy_t = (D(t)x_t + \sum_{i-1}^{k} E_i(t)u_i(t,y)x_t + F(t)u(t,y))dt + R(t,y)dv_t,$$

which without multiplicative noise, G(t,y) = G(t), may be interpreted directly as the traditional white-noise-driven bilinear system, BLS. Here, A(t), $B_i(t)$, C(t), G(t,y), D(t), $E_i(t)$, F(t), R(t,y), $u_i(t,y)$, i = 1, ..., k, are matrices of appropriate dimensions.

Estimation and control of BLSS, (2) and generalizations thereof, are studied by Mohler and Kolodziej (3), (4). Lipster and Shiryayev (1), (2) derive the finite-dimensional mean-square optimal state estimator for the case of (1) when initially the unmeasured state is conditionally Gaussian relative to the initial observation; i.e., $x_0|\mathbf{Y}_0$ is Gaussian where $\mathbf{Y}_0$ is the σ-algebra generated by the initial observations. Kolodziej rigorously extended this result to the vector case and considered a stochastic control problem (5). More recently, this work has been extended for the non-Gaussian initial condition for a large class of the scalar case of (1) in reference (6).

MODELS

Neutron Kinetics

The typical nuclear-fission process is described by a BLS of form (see reference (7))

$$\frac{dn}{dt} = \frac{[u_1(1-\beta)-i]}{T} n + \sum_{i=1}^{6} \lambda_i c_i,$$
$$\frac{dc_i}{dt} = \frac{\beta_i n}{T} - \lambda_i c_i, \quad i = 1, \ldots, 6 \quad , \tag{3}$$

where u_i is neutron multiplication; n is neutron population; c_i ($i = 1,\ldots,6$, for U-235) are a precursor populations (delayed neutrons); T is average neutron generation time; λ_i and β_i are decay constants and proportionality constants associated with the i-th group such that

$$\sum_{i=1}^{6} \beta_i = \beta.$$

For "zero"-power fission (i.e., negligible temperature increase), neutron multiplication is controlled directly. However, noise is particularly present as an added multiplier v, and as an added net neutron source, w (i.e., source minus loss). Hence, the BLSS model becomes

$$\frac{dn}{dt} = \frac{[(u_i+v)(1-\beta)-1]}{T} n + \sum_{i=1}^{6} \lambda_i c_i + w$$
$$\frac{dc_i}{dt} = \frac{\beta_i n}{T} - \lambda_i c_i, \quad i = 1, \ldots, 6. \tag{4}$$

Traditionally, n is measured and the process is a CLS. If v is a white-noise process, however, the solution to (4) does not exist in the usual sense. Then, it is sometimes convenient for analysis to use a stochastic integral formulation such as the Ito integral with the required correction term to relate the two integral formulations. Formally (4) can be written as

$$
\begin{aligned}
dn &= \left\{\frac{[(1-\beta)u_i-1]}{T} n + \sum_{i=1}^{6} \lambda_i c_i + \frac{1}{2}\left(\frac{(1-\beta)\ n}{T}\right)^2\right\}dt + \frac{(1-\beta)}{T} n\ dv + dw, \\
dc_i &= \left(\frac{\beta_i n}{T} - \lambda_i c_i\right)dt,\ i = 1, \ldots, 6.
\end{aligned}
\qquad (5)
$$

Here, v, w are Wiener processes with unity variance assumed for convenience. Neutron level (proportional to power) is a measured variable, so that with $u_1 = u_1(n,t)$, normal case, (5) is a special form of (1).

Heat Exchange

In heat-exchange dynamics, BLS arise if it is assumed that conduction is manipulated or that the convective coefficient is changed such as by altering coolant mass flow rate. e.g., consider an energy balance on a perfectly insulated cylinder (such as reactor core element) of average temperature T_c and mass heat capacity c_1 with negligible axial conduction. Coolant flows through the cylinder, and has average temperature T_g, mass heat capacity c_2, specific heat c_p and mass flow rate u_2. First, assume that the weighted average coolant temperature is given by

$$T_g = \frac{(T_i + \theta\ T_o)}{1 + \theta}$$

where θ is a constant, and T_i, T_o are inlet and outlet temperatures, respectively. Then, the energy balance yields

$$
\begin{aligned}
c_1 \frac{dT_c}{dt} &= h(u_2)\ (T_g-T_c) + n \\
c_2 \frac{dT_g}{dt} &= c_p\ (1 + \tfrac{1}{\theta})\ u_2\ (T_g-T_i) + h(u_2)\ (T_c-T_g),
\end{aligned}
\qquad (6)
$$

where heat-transfer times area, $h(u_2)$, is nearly linear in coolant mass flow rate u_2 (a control) with $h(.) \cong c_h\ u_2$. Now, (6) is a BLS if T_i is given (as is usual).

Sometimes, h(.) is not well defined, and correspondingly multiplicative noise may be introduced into the model (6). Also, random heat variations may occur so that (6) is approximated by the Ito-form of stochastic equations as again a special form of (1). Here, the presence of non-independent noise sources are noted which can add some complication. Also, it is necessary to measure both T_g and T_c (such as by thermocouples) for this form.

If a nuclear power reactor with temperature reactivity (normally negative neutron-multiplication feedback due to core expansion and fission cross-section contraction with temperature), is considered so that the first of (4) becomes

$$\frac{dn}{dt} = \frac{[(u_1 + v + u_T)(1-\beta) - 1]}{T} n + \sum_{i=1}^{6} \lambda_i c_i + w$$

where $u_T = c_T T_c + c_o$. Here, c_T, the temperature coefficient, and c_o are constants.

Again, the system in Ito form is a special case of (1) with both temperatures (T_c, T_g) measured. It might be noted that most of the random feedback multiplication occurs in the v term. By neglect of the random heat-transfer term, this would only require a measurement of n. The significance of such mathematical structures is discussed below.

Humoral Immune Process

Similar to nuclear fission and population equations, cellular division and differentiation may be derived as cascades of BLS. Depending on the measurements and noise structures, the system may take the form of (1). Similar structures arise for molecular binding. A good example of the combination of such processes occurs in the humoral immune system which leads to the generation of antibodies of molecular concentration x_3, for the destruction of alien (antigen), of molecular concentration y. Following reference (8) a stochastic model of the humoral system takes the following form:

$$\begin{aligned} dx_1 &= [a_1 x_1 + b_1(y) x_1] dt + g_1 dw_t , \\ dx_2 &= [a_2 x_2 + b_2(y) x_1] dt, \\ dx_3 &= [a_3 x_3 + a_{23} x_2 + b_3(y) x_3] dt, \\ dy_t &= [c y + h(y) x_3] dt + g_2 y_t dv_t. \end{aligned} \tag{7}$$

Here, the immune-complex dynamics have been neglected (8). x_1 is immunocompetent-cell concentration x_2 is plasma-cell concentration which leads to the generation of antibodies with concentration x_3. ai, $i = 1,2,3,a_{23}$, c and g_1, g_2 are constants. $b_1(y)$ involves probability of stimulation and differentiation and $b_2(y)$ is probability of differentiation. a_1, a_3 and c include life times and correction terms for multiplicative noise terms. h(y) is a molecular association term. w_t approximates the source rate of stem cells from bone marrow minus losses and the randomness associated with stimulation. v_t approximates the randomness associated with antigen growth.

STATE ESTIMATION

Estimation of system state is considered so that the mean-square error (m.s.e.) is a minimum. This starts from the basic relation of the conditional expectation.

$$m_t = \mathbf{E}\,[x_t | \mathbf{Y}_t],\ 0 \leq t \leq T.$$

$\mathbf{Y}_t$ refers to the "measurements", or, more precisely, the corresponding σ-algebra generated by y_s: $0 \leq s \leq t$.

If x_0 is conditionally Gaussian with respect to the initial observation y_0, then under certain broad assumptions it is shown in references (2) and (4) that the distribution $x_t | \mathbf{Y}_t$ is Gaussian and that the optimal m.s.e. filter is given by

$$dm_t = (F\,m_t + f)\,dt - K\,Q^{-1}\,[dy_t - (H\,m_t + h)\,dt],$$

$$dP_t = [F\,P_t + P_t\,F' + G\,G' - K\,K']\,dt, \qquad (8)$$

$$K = (G\,R' + P_t\,H')\,Q^{-1}, \qquad Q^2 = R(y,t)\,R'(y,t).$$

Here, the arguments generally are dropped for convenience; prime (') designates matrix transpose.

In reference (6), it is shown that a finite-dimensional filter may be derived for the case that x_0 is not conditionally Gaussian relative to y_0. More precisely, the optimal m.s.e. filter is derived for the scalar case with

$$dx_t = (f(t,y) + F(t,y)\, x_t)\, dt + g(t,y)\, dw_t + q(t,y)\, dv_t,$$
$$dy_t = (h(t,y) + H(t,y)\, x_t)\, dt + dv_t, \tag{9}$$

where f, F, g, q, h, H are nonanticipative functionals of y (i.e., $\mathbf{Y}_t$ measurable with $\mathbf{Y}_t = \sigma\text{-alg}\{y_s\colon 0 \le s \le t\}$, and w, v are independent Wiener processes. The conditional distribution of the initial states $P(a) = P(x_0 \le a \mid \mathbf{Y}_o)$ is given. Then, the minimum m.s.e. estimator is given by

$$
\begin{aligned}
dm_t &= (f + F\, m_t)\, dt + (q + P_t\, H)\, d\mu_t, \\
d\mu_t &= dy_t - (h + H\, m_t)\, dt, \\
P_t &= Q_t + R_t^{\,2}\, (I_t(2) - I_t(1)^2), \\
dQ_t &= (2\, F\, Q_t + g^2 + q^2 - (q + Q_t\, H)^2)\, dt, \\
x_0 &= \int a\, dP(a), \qquad Q_0 = 0, \qquad\qquad (10) \\
I_t(n) &= \frac{\int a^n \exp(a^2\, F_1 + a\, F_2)\, dP(a)}{\int \exp(a^2\, F_1 + a\, F_2)\, dP(a)}, \quad n = 1,\ 2, \\
dF_1 &= 0.5\, (R_t\, H)^2\, dt, \qquad F_1(0) = 0, \\
dF_2 &= R_t\, H\, (d\mu_t + I_t(1)\, H\, dt), \qquad F_2(0) = 0, \\
dR_t &= (F - H\, (q + Q_t\, H))\, R_t\, dt, \qquad R_0 = I.
\end{aligned}
$$

The characteristic function, e(z), of $x_t|\mathbf{Y}_t$ is given by

$$e(z) = \exp(iz\, (m_t - R_t\, I_t(1)) - 0.5\, z^2\, Q_t)$$

$$\frac{\int \exp(a^2\, F_1 + a\, (F_2 + iz\, R_t))\, dP(a)}{\int \exp(a^2\, F_1 + a\, F_2)\, dP(a)}.$$

Note here that in both cases

$$P_t = \mathbf{E}\,[(x_t - m_t)^2 | \mathbf{Y}_t],$$

the conditional variance, and μ_t is an innovation process.

The state estimators, which are presented here, with the appropriate assumptions may be applied to numerous CLS or BLSS including the above neutronic, heat-exchange and immunological process.

Tracking

For more highly nonlinear processes such as arise in the receiver (observation) model for target tracking, the above methodology may be adapted by a two-step nonlinear filter TNF. In this manner, the best BLSS observation fit is computed first. Then the state estimator is derived. For example, a typical second-order linear state model takes the form of

$$\begin{aligned} dx_1 &= x_2\, dt, \\ dx_2 &= -\alpha\, x_2\, dt + g\, dw_t. \end{aligned} \qquad (11)$$

A linear observation, with a product position correction term due to target motion may take the form of

$$\begin{aligned} dy_1 &= (x_1 + \frac{x_1 x_2}{c})\, dt + r_1\, dv_{1t}, \\ dy_2 &= x_2\, dt + r_2\, dv_{2t}. \end{aligned} \qquad (12)$$

It is assumed that w, v_1, v_2 are independent Wiener processes. α is a drag coefficient, c is acoustic velocity, and g, r_1, r_2 are standard deviations.

To fit the above control-approximation approach, the best bilinear observation approximation is obtained such that u minimizes

$$J(u) = E\,\{ \int_0^T (x_1 x_2 - u_t x_1)^2 \}\, dt. \qquad (13)$$

The filtering problem may be solved in two stages again. First, however, the standard Kalman-Bucy filter (KBF), $\hat{x}_1=\mathbf{E}[x_1|y_2]$, is obtained from the linear, y_2, observation. Then the nonlinear conditional-Gaussian filter, NCF, improves the estimate by using the y_1 observation.

In this manner, the KBF yields

$$d\hat{x}_1 = \hat{x}_2\, dt + \frac{P_3}{r_2}\, d\nu_t,$$

$$d\hat{x}_2 = -\alpha\, \hat{x}_2\, dt + \frac{P_2}{r_2}\, d\nu_t, \tag{14}$$

$$d\nu_t = \frac{1}{r_2}\,(dy_2 - \hat{x}_2\, dt),$$

where P_2, P_3 are the standard time-variant solutions to components of the Riccati equation. The NCF yields

$$dm_1 = m_2\, dt + \frac{Q_1}{r_1}\,(1 + \frac{u_t}{c})\, d\mu_t,$$

$$m_2 = -\alpha\, m_2\, dt + \frac{Q_3}{r_1}\,(1 + \frac{u_t}{c})\, d\mu_t,$$

$$d\mu_t = \frac{1}{r_1}\left(dy_1 - m_1\,(1 + \frac{u_t}{c})\, dt\right), \tag{15}$$

$$dQ_1 = \left(2\, Q_3 - (1 + \frac{u_t}{c})^2\, \frac{Q_1^2}{r_1^2}\right) dt,$$

$$dQ_2 = \left(Q_2 - \alpha\, Q_3 - (1 + \frac{u_t}{c})^2\, \frac{Q_1\, Q_3}{r_1^2}\right) dt,$$

$$dQ_3 = \left(-2\, \alpha\, Q_2 + g^2 - (1 + \frac{u_t}{c})^2\, \frac{Q_3^2}{r_1^2}\right) dt.$$

The optimal control, computed after filter stage 1, in order to obtain (15), is

$$u_L = \hat{x}_2 + \frac{2\hat{x}_1 P_3}{(\hat{x}_1^2 + P_1)}, \tag{16}$$

by solution of the Bellman-Hamilton-Jacobi stochastic equation.

A comparison of this TNF and the two-measurement KBF is given in Reference (9), and indicates the improved accuracy of nonlinear tracking by an order of magnitude in some cases. Improvement over that of the extended KBF was much less, but still typically 50% (10). Consideration of the arc tangent nonlinearity for bearings-only track results in a more accurate TNF than the extended KBF, but the improvement is not as pronounced.

REFERENCES

1. Lipster, R.S. and Shiryayev, A.N. Statistics of Random Processes, I - General Theory. Springer-Verlag, New York, 1977.
2. Lipster, R.S. and Shiryayev, A.N. Statistics of Random Processes, II - Applications. Springer Verlag, New York, 1978.
3. Mohler, R.R. and Kolodziej, W.J. IEEE Trans. Sys. Man. and Cyb. 10: 913-920, 1980.
4. Kolodziej, W.J. and Mohler, R.R. IEEE Trans. Auto, Cont. 26: 1048-1053, 1981.
5. Kolodziej, W.J. Conditionally Gaussian Processes in Stochastic Control, Ph.D. Dissertation, Oregon State University, Corvallis, OR, 1980.
6. Kolodziej, W.J. and Mohler, R.R. SIAM J. Cont. and Optimiz. 23: to appear, 1985.
7. Mohler, R.R. Bilinear Control Processes, Academic Press, New York, 1973.
8. Mohler, R.R., Bruni, C. and Gandolfi, A. Proc. IEEE 68: 964-990, 1980.
9. Mohler, R.R., Kolodziej, W.J., Brunk, H.D., and Engelbrecht, R.S. Statistical Signal Processing, (Eds. E. Wegman and J. Smith), Marcel Dekker, New York, 1985, pp. 253-264.
10. Halawani, T.V., Mohler, R.R., and Kolodziej, W.J. IEEE Trans. Acous, Sp. and Sig. Processing 32: 344-352, 1984.